What are species? What are the factors involved in their evolution? Dr Max King presents an up-to-date synthesis of theoretical, experimental and descriptive perspectives on speciation in higher organisms. The book provides a fresh insight into the processes involved in speciation utilizing the multi-dimensional databases now available. The author clearly and concisely analyses the most recent research in plant and animal populations, concentrating on the evolutionary processes, the role of chromosomes and the genetic mechanisms involved in speciation.

Species Evolution

the role of chromosome change

Species

Evolution

the role of chromosome change

MAX KING

Honorary Senior Research Fellow,
School of Genetics and Human Variation,
La Trobe University, Victoria, Australia

CAMBRIDGE
UNIVERSITY PRESS

CAMBRIDGE UNIVERSITY PRESS
Cambridge, New York, Melbourne, Madrid, Cape Town, Singapore, São Paulo

Cambridge University Press
The Edinburgh Building, Cambridge CB2 8RU, UK

Published in the United States of America by Cambridge University Press, New York

www.cambridge.org
Information on this title: www.cambridge.org/9780521353083

First published 1993
Reprinted 1994
First paperback edition 1995

A catalogue record for this publication is available from the British Library

Library of Congress Cataloguing in Publication data
King, Max.
Species Evolution : the role of chromosome change / by Max King.
p.   cm.
Includes bibliographical references (p.   ) and index.
ISBN 0-521-35308-4 (hc)
1. Species.   2. Variation (Biology)   3. Mutation (Biology)
I. Title
QH380.K56   1993
575.2–dc20   92-33870   CIP

ISBN 978-0-521-35308-3 hardback
ISBN 978-0-521-48454-1 paperback

Transferred to digital printing 2008

For Pamela

Contents

Preface

For me, the late 1970s and early 1980s provided one of the great insights into the scientific approach. After gaining my Ph.D. in the Genetics department at the University of Adelaide in 1976, I worked with Bernard John in the Department of Population Biology at the Research School of Biological Sciences (Australian National University), for some nine and a half years. Here, most of my research was on the population cytogenetics of a number of Australian grasshopper species, and a continuation of my studies on the structural and population cytogenetics of Australian amphibians and reptiles.

At that time, Australia was one of the great centres for cytogenetic research, and the Department of Population Biology boasted the presence of both Bernard John (the chair and later Director of the Research School of Biological Sciences) and Michael White (who had retired from the chair of Genetics at Melbourne University and continued his research at the Department of Population Biology). Since both of these strong-minded individuals had very different views about most things, debate and discussion on the pros and cons of any evolutionary issue or cytogenetic principle were possible at any time. The atmosphere created for research was inescapably productive. On more than one occasion, determined discussion during or after the weekly seminar resulted in the publication of a paper on some aspect or interpretation of evolutionary theory.

For me, the learning experience at ANU, and in other spheres too, included the appreciation that some scientists would simply not accept the possibility that a particular theory was either correct or incorrect, despite incontrovertible evidence to support or refute it. For example, some had established their own view on speciation and simply stood their ground. Presumably this was because they had published extensively supporting one concept and either felt that they would look silly, or – don't mention the thought – might be wrong, if they did otherwise. Virtually any evolutionary or cytogenetic issue can be debated, and if one has a pre-ordained position there seems to be little value in pursuing the matter. Unfortunately, speciation research appears to have attracted its fair share of dogmatists in the past and will undoubtedly attract them in the future. My own view is that intransigence is not what science is about. A theory is only as good as the evidence which supports it, and if the evidence is starting to go the other way, you look for a new theory. In this regard, the purpose of my using two quotations by Ernst Mayr, drawn from 1963 and 1982 publications, to introduce Chapter 9, was not to show that he was wrong in his earlier conclusion, but that he had the integrity to change his mind from the earlier to the latter. It would be nice to think that this was a more widespread trait.

One might simply ask, what hope do we have of producing a unified perspective on the modes of speciation when we start from a position of polarization on every single issue? It is not even possible to satisfy people with a definition of what a species is, let alone explain how it evolved from its congener. In writing this book, I have taken the approach that basic concepts must be explained in terms of their weaknesses and strengths, acknowledging the fact that the possibility for disagreement is endless.

Because of the vast body of published information, I have had to be particularly ruthless in my culling of the literature. In this regard, it is more important to provide the most comprehensive account of the action of, and evidence for, intricate mechanisms involved in speciation processes, rather than provide yet another loose overview. Although continuity and background information have also necessitated a broad coverage, such areas have been deliberately minimized. Thus, while a number of models for chromosomal speciation are undoubtedly sympatric in their origin and are considered in detail, sympatric speciation is not dealt with here in any detail. Equally, modes of speciation in microorganisms and asexual species are not considered. The literature on polyploidy and partheno-genesis is extensive and there are a number of excellent volumes which deal with these subjects satisfactorily. Chromosomal divergence is only one aspect of these processes and so, while not being irrelevant to the areas considered here, they are only briefly discussed. This book has been written about the two most common forms of speciation in bisexual plants and animals, the non-chromosomal forms of allopatric speciation and the processes of chromosomal speciation. But here again, while certain aspects of allopatric speciation are examined, emphasis is concentrated on chromosomal speciation and the processes which have enabled this to occur. I have had to be selective in the examples I have used and may no doubt be criticized on the use of too many to drive home a point; too few for clarity; or the use of those examples which fail to show the alternative view. In the case of the latter, I have tried at all times to present a balanced interpretation of the available data and to provide a series of possible conclusions where the data suggest that this was the case.

I recently read a book review in which a volume on meiosis was attacked for being written with a turgid style and for being far too sophisticated for the non-specialist reader. The book was the most comprehensive analysis of meiosis yet made. To me, there is no escape from this issue; the subject at hand is complex and to be dealt with fairly the complexity must be presented. This book is aimed at advanced students and evolutionary biologists. Even so, it is pretty difficult to present a theoretical perspective on the fixation of negatively heterotic rearrangements in founding populations as snappy reading. The fact of the matter is that repetitious examples are boring, but in some cases they are necessary. My only suggestion to the disillusioned reader is a detour into le Carré or Ludlum. I don't expect to be rushed for the film rights. However, I also think that the conclusions reached are compelling and are supported by an overwhelming amount of evidence.

I must acknowledge the assistance of Professor Peter Parsons who, when in the chair at La Trobe University, not only fostered my career in science, but suggested that I write this book. Many scientists have commented on various chapters of the manuscript and to them I am most grateful. Apart from Peter Parsons, Drs Graham Flannery and Neville White from La Trobe University and Dr Les Christidis from the Museum of Victoria were most helpful. My support at the Northern Territory Museum has been outstanding and in this regard I thank the Director Dr Colin Jack-Hinton.

Paul Horner, my technical assistant, has given me tireless help and I thank him for his efforts. I also thank Margaret Meshcherskij and Lorna Graverner for typing and Usha Dasari and Kim Cottee for their assistance in the library.

Max King

Copyright acknowledgements

In addition to those citations appearing in the text of this book, I further acknowledge the following individuals and organizations who have given specific permission for the use of copyright material:

Fig. 2.1 reproduced with permission from Sinauer Associates Inc., and A. R. Templeton, from *Speciation and its Consequences* 1989 ed. D. Otte and J. A. Endler.

Fig. 3.1 reproduced with permission of Allen Press, from *Evolution*, **13**, 389–404, 1959.

Fig. 3.2 reproduced with permission of Allen Press, from *Evolution*, **40**, 1171–84, 1986.

Fig. 3.3 reproduced with permission from Sinauer Associates Inc., and S. C. H. Barrett, from *Speciation and its Consequences* 1989 ed. D. Otte and J. A. Endler.

Fig. 4.1 reproduced with permission of Allen Press, from *Evolution*, **36**, 427–43, 1982.

Fig. 4.2 reproduced with permission of Routledge, Chapman and Hall Ltd, and E. Mayr, from *Evolution as a Process* 1954 ed. J. S. Huxley, A. C. Hardy and E. B. Ford.

Fig. 4.3 reproduced with permission of the University of Chicago Press, from the *American Naturalist*, **109**, 83–92, 1975.

Fig. 4.4 reproduced with permission of Oxford University Press, from *Genetics, Speciation and the Founder Principle* 1989 ed. L. V. Giddings, K. Y. Kaneshiro and W. W. Anderson. Also the Genetics Society of America from *Genetics*, **103**, 465–82, 1983.

Fig. 5.2 reproduced with permission of Springer Verlag, and B. John, from *Chromosoma* (Berl.), **52**, 123–36, 1975.

Fig. 5.3 reproduced with permission of Hodder and Stoughton Ltd, and B. John, from *Population Cytogenetics* Studies in Biology, **70**, 1976.

Fig. 5.4 reproduced with permission of the Indian Academy of Sciences, from *Journal of Genetics*, **24**, 405–74, 1931.

Fig. 5.5 reproduced with permission of Cambridge University Press, from *Animal Cytology and Evolution* 1973.

Fig. 5.6 reproduced with permission of Springer Verlag, from *Chromosoma* (Berl.), **88**, 57–68, 1983.

Fig. 6.2 reproduced with permission of Birkhäuser Verlag AG, from *Journal of Evolutionary Biology*, **3**, 133–37, 1990.

Fig. 6.3 reproduced with permission of the University of Chicago Press, from *The American Naturalist*, **137**, 379–91, 1991.

Fig. 6.4 reproduced with permission of the University of Chicago Press, from *The American Naturalist*, **137**, 430–42, 1991.

Fig. 6.5 reproduced with permission of the University of Chicago Press, from *The American Naturalist*, **137**, 287–331, 1991.

Fig. 6.6 reproduced with permission of the University of Chicago Press, from *The American Naturalist*, **137**, 349–58, 1991.

Fig. 7.1 reproduced with permission of the Institute of Genetics Lund, and I. Gustavsson, from *Hereditas*, **109**, 169–84, 1988.

Fig. 7.2 reproduced with permission of Academic Press, and C. A. Redi, from *Biological Journal of the Linnean Society*, **41**, 235–55, 1990.

Fig. 7.5 reproduced with permission of Springer Verlag, and Y. Rumpler, from *Chromosoma* (Berl.), **98**, 330–34, 1989.

Fig. 7.9 reproduced with permission of S. Karger AG, from *Cytogenetics and Cell Genetics*, **48**, 228–32, 1988.

Fig. 8.2 reproduced with permission of the Genetics Society of America, from *Genetics*, **105**, 969–83, 1983.

Fig. 8.3 reproduced with permission of J. Wahrman, from *Chromosomes Today*, **4**, 399–424, 1973.

Fig. 8.4 reproduced with permission of Allen Press, from *Evolution*, **30**, 831–40, 1976.

Fig. 8.5 reproduced with permission of E. Nevo, from *Acta Zoologica Fennica*, **170**, 131–36, 1985.

Fig. 8.7 reproduced with permission of Allen Press, from *Evolution*, **43**, 296–317, 1989.

Fig. 8.9 reproduced with permission of the Society of Systematic Biologists, from *Systematic Zoology*, **36**, 18–34, 1987.

Fig. 8.10 reproduced with permission of the Kew Conference Committee and J. B. Searle, from *Kew Chromosome Conference III* 1988 ed. P. E. Brandham.

Fig. 9.2 reproduced with permission of the University of Chicago Press, from *The American Naturalist*, **87**, 343–58, 1953.

Figs. 9.3 and 9.4 reproduced with permission of the Society of Systematics Biologists, from *Systematic Zoology*, **27**, 285–98, 1978.

Fig. 9.5 reproduced with permission of Allen Press, from *Evolution*, **10**, 126–38, 1956.

Fig. 10.1 reproduced with permission of Springer Verlag, and B. John, from *Chromosoma* (Berl.), **94**, 45–58, 1986.

Fig. 10.2 reproduced with permission of Routledge, Chapman and Hall Ltd, and B. John and G. Miklos, from *The Eukaryote Genome in Development and Evolution* 1988. Also with permission from Macmillan Magazines Ltd, from *Nature*, **284**, 426–30.

Fig. 10.3 reproduced with permission of Routledge, Chapman and Hall Ltd, and B. John and G. Miklos, from *The Eukaryote Genome in Development and Evolution* 1988.

Fig. 10.7 reproduced with permission of the American Association for the Advancement of Science and D. D. Shaw, from *Science*, **220**, 1165–67, 1983.

Fig. 10.8 reproduced with permission of Routledge, Chapman and Hall Ltd, and B. John and G. Miklos, from *The Eukaryote Genome in Development and Evolution*, 1988.

1

Introduction: genes, dreams and structural rearrangements

Post-mating effects, especially as observed in hybrid sterility and 'hybrid breakdown' are largely incidental to the speciation process. They do not appear to
serve to actively reinforce reproductive isolation . . . I feel that a de-emphasis on
post-mating isolation in animals may serve as a major step in the unification of
speciation theory in plants and animals.

(Carson, 1985, pp. 387–88)

Of all the mechanisms which have gained man's attention, those responsible
for producing the enormous biological diversity which surrounds us have been
paramount. One needs to look no further than the historical record to see that
not only was the cataloguing of plant and animal species and their variation a
significant pastime for our predecessors, but the origins of these forms were of
the greatest importance.

Some of the most outstanding examples of biological cataloguing are the
20 000-year-old cave paintings of Australia's aboriginal tribes. These paintings
depict the diversity of species which were hunted, or encountered during aboriginal man's everyday existence. They also show forms which were purely imaginary
and which were manifestations of phenomena which could not be readily
explained; among these the lightning man, *namarrgon*, is commonly seen.

One of the features of aboriginal rock art is an intense interest in the internal
structure of organisms. While such characteristics were undoubtedly a byproduct
of a definite gastronomic bent, there is little doubt that a commonality of form
between species was also recognized. Thus, reproductive organs, gastro-intestinal
systems, circulatory systems, muscle blocks, tendons and bones can be seen in
their X-ray art of vertebrates. In both aboriginal ceremony and myth, many of
these structures were antecedents of new biological forms; not a far cry from
Adam's rib. Equally, spiritual figures portrayed in ancient rock art, such as the
rainbow serpent, were implicated in the creation of new species. Here the interaction between this mystical reptile and the land forms it encountered on its

travels, initiated both the speciation process and spread the populations of man and other organisms. The oral history of these people portrayed in legend and ceremony confirms this creationism in its purest form; from the earth.

Yet, there is little difference between this and the formalized creationism which was introduced into the structured religions, be they Hellenism, Buddhism or Christianity, some 18 000 years later. Most of us are familiar with Christian creationist views on the origin of plant and animal species. The myths of the dreamtime exalted in aboriginal rock art and Old Testament views on the origins of life, both share a necessity for the intervention of an outside force to explain what could not be readily understood. While the creationists believe in a six-day miracle, at least the aboriginal dreamtime was temporally sequential, depending on the travels of the rainbow serpent.

The fact that these views are still being intoned and taken up by a growing number of people, tells us that science has failed to get its message across to a large section of the community: those who depend on faith and the literal interpretation of analogies as their guide to survival. Thus, the very strength of evolutionary theory, the vast database encompassing so many fields of biology, has become its greatest weakness. The fragmentation of approach to that common theory has resulted from an array of new techniques, each of which has established itself as an independent entity with its own discordant views on evolutionary processes and an extensive literature to support it. It might be seen as a sad comment that in the two hundred years since Linnaeus first described species as biological entities, there is still no universally accepted definition of what a species is. Nevertheless, this is a natural consequence of the fragmentation of disciplines and the particular requirements which each science has for a species definition. Any army would have considerable difficulty in defending itself when lines of communication are so overextended.

In the years which have elapsed since 1978 when Michael White wrote *Modes of Speciation*, a vast body of data has been published on every aspect of speciation theory. Among the many contributions, the most valuable, in terms of our understanding of the genetics of speciation, have come from two areas. The first of these involves the study of fertility and viability effects of structural chromosomal rearrangements on interracial and interspecific hybrids. This type of information can provide a direct assessment of the role of chromosome change in speciation. The second area includes the application of biochemical and molecular techniques which have only recently been developed, or applied to problems of plant and animal speciation. Multidisciplinary investigations can provide a most valuable perspective into the evolution of particular species complexes. For example, a lineage of chromosome races or species may be investigated chromosomally with standard preparations, C, G, Ag, or fluorescence banding with numerous fluorochromes using chromosomes derived from lymphocyte cultures. Liver and other tissues may be analysed electrophoretically for 60 presumptive loci, and the mitochondrial DNA (mtDNA) or ribosomal DNA (rDNA) may be sequenced or mapped with restriction endonucleases. Similarly, immunogenetic techniques such as the microcomplement fixation of serum albumin, which requires just a few drops of blood, can ascertain the relationships between these forms and establish the time of their evolutionary divergence. With any luck at all, studies on the comparative morphology of members of this complex can then be made. A suite of techniques providing alternative or corroborative phylogenies for the evolution of a species complex can give a direct assessment on both the degree and timing of genetic divergence and an array of perspectives on the evolution of that complex. Moreover, since all of these techniques can be applied to samples obtained from the same specimens, the impact of the findings is maximized. Comprehensive data are indispensable, and when independent data sets are corroborative in their findings, the conclusions which may be reached are often indisputable.

Despite the substantial body of new and exciting cytogenetic, biochemical and molecular data, and the accrued information of a hundred years of plant and animal cytogenetic research, *Modes of Speciation* was the last published volume which provided an insight into chromosomal speciation theory. Numerous recently published books or conference proceedings, have either downgraded the significance of chromosomal speciation, or simply ignored it in favour of speciation by genetic differentiation (see the introductory quotation to this chapter). Indeed, the volumes presented by this essentially North American coterie of like-minded geneticists would lead the reader to think that much of the data presented in this book did not exist. Consequently, I have attempted to correct this imbalance by providing a broad and integrated view on the concept of chromosomal speciation and the overwhelming amount of evidence which now supports it.

One of the recent developments associated with increased emphasis on genetic and molecular research into speciation, has been an amplified attack on the biological species concept (BSC). This concept remains as the most used species definition despite having been under relatively constant criticism since its introduction. The more recent arguments used against the BSC originated from two core areas. First, the application of biochemical and molecular techniques has revealed the existence of numerous genetically distinct populations within species. Many systematists with a cladistic interest feel that distinct populations of this type should be recognized as species. Consequently, there is a push from this direction to recognize evolutionary and phylogenetic species concepts. The second area of attack on the BSC was made by those who view species as having evolved by purely allopatric means, involving the gradual accumulation of genetic differences. These authors have argued against the concept of chromosomally induced speciation and the likelihood that such changes can form reproductive isolating mechanisms. They also consider such mechanisms to be incidental to the speciation process. Since this is the very basis of the BSC and some evolutionary species concepts, this can be viewed as a significant area of conflict. In the second chapter of this book, species concepts are considered and the question of the most appropriate species concept is addressed.

In the third chapter, particular aspects of the process of allopatric speciation in isolated populations are examined. The key question as to how reproductive isolation can be attained by genic differentiation is approached from three perspectives. First, a basis for comparison is made by defining our understanding of what reproductive isolation is and by describing the nature of genic variation within and between speciating lineages. The second approach has used the large number of hybridization and genetical analyses in *Drosophila* to establish a relationship between genic and reproductive isolation. The third perspective briefly examines the possibility of speciation with major genetic divergence and an apparent absence of reproductive isolation. This is contrasted with rapidly attained reproductive isolation in some flowering plants.

A contrasting form of allopatric speciation involving the revolutionary conditions associated with founder populations is considered in the fourth chapter. Not only do these conditions provide for rapid genetic differentiation, but they also introduce the likelihood of the fixation of chromosomal rearrangements which can act as reproductive isolating mechanisms. The concept of the founder population is most important to the possibility of chromosomal speciation. It is not surprising that the founder effect is itself a matter of debate. The different models of founder effect and related concepts are discussed and the predictions, in terms of population structure and impact on genetic variability, are assessed.

One of the major problems facing evolutionary biologists has been the appreciation of what types of chromosomal rearrangements are associated with speciation. Too many attacks have been made on chromosomal speciation concepts by those who are unaware of a distinction between balanced polymorphisms and negatively heterotic rearrangements. In the fifth chapter of this volume, those chromosomal rearrangements which may be and also those which cannot be involved in the speciation processes are considered. In the second part of this chapter, theoretical studies which have used chromosomal databases encompassing a considerable variety of structural rearrangements, and which have been widely quoted as supporting or refuting chromosomal speciation hypotheses, are also examined.

The sixth chapter is subdivided into two basic areas which explain aspects of the fixation of chromosomal rearrangements in isolated populations. The first of those is to consider factors responsible for the predominance and high incidence of particular forms of chromosomal rearrangements in chromosomally speciating complexes, such as mutation rate, the direction of change and meiotic drive. The second section considers some of the theoretical models which have been advanced to explain how negatively heterotic rearrangements can reach fixation in isolated populations.

The great body of evidence derived from chromosomally speciating lineages is presented in Chapters 7 and 8. In the former, the fertility and viability effects of structural rearrangements are examined in two situations. The first of these considers spontaneous mutations in domesticated species. The second, and largest, section involves the consideration of natural and laboratory hybrids between members of chromosomally speciating lineages. The impact of chromosomal rearrangements on the fertility and viability of these hybrids is described in considerable detail.

This theme is followed in Chapter 8 where complexes of species which have been examined both chromosomally and genetically by either protein electrophoresis or some form of molecular analysis are compared. A simple framework for comparison has been constructed, where lineages of species are categorized as ancient or recent. The simplest of questions is asked. In these speciating complexes, does the evidence suggest whether the chromosome race has been formed before or after genetic differentiation? That is, an attempt is made to determine the primacy of chromosomal divergence in an allopatric population, and thus establish chromosomal speciation.

Chapter 9 examines the major models which have been advocated for chromosomal speciation and discusses their relative advantages and disadvantages. These are divided into three categories, based purely on the proposed area of origin. That is, whether the speciation event occurred within or outside of the parental species distribution, or within the hybrid zone. Twelve different models for internal and external chromosomal speciation are critically examined. Similarly, three different modes of speciation in hybrid zones are considered.

The development of molecular techniques and their application to speciation research has greatly changed our understanding of that process and previously unheard of evolutionary mechanisms have been described. Chapter 10 discusses the major areas of molecular research with a view to explaining existing cladogenic processes. Two questions are asked. First, can molecular turnover mechanisms provide processes by which indistinguishable populations can establish postmating isolating mechanisms which are sufficiently powerful to enable speciation to occur? Second, have molecular mechanisms been detected which can enhance the formation and fixation of chromosomal rearrangements, or genetic divergence, thus providing additional support for existing modes of speciation? Answers to these questions were developed through the analysis of the pattern of concerted evolution and the process of molecular drive. The evolutionary impact of genomic turnover mechanisms is considered.

The eleventh and last chapter provides a résumé of the concepts and conditions necessary for chromosomal speciation and touches briefly on the analysis of hybrid zones, on the role of chromosomal evolution in macroevolution, and the evolutionary significance of chromosomal speciation.

Hopefully, this volume will correct the perceived imbalance in speciation theory which currently exists. Chromosomal speciation remains as one of the two most significant forms of speciation and, as the growing body of evidence now shows, it is one of the most common.

2
The species – what's in a name?

The unit of evolution is the terminal taxon, the isolated interbreeding population which is an objective reality. All living beings belong to a terminal taxon, but whether or not a given species is a terminal taxon is unpredictable, being dependent on future discovery.

(Løvtrup, 1979, p. 388)

Perhaps the short quotation from Løvtrup (1979) which introduces this chapter best sums up the dilemma faced by biologists and does so in the simplest possible way. This most basic of biological problems, that of defining what a species is, appears to be no more resolvable today than it was two hundred years ago. Indeed, it could be argued that the situation is deteriorating, since with the development of sophisticated biochemical and molecular techniques and their application to biological populations, the nexus between a morphologically recognizable species and that isolated, interbreeding population which is Løvtrup's 'terminal taxon' is being forced further and further apart.

It is no exaggeration to say that the history of science is littered with discarded species concepts. This might well be due to the fact that our perception of what a species is necessarily changes with the additional knowledge which we have gained. While there is little doubt that Agassiz' (1857) view of a species as 'thoughts of the creator which are real' would gain little acceptance today, at least in the scientific community, it might be surprising that Darwin's own view of a species may suffer the same fate. Darwin regarded species as open systems with fluid borders which could only be subjectively delimited. The different species concepts recognized today all involve closed systems with what are at least generally definable borders.

It is not the intention of this chapter to provide an historical view of what a species is. Mayr (1963, 1982a), Dobzhansky (1970), Ghiselin (1974) and Slobodchikoff (1976) can provide the reader with a surfeit of information on this subject. It is more important to understand the species concepts which are either in current use, or which have at least been proferred as usable models. Not all of these will be included here since the aim of this chapter is to provide a working basis for this volume. Thus, an attempt is made to define the biological species concept and to integrate the ramifications of criticisms made against this model. Equally, the validity of certain possible replacements for the BSC are critically examined, as are alternative approaches such as the evolutionary species concept and derivations of this.

2.1 The Linnaean species, morphology and systematics

It is no idle comment to say that Linnaeus (1707–78) was the father of modern systematics. He introduced the facility of binomial nomenclature, and the view that every organism must belong to the lowest taxonomic entity, the species. Nevertheless, while Linnaeus put forward the basic framework for classification and used it to categorize vast numbers of organisms, he was also a creationist who maintained that biological species were real entities created by God.

> There are as many species as the infinite being created diverse forms in the beginning, which, following the laws of generation, produced as many others but always similar to them: Therefore there are as many species as we have different structures before us today.
>
> (Linnaeus, 1751, in Mayr, 1982a, p. 258).

A consequence of this attitude is the belief that species are well-defined and constant in form. While this is out of tune with both evolutionary theory and observed variation, there is little doubt that if man is to hope to catalogue the organisms present around him, he can only reasonably describe those entities which are perceived as being different. Thus, the vast majority of species which were described during the drive to categorize the world's fauna, which occurred in the nineteenth and early twentieth centuries, relied on the morphological species concept. Indeed, most species described by today's taxonomists are also morphological species described under the rules of the International Commission of Zoological Nomenclature or its botanical equivalent.

The date 1 January 1758, was arbitrarily fixed by the zoological code as the starting point of zoological nomenclature, for this date coincided with the publication of Linnaeus' *Systema Naturae* (10th edition) and Clerck's *Aranei Svecici*. Zoological nomenclature is the system of scientific names applied to taxonomic units of animals known to occur in nature. The code regulates the naming of taxa in the family, genus and species groups. Names regulated by the code are attached to a name-bearing type specimen, and in the case of the species this is the holotype. The code was introduced as a means of regulating nomenclature following the confusion which occurred after the introduction of Linnaeus' binomial naming system and was simply a means of regulating priority and removing synonymy.

Two of the definitions which have been used to define the morphospecies are 'established by morphological similarity regardless of other considerations' (Cain, 1954), and in a parody Mayr (1963, p. 31) suggested that:

> Natural populations considered by general consent to be species are morphologically distinct. Morphological distinctness is thus the criterion of species rank. Consequently, any natural population that is morphologically distinct must be recognized as a separate species.

With open-ended definitions of this form the morphological species concept posed a suite of typological problems. In some cases, polymorphic variants within a species have been defined as species; in others different sexes of a species have themselves been defined as species; and further, morphologically different populations have been regarded as species.

Mayr (1963) attacked the use of morphospecies because the concept ignores the primary role of reproductive isolation and concentrates on the secondary role of morphological differences. He argued that the vulnerability of the purely morphological species concept in sexually reproducing species can be shown by:

1 The presence of conspicuous morphological differences among conspecific individuals and populations (intraspecific variation). That is, where there are often greater differences between individuals, or populations, than between related species.
2 The virtual absence of morphological differences among some sympatric populations (sibling species), that otherwise have all the characteristics of good species, being reproductively isolated from each other.

Nevertheless, despite these anomalies the morphological species remains as the baseline for the taxonomic description of our flora and fauna. They must remain so for they are tangible, recognizable units which can be diagnosed by their innate form. However, most modern biologists also place the species they deal with in the context of their biological distinctiveness. Thus, while such biological species may not be described as taxonomic entities, they are sometimes described as biological entities.

2.2 Biological species concepts

2.2.1 The biological species concept (BSC) (Mayr, 1942, 1963)

Any definition of a species must cope with the fact that speciation is an ongoing process and, as such, the analysis of existing species at any instance of time must reveal a continuum of variation, ranging from species with profound morphological differences between them, through every stage of differentiation, to those populations which are beginning to diverge and are showing the earliest stages of doing so. Traditional systematists who only recognize morphological species because of the self-imposed limitations of the Linnaean binomial system, would regard all of these entities, including cryptic and sibling species, as intraspecific variation. Such an approach ignores the biological reality of much of the variation we encounter.

The emergence of the biological species can be traced back to the early nineteenth century and this has been done most eloquently by Dobzhansky (1970) and Mayr (1982a). The key decision which created the BSC was the recognition by Dobzhansky (1937) that the process responsible for species formation was the development of reproductive isolating mechanisms. A host of definitions have been produced based on this most fundamental character:

> Species are groups of actually or potentially interbreeding natural populations which are reproductively isolated from other such groups
>
> (Mayr, 1942, p. 120).

['actually or potentially' was deleted from Mayr's 1969 definition].

and

> That stage of the evolutionary process at which the once, actually or potentially interbreeding array of forms becomes segregated in two or more separate arrays which are physiologically incapable of interbreeding
>
> (Dobzhansky, 1937, p. 312).

and

> Species are ... systems of populations; the gene exchange between these systems is limited or prevented in nature by a reproductive isolating mechanism or perhaps by a combination of several such mechanisms
>
> (Dobzhansky, 1970, p. 357).

and

> The sum total of the races that interbreed frequently or occasionally with one another, and that intergrade more or less continuously in their phenotypic characters, is the species
>
> (Grant, 1963).

White (1978a) accepted the basic concepts advanced by Mayr (1963) and Dobzhansky (1970), but also emphasized in the strongest terms that a species 'was at the same time a reproductive community, a gene pool, and a genetic system'.

All of these definitions involve the same basic principles, but that which is almost universally used and most often referred to as the BSC is Mayr's (1942) definition.

Biological species have the following characteristics (Mayr, 1963):

1 Species are defined by distinctness rather than by differences.
2 Species consist of populations rather than unconnected individuals.
3 Species are not defined by the fertility of individuals but by the reproductive isolation of populations.
4 Species are reproductive communities in which individuals of animal species recognize potential mates for reproduction.
5 The species is an ecological unit that, regardless of the individuals composing it, interacts as a unit with other species with which it shares the environment.
6 The species is a genetic unit, a gene pool, whereas the individual is a temporary vessel holding a portion of the gene pool for a short time.

The BSC was proposed as a concept which would apply to sexually reproducing species and it had no application to asexually or uniparentally reproducing organisms.

The critics of the biological species concept have been many and varied. Most of the earlier criticisms centred on the difficulty of applying the BSC as a taxonomic tool. For example, Grant (1957) in a survey of 11 Californian plant genera found that discrete recognizable species were a small fraction of the forms analysed, most of which were problematical and ill-defined. Similarly, Ehrlich (1961) analysed nearctic butterflies and commented that the presence of clearly defined species was a myth, since over one half of the genera resisted partitioning. Even when presumptive species were sympatric, partial interbreeding made a clear-cut decision on their status impossible. Ehrlich opted for the abandoning of the BSC and the adoption of a numerical approach to taxonomy.

The BSC was attacked by Simpson (1961) on its failure to cope with temporally sequential species and uniparental organisms (despite the fact that these were excluded from the concept). Simpson also pointed to the difficulty in assessing the degree of reproductive isolation between species. That is, while some taxonomists insist on absolute and permanent reproductive isolation between species, most hold the view that 'good' species are present if the hybrids produced are infertile. The arbitrary nature of the BSC in determining the degree of sterility was also objected to by Sokal and Crovello (1970), who emphasized that if a hybrid is produced in nature from two species and there is any backcrossing at all, then by a strict application of the BSC the two parents belong to the same species, even if such hybrids appear in only a small part of the range of the species. They suggest that this approach is rarely taken, since investigators establish an arbitrary level of interbreeding which they will tolerate before assigning plants or animals to the same species. Sokal and Crovello also pointed out that the BSC was typological because it was defined by strict genetic criteria which were rarely tested and which were not necessarily met by the constituents. Sokal and Crovello argued that a purely phenetic approach to the definition of a species was desirable.

Many of the critics of the BSC adopted their position because of an unwillingness to accept that reproductive isolation should be used as a criterion for a species definition. In some cases this stance was taken because definitions were incompatible with cladistic theory, or had grey areas in this regard (Wiley, 1978; Cracraft, 1983; Frost and Hillis, 1990). For example, Frost and Hillis (1990) found it difficult to use the BSC because in their view reproductive compatibility among populations was a shared primitive feature and they could only make meaningful conclusions using shared derived characters. This would appear to be an ill-considered view, since reproductive isolation, the feature which distinguishes derived species, is a derived character. If one considers species as interbreeding populations, reproductive compatibility is a shared primitive feature, but reproductive isolation of a population from other populations is a derived character.

Paterson (1985) made a fundamental assault on the BSC which he termed the 'isolation' concept. Paterson pointed to the fact that the BSC was basically a negative concept involving the formation of isolating mechanisms. In this respect he regarded the BSC as being misleading, since the term 'isolating mechanism' implies that isolation is the evolutionary function of those traits which define a species. Moreover, he reasoned that because allopatric speciation is the predominant cladogenic process, all differences which are established between species are achieved in isolation. Paterson argued that a logical extension of this point is that those phenotypes which are responsible for isolating mechanisms between species could never have been functioning as isolating mechanisms when they were established, because parental and daughter populations were formed in isolation. Consequently, isolating mechanisms are a product of isolation, rather than being a causal mechanism responsible for the production of new species in isolation. Paterson also argues that those who support the BSC have been seduced into thinking that isolation has been selected for despite logical and mechanistic difficulties involved in such a process. These include the inherent difficulty in selecting for sterility and the numerous models which have been generated to show that this is so. Paterson asserts that sterility can only be acquired as an accidental consequence of other changes.

The arguments presented by Paterson (1985) on selection for isolation, were largely supported by Templeton (1987, 1989), and are dealt with in Section 2.2.4.

2.2.2 The biological species concept mark II (Mayr, 1982a)

In an attempt to overcome some of the more trenchant criticisms of the BSC, Mayr (1982a) presented a more descriptive definition of a biological species: 'A species is a reproductive community of populations (reproductively isolated from others) that occupies a specific niche in nature' (Mayr, 1982a, p. 273).

Clearly, the introduction of 'niche' has been used to broaden the original BSC into a more generalized concept and thus account for asexual species: 'speciation is not completed by the acquisition of isolating mechanisms but requires also the acquisition of adaptations that permit the coexistence with potential competitors' (Mayr, 1982a, p. 275). Further, Mayr reasoned that the major biological meaning of reproductive isolation is that it provides protection for a genotype adapted for the utilization of a specific niche. Undoubtedly, these changes were designed to allay some of the criticisms which have been directed at the BSC. Nevertheless, the addition of qualifiers such as 'reproductive community' and the concept of 'niche' have introduced a series of unquantifiable characteristics. This step has not been accepted by many biologists, and in a sustained attack Hengeveld (1988) laid bare the problems with the revised BSC. Hengeveld attacked this definition from four directions:

1 The concept of the niche in ecology is typological, and on the basis of Mayr's own views on typological evidence, this should not be used.
2 There are fundamental problems in defining the concepts of 'niche' and 'population'.
3 Since processes operate on different spatio-temporal scales, clear-cut ecological discontinuities cannot be drawn between species.
4 Including the concept of 'niche' in species definitions would restrict such definitions to animal species only.

To paraphrase Hengeveld: adopting the concept of niche within biological species definitions is not recommended.

2.2.3 The recognition species concept (Paterson, 1978, 1985)

In 1978 Paterson produced what must be regarded as a highly contentious view of the species, although it has had limited support (Carson, 1985; Spencer *et al.*, 1986, 1987). What is remarkable is the time taken for such a view to surface, after the general acceptance of the biological species concept for some 50 years.

Paterson introduced his recognition concept because of a difficulty in reconciling characteristics of mosquito species evolution within the constraints of a biological species. In essence, the recognition concept proposes that sexual reproduction is fundamental and that species are an incidental consequence to the evolution of sex. That is, species arise by a fertilization system adapting to a new habitat in order to ensure effective fertilization in that new habitat.

Paterson's model is based on fertilization mechanisms in biparental organisms (to which it is restricted), and the interactions between mating partners. The essential features of the recognition species concept are as follows:

1 The members of a species share a specific mate recognition system (SMRS) to ensure effective syngamy within a population of organisms occupying their preferred habitat.
2 The characters of the SMRS are adapted to function efficiently in this preferred habitat.
3 A new species arises when all members of a small, isolated subpopulation of a parental species have acquired a new SMRS. This facilitates syngamy under the new conditions and makes effective signalling impossible between daughter and parental populations. Thus, speciation is the incidental consequence of adaptation to a new environment.

Paterson describes the specific mate recognition systems as signal and response interactions. They include such fundamentals as pollen/stigma interactions and sperm/ova interactions, and in many respects they are analogous to antibody/antigen interactions. SMRS interactions do not include those between flowers and pollinators, which must be regarded as vector systems.

Paterson (1985, p. 25) defined a species as 'The most inclusive population of individual biparental organisms which share a common fertilization system'. It is axiomatic that such a fertilization system determines:

1 The limits of gene exchange.
2 That members of a species mate positively assortatively through the functioning of their SMRS.
3 That effective syngamy occurs within a population of a species occupying a preferred habitat and this is ensured by the SMRS.
4 That limits to the field of gene recombination exist.

It is fundamental to the recognition concept that the SMRS not only ensures mating, but that it also prevents hybridization with other species. That is, when mating recognition occurs, hybridization is impossible. Paterson regards parapatrically distributed allopatric species which form a narrow hybrid zone between them as being conspecific, since anything that hybridizes cannot be a separate species. The systematic implications of this concept are overwhelming.

It is now worth taking a closer look at some of the problems associated with this view of a species. Butlin (1987) directed his attack on the concept solely at the hybridization standpoint, for the logic of regarding two species as being conspecific if they produce a hybrid with zero fitness is hard to justify. Butlin felt that the recognition concept was unsuitable since it defines species by an unknown future potential. Equally, it ignores those many species which have a stable coexistence in parapatry. These species do not exchange genes and therefore remain distinct, or diverge, despite having overlapping or abutting distributions. Butlin (1987, p. 462) made a decisive comment 'Species should be defined by the absence of gene flow, whatever the characters responsible for its prevention.'

In an avid defence of the recognition concept Spencer et al. (1987, p. 960) argued 'that post-mating isolation should be relegated to the same (intraspecific) category as geographical isolation. The effects are the same: They both provide a temporary and leaky genetic independence quite different from the permanent independence resulting from speciation' and further 'Indeed, the continual gene flow across most, if not all, hybrid zones (Barton and Hewitt, 1981) seems to us to be good evidence that such populations should be considered conspecific. In fact, the amount of gene flow between such populations is certainly greater than between populations that have no post-mating barriers and occur on distant oceanic islands'.

One of the most common means of speciation in flowering plants is by the process of polyploidization, the duplication of the whole genome. The attainment of polyploidy is often regarded as being instant speciation, since the diploid ancestor and its polyploid daughter species are reproductively isolated by meiotic problems, and either total sterility or semisterility are caused by the asymmetrical ploidy of F1 hybrids. Paterson (1987) regards autopolyploids as being conspecific, since reproductive isolation between them and their direct ancestor is sometimes not complete, and because the genes of the diploid are simply the same as the polyploid although varying in dose. By the same token, he could not validly recognize allopolyploids, because they are the product of hybridization and thus the parental species would have also been conspecific because recognition systems had operated. A large proportion of the plant kingdom and a number of animal species currently regarded as good biological and taxonomic species would therefore be regarded as conspecific with diploid ancestors under Paterson's definition.

This would be so despite the absence of gene flow between them and fixation of substantial morphological and/or genic differences.

The recognition concept of speciation is both divisive and biologically inexact. Its ramifications are profound, for many of the world's species are not sympatric in distribution, but are either completely allopatric, or allopatric but with a parapatric boundary between species. Nevertheless, they are morphologically recognizable entities which show either a restriction, or complete absence, of gene flow with their parapatric neighbours, the level of this depending on the degree of hybridization in the parapatric contact zone. The recognition concept would lead to the complete taxonomic destabilization of groups where any degree of hybridization occurred, for they would be regarded as being conspecific. Indeed, the horses provide the best example of this dilemma. The seven extant species are allopatrically distributed and in certain situations produce hybrids which are quite viable (mules and hinnies from donkey × horse matings), yet generally infertile (see Section 7.3.4). Does this mean that the many zebras, donkeys and horses are all part of the one species because gametes recognize one another and produce occasional viable zygotes? The recognition concept would suggest that all of the species are conspecific because they form sterile F1 hybrids. The same argument can be applied to the many species of *Drosophila* which can be artificially hybridized. For example, 266 hybridizations between 191 *Drosophila* species were reported by Bock (1984). The consequence of the recognition concept would be to insist that these species were also conspecific.

The legion of inconsistencies present in the recognition concept renders it unacceptable as a usable species definition.

2.2.4 The cohesion species concept (Templeton, 1989)

Templeton (1989) argued that isolation was an irrelevant function in the process of speciation, since it is a negative phenomenon which could not be selected for. He suggested that it was more important to focus attention on the functions of pre-mating mechanisms which facilitate reproduction. Templeton also argued that the isolation function could arise as a byproduct of other reproductive functions, but in general isolation was not an active part in the process of speciation. Consequently, isolating mechanisms were a misleading way of thinking about the process of speciation. Paterson (1978, 1985) had focused attention on the positive function of these mechanisms in facilitating reproduction. That is, a species is a field of recombination and the limits of this field are fertilization mechanisms. Templeton (1989) regarded the stance adopted by Paterson as being superior to that of the BSC, despite the profound criticisms which have been directed at Paterson's views.

In attempting to develop the same approach, Templeton defined a species in terms of genetic and phenotypic cohesion as 'the most inclusive population of individuals having the potential for phenotypic cohesion through intrinsic cohesion mechanisms' (Templeton, 1989, p. 12).

This concept was drawn from the evolutionary, BSC and recognition concepts and has a mechanistic focus like the BSC, although the emphasis is on cohesion mechanisms rather than isolating mechanisms. The cohesion mechanisms that define species status are those that promote genetic relatedness and determine the population boundaries for the actions of evolutionary forces. Templeton defined two basic classes of cohesion mechanisms and he termed these genetic exchangeability and demographic exchangeability. Genetic exchangeability includes those factors which define the limits of the spread of new genetic variants through gene flow. These may act either by promoting identity by gene flow, or preserving identity through lack of gene flow (i.e. isolating mechanisms). In contrast, demographic exchangeability includes those factors which define the fundamental niche and determine the limits of spread of new genetic variants through genetic drift and natural selection. That is, the limits of replacement or displacement of alleles within a population (Fig. 2.1).

A number of advantages of the cohesion concept over the recognition and BSC were perceived by Templeton (1989) and these include:

1 A range of cohesion mechanisms define a species, rather than gene flow which is the major component of the alternative models.
2 The cohesion concept can be applied to a range of organisms which have diverse reproductive strategies and lifestyles.
3 A 'good' species could be defined as one with distinct levels of genetic and demographic exchangeability. Templeton regarded this as an important point since the boundaries defined by genetic and demographic exchangeability are different, and the other species concepts only recognize the former of these.
4 The cohesion concept clarifies the evolutionary significance of sympatric models of speciation, for the evolution of demographic non-exchangeability triggers the speciation process.
5 Since a broad set of 'microevolutionary' processes are involved in a species definition, natural selection can be dealt with as a general principle rather than by having to refer to the effects on gene flow.

The question is, of course, is the cohesion concept sufficiently different from the BSC to make it worthwhile? While emphasis has been placed on characteristics which make a species a cohesive unit, the majority of these have been included in the BSC, or are at least tacitly accepted as being a part of that concept. That is, those characters loosely termed as demographic exchangeability; including drift and natural selection and the realizable boundaries of a population.

In many respects the cohesive concept is purely an optimistic rewriting of the BSC, with an emphasis on those factors which hold a species together rather than those which isolate it from the next. But even so, the overriding emphasis on gene flow and the factors which define the limits of the spread of new genetic variants (i.e. genetic exchangeability) gives one the impression that this is the BSC in a new guise. All factors which are defined as promoting identity, such as fertilization systems, developmental systems and the processes referred to as demographic exchangeability, are cohesive forces, whereas, isolating mechanisms promote identity by the absence of gene flow and are therefore divisive. Indeed, isolating mechanisms are listed as one of the two key mechanisms in promoting genetic identity by preventing gene flow. By implication, isolation mechanisms can be significant as a major cladogenic mechanism. There seems to be no worthwhile difference between this concept and that of the biological species. Indeed, Templeton asserted, when talking of sympatric speciation, that it is 'the evolution of demographic non-exchangeability that triggers the speciation

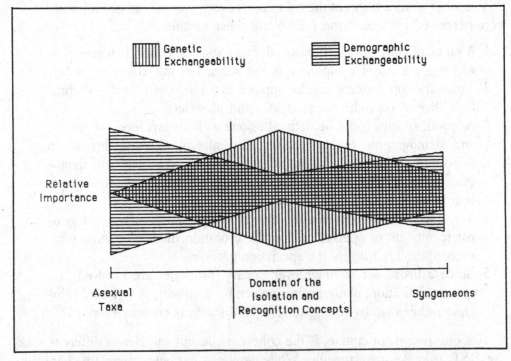

Fig. 2.1. Templeton (1989) provided this diagram to demonstrate his view on the relative importance of demographic and genetic exchangeability in the cohesion species concept. (Redrawn from Templeton, 1989.)

process'. This reads remarkably like the absence of gene flow and the presence of isolation.

A series of significant criticisms of the cohesion concept were made by Endler (1989), who attacked the premise that gene flow holds species together by homogenizing allele frequencies and co-adapted gene complexes. He suggested that these mechanisms have been discredited by the isolation by distance concept and the fact that homogenization only occurs if alleles are selectively neutral. Endler also argued that this concept was operationally unusable and can degenerate into a phenetic species concept, since deciding what a species is, is a most difficult process with a cohesive definition.

2.3 Evolutionary species concepts

'Species do evolve, and almost always do so gradually. Among evolutionary species there cannot possibly be a general dichotomy between free interbreeding and no interbreeding. Every intermediate stage occurs, and there is no practically definable point in time when two intraspecific populations suddenly become separate species' (Simpson, 1961, p. 152).

Evolutionary species concepts were introduced because the biological species concept could not be applied to temporally sequential species or uniparental organisms. In its original form, the evolutionary species concept was proposed as a palaeontological concept by Simpson (1961). However, subsequent derivative definitions changed the intent and validity of Simpson's view to the extreme.

2.3.1 Simpson's (1961) evolutionary species concept

Simpson felt that it was necessary to have a broader definition that related the genetical species (his term for the biological species) directly to the evolutionary process, and which would thus include applicability to fossil material. He proposed the following broad definition:

> An evolutionary species is a lineage (an ancestor–descendant sequence of populations), evolving separately from others and with its own unitary evolutionary role and tendencies

> (Simpson, 1961, p. 153).

Since this concept was designed to cope with palaeontological sequences, it viewed species as temporal lineages, the constituents of which changed with time. Thus, a name could be fixed to a phenotypically distinct form within a lineage.

Another name could be fixed to a subsequent form in the same lineage at a later time. These 'successive species' were a direct reflection of phyletic evolution, an approach which has not been accepted by today's phylogeneticists, who regard speciation as a dichotomous event.

Since biological species reflect morphological divergence between lineages at any one time, Simpson reasoned that at any point in time there would be a correspondence between biological and evolutionary species in biparental organisms (see Fig. 2.2).

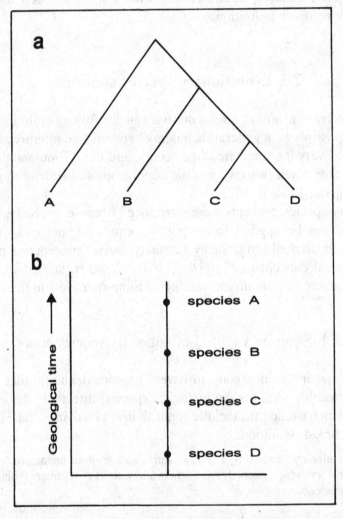

Fig. 2.2. a. A lineage of biological species showing that speciation is accompanied by a dichotomous branching event. b. The evolutionary species concept in Simpson's (1961) terms, suggested that a single lineage could change from species to species through geological time.

An additional aspect of the evolutionary species concept was the possibility that two species could interbreed without losing their distinctive evolutionary roles. Most significantly, the amount of interbreeding allowable was precisely that much as does not 'cause their roles to merge'. Clearly, since uniparental populations have evolutionary roles, these were included under the definition of an evolutionary species. Simpson equated 'roles' with 'niches', and a niche was seen as a way of life, or relationship that a population of animals had to the environment.

It is of some interest that Mayr (1982a) criticized the incorporation of the 'unitary evolutionary role' into a species definition, since such a character was immeasurable. Mayr (1982a) had himself incorporated the concept of the 'niche' into his revised definition of the BSC and met with a similar level of criticism from Hengeveld (1988).

Simpson also emphasized that different genetical species which lacked any determinable anatomical or ecological distinction should be regarded as single species, since they did not have a definable separate evolutionary role. Polytypic and sibling species were therefore abolished under the evolutionary species concept, although Simpson went to some length to argue for the inclusion of subspecies. Nevertheless, even though the deme might be the basic population unit, Simpson firmly rejected any possibility that these should enter into classification or nomenclature.

The first of the two major criticisms which Mayr (1982a) made of Simpson's concept, centres on the fact that it is basically a typological description and ignores the possibility of cryptic and polytypic species. Secondly, he pointed out that evolutionary species definitions in general, minimize factors responsible for the causation and maintenance of discontinuities which occur between species, and concentrate on problems of how to delimit multidimensional species taxa. While these criticisms reflect Mayr's own special areas of interest, the most fundamental criticism is that of multiple species occurring in a simple non-branching lineage.

2.3.2 Wiley's (1978) evolutionary species concept

In an attempt to resurrect and defend the species concept of Simpson (1961) as best suited for dealing with the species and its origins, Wiley (1978, p. 18) changed the definition of the evolutionary species to:

> A species is a single lineage of ancestral descendant populations of organisms which maintains its identity from other such lineages and which has its own evolutionary tendencies and historical fate.

This new definition has the following advantages (Wiley, 1978):

1 It implies that the species is the most inclusive unit of evolution.
2 It does not imply that species must change; a species 'maintains its identity' rather than 'evolves'.
3 'Identity' in the context of this definition means individual identity and does not infer either stasis or change in morphology.
4 Species are thought of as individuals rather than classes (a reference to the theories of Ghiselin (1974)).
5 Species are historical, temporal and spatial entities.
6 Whether a group of organisms is or is not a species becomes an hypothesis to be tested. Evidence which can be used to test this proposition is derived from genetic, phenetic, spatial, temporal, ecological, biochemical and/or behavioural sources.

Wiley (1978) then delimited the following corollaries which were implicit in his definition.

1 All organisms, past and present, belong to some evolutionary species.
2 Separate evolutionary lineages (species) must be reproductively isolated from one another to the extent that this is required for maintaining their separate identities, tendencies and historical fates.
3 The evolutionary species concept does not demand that there be morphological or phenetic differences between species, nor does it preclude such differences.
4 No presumed separate, single evolutionary lineage may be subdivided into a series of separate ancestral and descendant species.

It is noteworthy that Wiley (1978), like Simpson before him, argued that the evolutionary species concept included asexual species complexes. Moreover, Wiley concluded that in both allopatric demes and asexual species the absence of divergence between entities requires that they should be grouped into single species.

While both Simpson's and Wiley's definitions are superficially most similar, a close examination of the corollaries to Wiley's concept, changes our appreciation greatly. Thus, Wiley reasoned that all species past and present belong to the same evolutionary species and that no lineage may be subdivided into ancestral and descendant species, whereas, Simpson went to some lengths to define successive species within lineages, which should have morphological differences between them, at least as great as sequential differences among contemporaneous species of the same group, or closely allied groups. Second, Wiley argued that his evolutionary species concept did not demand morphological or phenetic differences between species, whereas Simpson was quite clearcut in the necessity for defining such characters in an evolutionary species.

It is clear that Wiley had in many respects made these changes to comply with our understanding of more recent evolutionary theory and the application of modern practices and principles (phylogenetic classifications are incompatible with phenetic classifications). In addition, he had removed the 'unitary evolutionary role', ostensibly to alleviate the criticism associated with niche descriptions. Nevertheless, the inclusion of morphologically undefined species (which alleviated Mayr's criticism of the neglect of sibling species and polytypism), greatly changed Simpson's original concept of an evolutionary species and can only result in confusion. Indeed, this step opened a Pandora's box of nomenclatorial problems. Naming morphologically indistinguishable species, could not only lead to taxonomic mayhem, but could also further erode the correspondence between biological and evolutionary species which Simpson (1961) cited as a reality at any point in time.

It is noteworthy that Hecht (1983) and Hecht and Hoffman (1986) expressed the view that none of the criteria for the recognition of evolutionary species could be unequivocally used. Not only is this due to the gaps in the fossil record, which make it impossible to distinguish between cladogenic events and migration, it is also due to the underlying assumption of the evolutionary species concept, that extinct species are comparable to extant morphospecies. The premise that morphospecies approximate actual biological species is without foundation.

Mayr (1982a) attacked the typological approach adopted by cladists such as Wiley, who regard species as the segment of any phyletic lineage between two branching points and ignore reproductive isolation and its significance.

> The principal weakness of so-called evolutionary species definitions is that they minimize (if not ignore) the crucial species problem, the causation and maintenance of discontinuities between species, and concentrate instead on the problem of how to delimit multidimensional species taxa. Yet they do not even meet the limited objective of how to delimit such open-ended systems
>
> (Mayr, 1982a, p. 294).

2.3.3 The ecological species concept (Van Valen, 1976)

Van Valen (1976) proposed his ecological species concept because of what he perceived as a discordance between the biological species concept and particular organisms such as the American oak trees (*Quercus*). Van Valen described his view of the species as being 'radical', and modified Simpson's evolutionary species to fit in with his own requirements. Three overriding principles governed Van Valen's view of the species: first, that genes were of minor importance in evolution and should only be considered as molecules; second, that the control of evolution was ecological and under the constraints of individual development; and, third, selection acts primarily on genotypes.

Thus, Van Valen's ecological view of an evolutionary species was defined as:

> A lineage (or a closely related set of lineages) which occupies an adaptive zone minimally different from that of any other lineage in its range and which evolves separately from all lineages outside its range
>
> (Van Valen, 1976, p. 233).

Van Valen also added the following qualifications:

1 A population was a group of individuals in which adjacent individuals at least occasionally exchanged genes reproductively and did so more frequently than with individuals outside the population.
2 Lineages were closely related if they had occupied the 'adaptive zone' since their latest common ancestor. If that zone had changed, they were closely related if new adaptations had been transferred among lineages rather than originating separately in each.
3 An 'adaptive zone' was some part of the resource space together with whatever predation and parasitism occurred on the group considered. It was part of the environment and existed independently of any inhabitants it might have.
4 'Range' was both geographic and temporal.
5 The 'occupation' of an adaptive zone was a difference in population density; a species could occupy more than one adaptive zone.
6 Reproductive isolation was of minor evolutionary importance and needed little consideration.

One of the major problems associated with Van Valen's view of a species is the difficulty in quantifying such characteristics as 'adaptive zone' and 'range'. On these grounds alone such a definition would appear to be largely unworkable. Van Valen's ideas on the significance of genic change and reproductive isolation appear to be unorthodox.

Two criticisms of this concept were made by Wiley (1978). First, species do not have to occupy minimally different niches or adaptive zones from other species to be regarded as species: two species can occupy the same niche. Second, on the basis of Van Valen's definition, one could argue that a species forced to extinction in a particular habitat as a result of interspecific competition, was not really a species at all. This may occur despite the fact that the separate species were reproductively isolated.

2.3.4 The phylogenetic species concept (Cracraft, 1983)

A perceived failure of the BSC to resolve the pattern and process of taxonomic differentiation and its emphasis on reproductive isolation led Cracraft (1983) to propose his model. He reasoned that if reproductive isolation was not regarded as the central issue in taxonomic differentiation, then differentiation was the primary act and interbreeding was secondary. A number of other similar concepts have been proposed (Rosen, 1978; Donoghue, 1985; Echelle, 1990); however, Cracraft's model will be discussed in detail here.

Cracraft (1983) argued that taxonomic units are recognized by the possession of unique phenotypes, and that these are not necessarily morphological, but may also include recognizable biochemical, physiological or behavioural attributes. It is these intrinsic characteristics which prevent interbreeding with other taxa. Clearly, evolutionary taxonomic units of this type may be equivalent to biological species in some cases, but these cases would only occur when species are monotypic. It is much more likely that a biological species would contain two or more evolutionary taxonomic units.

To overcome this problem Cracraft proposed that a species should be defined from the perspective of the results of evolution, rather than the processes which produce those results. Consequently, 'A species is the smallest diagnosable cluster of individual organisms within which there is a parental pattern of ancestry and descent' (Cracraft, 1983, p. 170).

Clearly, the phylogenetic species concept eliminated any reference to reproductive isolation from other species-level taxa. Species are recognized strictly in terms of their status as diagnosable evolutionary taxa. Thus, two sister taxa could broadly hybridize and still be considered as species if each was diagnosable as a discrete taxon. Cracraft emphasized that the phylogenetic species concept did not deny the importance of reproductive cohesion or disjunction, but claimed that the inclusion of these criteria in a species definition impeded our understanding of that relationship.

Cracraft provided the following qualifications for his definition of a phylogenetic species:

1 Although most species would be defined by uniquely derived characters, these cannot be included in the species definition, since it would not be possible to recognize ancestral species.
2 Species must be diagnosable from all other species.
3 Diagnostic characters must be passed from generation to generation and must be taken to define a reproductive community.
4 A species definition must have some notion of reproductive cohesion,

of parental ancestry and descent, although this is not predicated on reproductive disjunction.

Cracraft raised what he perceived to be problems with the phylogenetic species concept and chief among these was the recognition of many more species-level taxa, and the elevation of some subspecies to species. Second, Cracraft concluded that isolated populations or demes which were recognized by discrete biochemical characters should be recognized as species if these were distinct units of evolution. Third, he believed that it would be necessary to revise much of our current species-level systematics.

The fundamental weakness of a concept which recognizes any character as being suitable for diagnosis rather than just morphological characters, is the limit to which this might be pursued and the value of the species defined. For example, let us consider a species which is morphologically uniform throughout its distribution. Suppose that a chromosomal analysis of populations had been made and that this determined that two chromosome races were present. The phylogenetic species concept would recognize these chromosome races as species. If a subsequent electrophoretic analysis of soluble proteins was made on populations within one of the chromosome races and a series of populations could be defined by the possession of fixed differences at certain electrophoretic loci, these populations could also be regarded as phylogenetic species. To further extend the limits, a mtDNA restriction endonuclease analysis could be performed on one of these electrophoretically distinguishable populations, and further subdivide it into diagnostically discrete lineages which could also be defined as phylogenetic species. Presumably, this analysis could be further extended to DNA sequence analysis of the nuclear genome, and yet further subdivisions would be possible. Clearly, these are all recognized as basic evolutionary units, although anything that could be morphologically recognized as a species was dispensed with five levels of resolution ago.

The question may be raised as to what units would be recognized when there was disagreement between some of the techniques used to define phylogenetic species. For example, when a widely distributed species of rodent such as *Peromyscus maniculatus* (see Section 8.3.2 for a detailed discussion) had numerous subspecies defined by morphology, mtDNA analysis suggested that many of these were conspecific and others subdivided, whereas other diagnosable entities were comprised of populations of several subspecies. There are a number of other examples where biochemical and molecular techniques are in absolute disagreement with morphological species, or with each other.

One of the great problems associated with the phylogenetic species approach is the inconsistency which it creates within morphologically recognizable species.

The great majority of species would remain as morphological species because of their low population numbers or rarity. On the other hand, certain species would be examined with every technique which was developed for molecular or bio-chemical research, because they are readily available as laboratory animals. This is the case today, and there is little doubt that such a scenario will operate in the future. The phylogenetic species concept has the potential to create an enormous imbalance between taxa, if each of the detected entities are to be recognized.

The four evolutionary species concepts described above appear to have little to offer to the working biologist. They are specialized to the extreme and are similar to each other in name alone. Sokal and Crovello (1970, p. 146) made the following observation on Simpson's (1961) evolutionary species concept: 'This is so vague as to make any attempt at operational definition foredoomed to failure'. These comments may well have a wider application.

2.4 Concluding remarks

Our requirements for a species concept are essentially bidimensional. First, some-thing that will satisfy the community in general and meet its desire that a name should be placed on any animal or plant species which is encountered. This is not only an essential prerequisite for cataloguing our flora and fauna, but is critical for the control of insect pests and disease, and for the conservation and utilization of plant and animal resources. A species without a name is essentially unprotected by our legal system. On the basis of this simple requirement the morphologically recognizable species defined by a Linnaean binomial is solidly entrenched. This situation will continue: the taxonomic species remains immutable.

Our second requirement is for a species concept that will satisfy the very different needs of the scientific community. As we have seen, this is where the trouble starts. Palaeontologists, geneticists, systematists and ecologists have very different approaches to what they perceive a species to be. This is a not only a product of scientific training, it is also a byproduct of the techniques used in everyday research endeavours. The basic dichotomy between biological and evol-utionary species concepts is, in a simplistic sense, a reflection of these differences.

Thus, Simpson (1961) introduced the evolutionary species concept because he was a palaeontologist who had to deal with both past and present faunas, whereas the BSC only deals with the present. Wiley's (1978, 1981) modified evolutionary species concept is an essentially cladistic approach, where species are lineages of ancestor/descendant populations, and it attempts to bridge the gap between evolutionary process and pattern of descent. Nevertheless, Wiley (1981) was dia-metrically opposed to the view that phyletic evolution within a lineage could cause speciation, whereas this was the benchmark of Simpson's model; that is,

Simpson's recognition of allochronic species (successional species within a single lineage).

The version of the ecological species concept proposed by Van Valen (1976) is a proclaimed radical view which is unquantifiable because of the incorporation of 'adaptive zone' and 'range' in the definition. Nevertheless, this view was a product of Van Valen's particular research activities and requirements, and his dissatisfaction with the BSC. Indeed, discussion of this theory was included because it demonstrated the relationship between species concepts and research interests.

The phylogenetic species concept proposed by Cracraft (1983), with its definition which regards any diagnosable difference as a valid specific character, must be regarded in a similar vein. Thus, a morphologically recognized species may be regarded as a composite of subtly defined cryptic species each of which has equal status. Such a definition takes the systematic approach to an extreme which would appear to be unworkable. It is noteworthy that Cracraft's definition appears to have been derived from Wiley's evolutionary species concept, which itself incorporated a range of diagnostic characters other than morphological differences. Nevertheless, where Wiley maintains the significance of the reproductive isolation between species (corollary 2), any pretext to following this criteria is dropped in the phylogenetic species concept. Yet the very basis for maintaining the integrity of genetically distinct allopatric populations is the absence or prevention of gene flow. They are reproductively isolated even if only by geographic proximity.

Both evolutionary and phylogenetic species complexes fail to distinguish between species and lineages and in reality are a description of lineages. Bock (1986, 1989) was highly critical of such definitions, suggesting that they result from a failure of the authors to distinguish between theoretical definitions and objects in nature. In fact, they are an easy way out for cladists who wish to define any recognizable genetic entity as a species.

Wiley (1981) made the compelling point that the reason for the great popularity of the BSC is that biologists perceive entities fitting this concept in nature. The fact is that many biologists dealing with the systematics of plants and animals use the BSC in conjunction with the descriptive morphospecies. For while it is true that not many biological species have been described which are cryptic with their sibling species (for they may not have any characters which diagnose them), they are still regarded as an entity and perceived as such because of their reproductive isolation. Indeed, the key character which separates all morphospecies is their reproductive isolation, and they are continuously observed to retain this.

It would appear that there is no good reason to abandon the BSC, a concept with an enormous coverage. The fact that it does not include asexual species would seem to be irrelevant, for this was known at its inception. In many respects it is analogous to the morphological species which ignores cryptic forms. One does not simply abandon a widely accepted system because of a few exclusions; one sets up special-case definitions for these. In many respects the BSC does just this for cryptic species.

Perhaps this point may give us some insight to recently proposed biological alternatives to the BSC: the recognition and cohesion species concepts. It is of significance that both Paterson and Templeton, the authors of the respective concepts, are proponents of the view that the vast majority of instances of speciation are of a purely allopatric origin. It is noteworthy that both authors have in the past argued fervently against post-mating isolating mechanisms and in particular against the possibility of chromosome change playing a role in speciation (Paterson, 1978, 1982, 1985; Templeton, 1980, 1981, 1982). Unaccountably, and despite this stance, Templeton's cohesion species concept includes post-mating isolating mechanisms as an important factor in speciation.

It would appear that both recognition and cohesion species concepts are a byproduct of the positions which Paterson and Templeton have adopted. Thus, if it is argued that genic changes are responsible for speciation and that chromosomally induced post-mating isolating mechanisms do not play a role in speciation, one can hardly accept a definition for biological species which is based on reproductive isolation. The logical alternative is to advocate reproductive cohesion and specific mate recognition systems. Nevertheless, the recognition concept denies the possibility of any form of hybridization between good species and produces a completely untenable situation, with profound effects on taxonomic nomenclature. Indeed, the argument has been advanced by Hecht and Hoffman (1986) and Templeton (1989) that the recognition concept and BSC are two sides of the same coin. This is a little difficult to accept when the narrow view of a species espoused by Paterson (1985, 1987) is considered.

One of the fundamental criticisms of the BSC advanced by both Paterson (1985) and Templeton (1989) centred on selection for reproductive isolation. Both authors seemed to have the opinion that proponents for reproductive isolating mechanisms also supported the view that these mechanisms were selected for (see Sections 2.2.1 and 2.2.4). Nothing is further from the truth. In the case of genetically induced reproductive isolation, chance mutation in a founding population, or the gradual accumulation of accrued genic differences, may result in significant reproductive differences being established between parent and daughter populations. These would render a hybrid sterile when parent and daughter populations hybridized. In both cases, selection has not been for any reproductive isolating function of the genic differences established. The degree of reproductive isolation is purely a function of the degree of reproductive differentiation and the accompanying loss of reproductive compatibility. In the case of negatively heterotic chromosomal rearrangements, these would have to have a degree of sterility associated with them when initially formed in the population isolate. They are negatively heterotic and are *selected against* as heterozygotes. Such rearrangements can only reach fixation if the population dynamics in a founder situation lead to a chance fixation; if they were established as homozygotes (which is most unlikely); or if they were 'driven' to fixation by meiotic drive (see Section 6.4). The latter is either a product of cellular mechanics or a complex genetic system and is not selection. If the rearrangements happen to reach fixation, the same level of sterility would be present in the hybrid between parent and daughter species that was present at the time of induction as a structural heterozygote. However, in such a situation meiotic drive would be absent. Both genic and chromosomally induced reproductive isolating mechanisms are not selected for as such. To claim otherwise would appear to be a misrepresentation of the facts.

The great body of data presented in this book not only demonstrates the primacy of chromosome change in establishing divergence between species, it also shows the significance of these differences in establishing profound reproductive isolation. It is not reasonable nor realistic to ignore the many examples of reproductive isolation induced by chromosome changes in otherwise undifferentiated populations. Since chromosomal speciation is one of the major modes of speciation and it is based on reproductive isolation, the only possibility is to recognize a species definition incorporating this process, i.e. the BSC. The key to the success of speciating complexes which have evolved by allopatric or other means, is the prevention or restriction of gene flow between parental and daughter species. The fixation of multiple, or major gene differences, or structural chromosomal rearrangements which can effectively act as post-mating isolating mechanisms by preventing gene flow, enforce reproductive isolation. The most functionally obvious difference between biological species is the attainment of reproductive isolation.

3
Genic differentiation, reproductive isolation and speciation in allopatric populations

> How much of the genome is involved in the early steps of divergence between two populations, causing them to be reproductively isolated from each other? We do not know. Mayr writes of a 'genetic revolution' in speciation, but we cannot put quantitative limits on this revolution (which may after all turn out to be only a minor reform) until we begin to characterise the genetic differences between populations at various stages of phenotypic divergence.
>
> (Lewontin, 1974, p. 160)

The classical view of allopatric speciation suggests that isolated but undifferentiated populations of a species, can gradually speciate over time by the independent accumulation of genetically based differences. The attainment of reproductive isolation in such a situation is the key to allopatric speciation. Nevertheless, the precise means by which this occurs remains as the most controversial aspect of this mode of speciation. In this chapter, a number of perspectives are provided on the evolution of genetic variation within and between species; between central and marginal populations; on the degree and nature of genic differentiation associated with speciation; and on the means of attaining reproductive isolation in allopatric speciation.

3.1 Reproductive isolation: dichotomous views

Attitudes to what constitutes reproductive isolation between species differ markedly between individual authors. Thus, Futuyma and Mayer (1980) regard species as being

. . . a group of populations whose evolutionary pathway is distinct and independent from that of other groups' . . . a distinct and independent evolutionary path is achieved by the group's reproductive isolation from other groups . . . We consider groups to have achieved full species status, then, if they are or could be (given the opportunity) truly sympatric without losing their separate identities through interbreeding

(p. 256).

This definition will not countenance the possibility of hybridization between species. That is, neither restrictions of gene flow by extrinsic barriers or isolating mechanisms satisfy the above definition: 'biological barriers to gene flow must be virtually impermeable' (p. 256).

The views promoted by Paterson (1986) are little different. His specific mate recognition system insists that gametic recognition defines conspecificity. Thus, if gametes recognize each other to the degree that syngamy occurs; regardless of the situation, those species must be conspecific. Hybridization between species is therefore an impossibility.

Under both Futuyma and Mayer and Paterson's definitions, a large number of organisms which are recognized as taxonomic species would be considered to be conspecific.

Patton and Smith (1989, p. 290) use the following operational definition of a species and have a less absolute view on reproductive isolation.

1 Contact populations that do not form hybrids and that, therefore, are both reproductively and genetically isolated are considered separate species.
2 Contact populations that hybridize, but for which hybrid class individuals are limited to the F1 generation, are genetically if not reproductively, isolated, and are considered separate species.
3 Contact populations that hybridize and produce a full array of hybrid class individuals in expectation of random mating within the zone of contact are considered conspecific.

This definition is far more realistic in terms of systematic reality than are the absolutist versions. Parapatric populations which are to all intents and purposes reproductively isolated because they only produce sterile F1 hybrids, are just as isolated as those species which fail to produce F1 hybrids, since there is an absolute barrier to gene flow.

The weakness of Patton and Smith's (1989) definition is that it does not consider F2 and backcross hybrids. If F1 hybrids can mate and produce F2 and backcross progeny which are absolutely sterile, there is no meaningful difference between these and crosses that produce sterile F1s. The difference is that by including F2 and backcross sterility in a definition, we include the possible ramifications of recombinational effects on the genome (see John and King, 1980) which are only realized in the F2 generation and would otherwise be ignored. Indeed, this is a sad reflection on many earlier studies examining the effects of hybridity, which only produced hybrids to the F1 level. If F1 hybrids were normal, or had only partially impaired fertility, they were used as an indication of an absence of reproductive isolating mechanisms. We now know that many of the more profound reorganizational effects are only seen in F2 and backcrosses (see Shaw and Coates, 1983).

Undoubtedly the definition provided by Key (1981) is the most realistic in terms of effective isolation and it is the view followed in this study:

> Reproductive isolation is the relationship between two populations that do not hybridize in the field, although in contact with each other, or, if they do, whose F1 hybrids leave no progeny of reproductive age, i.e. are infertile interse and in every backcross
>
> (p. 455).

This definition provides a realistic interpretation of reproductive isolation since the possibility of gene flow is obliterated at the backcross or F2 level.

One of the great problems associated with Futuyma and Mayer's (1980) and Paterson's (1986) definitions, is that they are biologically and systematically inappropriate. For example, by ignoring other than sympatric, non-hybridizing species, these authors disregard all species that occur in allopatric populations and happen to form hybrids in the laboratory, or that occur in allopatric populations which form hybrid zones. The great majority of species recognized by systematists are allopatrically distributed and many of these have parapatric contact zones with their ancestors. If none of these forms is recognized, we are in danger of having a species definition which fails to coincide with biological reality and by that I mean biological diversity. If one recognizes only the ten sympatric species, when 2000 equally distinct allopatric or parapatric species are present, the value of that species definition diminishes. Consequently, a definition of reproductive isolation such as Key's, has a far greater correspondence to systematic and biological reality, although still recognizing the genetic integrity of species.

The title of this section included the term 'dichotomous views' and in reality the few views discussed above are essentially dichotomous. Futuyma and Mayer (1980) and Paterson (1986) believe that species do not form hybrids with other species. Key (1981) and Patton and Smith (1989) argue that good species can form hybrids, but that gene flow is prevented by the sterility of these hybrids. The absence of gene flow between species indicates that they are reproductively isolated, whether they hybridize or not.

3.2 Genic variation

The most widely used biochemical method for estimating genic divergence within and between species has been protein electrophoresis. Most of this work has centred on those soluble proteins which catalyse chemical reactions (enzymes), in particular on the different forms of certain enzymes (isozymes), and on the allelic variation at enzyme loci (allozymes). It is not the intention of this study to explore the very considerable literature which now defines the extent of electrophoretic variation within and between populations of species from different taxa. There are many publications which do just that and it is to these that the reader is referred (Lewontin, 1974; Ayala, 1975, 1982; Nevo, 1978; Thorpe, 1983; Nevo *et al.*, 1984; Richardson *et al.*, 1986). Suffice it to say that there are now many hundreds of species which have been examined with the electrophoretic technique, and these have provided us with an extensive estimate of genic variation in plants and animals. The two most commonly used parameters are polymorphism (P) and heterozygosity (H). Ayala (1982) defined the former as the proportion of loci found to be polymorphic in the sample, whereas H estimates the average frequency of heterozygous loci per individual. It is now known that between 20 and 50% of loci are polymorphic in most organisms, and in vertebrates this range is from 15 to 30%. Heterozygosity levels range from 4 to 8% for vertebrate groups and 6 to 15% for invertebrates. Ayala (1982) indicated that there was both considerable heterogeneity in the amount of genetic variation among organisms within a group, and in some instances a considerable degree of genetic variation among local populations of the same species.

On a second level, measurements of electrophoretic similarity can be estimated between populations or species by using techniques such as genetic identity (I) or genetic distance (D) (Nei, 1972). In what must be the largest comparative study of its type, Thorpe (1983) utilized 7000 I distance values to distinguish between conspecific populations, 900 for comparing congeneric species and 160 for confamilial genera. Thorpe found that conspecific populations generally had an I value greater than 0.9, whereas congeneric species range from $I = 0.25$ to 0.85. Indeed, most mammals, reptiles, amphibians, fish, invertebrates and plants fell within these limits, but birds provided an exception, having reduced levels of biochemical evolution. The data provided overwhelming evidence that the molecular structure of proteins diverges with time following reproductive isolation. Equally, the amount of biochemical divergence within a species or genus will be broadly similar in most vertebrates, invertebrates and plants. Baverstock and Adams (1987) reinforced this finding when comparing both electrophoretic and microcomplement fixation data on a range of marsupial, rodent and reptile species. However, this does not mean to say that an exceptional genus such as *Drosophila* might not exhibit higher levels of genic variation than its confamilials (Nevo, 1978).

White (1978a) was critical of both the nature and the value of the electrophoretic technique. He felt that it provided a considerable underestimate of genetic variability within and between species, since as little as 50% of the genic variation present was detected. This was based on the assumption that electrophoresis only detected those nucleotide substitutions which changed the net electric charge and could not detect those which have no effect on charge. However, Ayala (1982) pointed to the fact that cryptic electrophoretic variation has been detected by sequential electrophoresis, heat denaturation, urea denaturation, monoclonal antibodies and peptide mapping. The few studies of this type which have been made suggest that cryptic variation is minor and that 'conventional electrophoresis may actually detect most of the protein variation present in natural populations. An increase of 20 percent in the effective number of alleles is not trivial, but it is not likely to have drastic consequences for most evolutionary considerations' (Ayala, 1982, p. 74).

It is clear that electrophoresis has provided a most valuable technique in establishing whether a species exists, yet its inconsistencies are legion. Richardson *et al.* (1986) suggest that allopatric populations with differences at more than 20% of their loci can be regarded as separate species on these data alone. Nevertheless, electrophoretically detected variability may be minimal between species. Carson (1982a) pointed to *Drosophila heteroneura* and *D. silvestris* as being a complex where morphologically, behaviourally and chromosomally divergent species are electrophoretically indistinguishable. He continued: 'In my opinion, electrophoretic similarity between species is less of a measure of overall genetic similarity than it is of the irrelevancy of this type of variation to adaptive evolution. Indeed, much of it may be neutral to selection' (Carson, 1982a, p. 429). While this statement may be regarded as a little extreme, Carson's view is in part supported by a mtDNA analysis of this complex which distinguished a number of forms not resolved by electrophoretic analysis (DeSalle *et al.*, 1986a, b).

The intrinsic value of electrophoretic analysis lies in the discovery of genic differences within speciating complexes which are not readily resolvable by morphological analyses. These differences may involve variation in polymorphism frequency, or the fixed differences which can occur between populations, subspecies, species or higher taxa. Adams *et al.* (1987) indicated 'that an effectively fixed difference occurs at a locus when the two taxa share no alleles at that locus, or any shared alleles occur at a frequency of less than 5% in one of the taxa', and they argued that only one of these markers was required to indicate the presence of a cryptic species in sympatry. Fixed electrophoretic differences may be as valuable as any morphometric character in such situations. The bat genus *Eptesicus* provides one of the best examples of the discovery of multiple cryptic species by electrophoretic analysis. Here, Adams *et al.* (1982, 1987) found a complex of nine electrophoretically recognizable species, in what had, up until 1975, been regarded as a single species. Of the 35 loci analysed, some forms expressed fixed differences at up to 29% of these loci; a truly extraordinary level of divergence.

A great deal of attention has been devoted to whether genetic changes of this type are adaptive or neutral in terms of their selective value. The only comment I would make on this issue is that genetic variants which may be neutral today may not have been neutral when they were initially formed. There is no reason why adaptive characters should not be transient in their status. The selectionist/ neutralist debate is adequately summarised by Kimura (1983, 1986), and will not be dealt with here.

Much of the research effort over the past 20 years has been on the analysis of what are termed structural loci. That is, those genes which are involved in the production of soluble proteins and which occupy the largest portion of the coding genome. Ayala (1975), Bush (1975), Wilson (1975), Wilson *et al.* (1975), Gould (1977), Templeton (1980), Mayr (1982b) and numerous other workers have made the distinction between these structural genes (which determine the synthesis of proteins), and a second component of the genome made up of regulatory genes. It is argued that such regulatory genes are evolutionarily significant because they control the action of structural genes and are responsible for major life processes such as development and reproduction.

Much of the emphasis on regulatory genes as a target for evolutionary activity was initiated when some of the comparative studies analysing structural loci, either failed to show differences between related species, or demonstrated a concordance between electrophoretic and lineage divergence. In some respects they provided a means of sustaining hypotheses which were not standing up to examination. However, the regulatory gene concept has recently been roundly criticized, primarily because it was drawn from studies on bacteria where both types of genes could be readily detected, and applied to eukaryotes where they could not be so easily defined (Charlesworth *et al.*, 1982; Grant, 1985; Miklos and John, 1987). Thompson (1987) pointed to the possibility that the regulator at one locus was itself likely to be a structural gene, or to require complexing with gene products to become functional in the repression of genic activity. Similarly, Grant (1985) described the situation in higher plants where there are a variety of controlling genes (inhibitors, minus modifiers, oppositional gene systems), and some of the structural genes, such as inhibitors, are also controlling genes. Thompson (1987) suggested that the significance of regulatory gene mutations in evolution has been overemphasized, because many of the regulatory genes modify coordinated gene activities, and the effects of mutation are so extreme that they would cause chaotic disorganization of the organism. Consequently, viable mutants of this type will be exceedingly rare and generally infertile. Clearly, the dichotomy between structural and regulatory genes has severe limitations which have invoked a significant level of opposition. This feature should be borne in mind when that terminology is used in this and following chapters.

3.3 Differentiation of central and peripheral populations: inversion polymorphism and genic effects

The considerable variation in genetic polymorphism, heterozygosity and fixed differences between species, provided impetus for the analysis of variation within species, and, in particular, the examination of possible differences between central and peripheral populations. This approach was the response to a series of most intensive analyses of chromosome inversion polymorphism in a number of *Drosophila* species.

The pioneering studies of da Cunha and Dobzhansky (1954) provided the first substantive evidence for a differentiation of central and marginal populations within a species distribution. These authors examined paracentric inversion sequence polymorphisms in populations of *Drosophila willistoni* in Central and South America. They found a reduction in the level of heterozygosity in peripheral populations when compared to that found in the centre of a species distribution (Fig. 3.1).

Subsequently, da Cunha *et al.* (1959) increased the number of the inversion sequences which were scored to include 50 different polytene blocks. They proposed that chromosomal polymorphism for these inversions was adaptive and reached three distinct conclusions:

1 Populations of the central part of the geographic distribution of a species tend to be more polymorphic than those on the periphery of the distribution or in peripheral pockets.
2 Populations which occupy rich and diversified environments were more polymorphic than those in marginal or submarginal environments.
3 Populations which face the competition of closely related species are more monomorphic than those which monopolize an environment.

At about the same time, Carson (1949, 1955) was analysing inversions in *Drosophila robusta* in eastern North America. Here, too, the percentage of individuals heterozygous for inversions was highest in the central populations. Carson (1955) proposed that crossing-over would occur less frequently in those populations in which individuals were heterozygous for inversions, than in those where

individuals were chromosomally monomorphic. The latter would by necessity
have a higher incidence of recombination, since inversion heterozygosity modifies
the location of chiasmata. In the highly polymorphic central populations, individuals would be more vigorous because of the heterotic buffering conferred by
structural hetrozygosity. Moreover, the long-term adaptation to a range of diverse

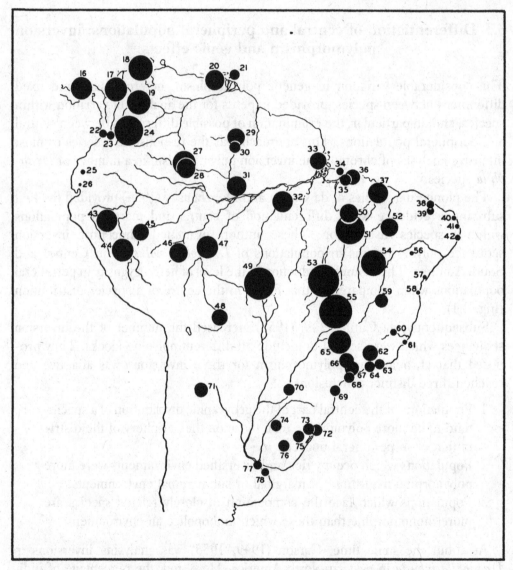

Fig. 3.1. A map showing the distribution and incidence of chromosome inversion polymorphism in *Drosophila willistoni* from South America. The diameter of the black circles is
proportional to the mean number of heterozygous inversions per inversion female. (Redrawn
from da Cunha *et al.*, 1959.)

niches in central populations would result in genomic stabilization. In contrast, the homozygous forms encountered in marginal populations would have a greater potential for adaptation when exposed to a new and unpredictable environment, because of their higher level of recombination and consequently the possibility for substantial genomic change.

Brussard (1984) summarized the consistent pattern of central/peripheral reduction in inversion heterozygosity encountered in most species of *Drosophila* analysed. He suggested that to minimize the production of individuals with extreme phenotypes, linkage arrangements which combine alleles of opposite effects would be favoured. In the centre of the distribution, linkage disequilibrium would be promoted by those epistatic interactions and inversions which stabilize the genome. Unpredictable and stressful environments characteristic of more peripheral populations would favour more extreme phenotypes. Free recombination would become increasingly important and inversions which retard recombination would be selected against in these marginal habitats.

The models which have been produced to explain the clines in inversion polymorphism in many *Drosophila* species, have not successfully explained the dramatic third chromosome inversion polymorphism differences which have been documented in *D. pseudoobscura* and *D. persimilis* by Dobzhansky and his colleagues (Dobzhansky and Epling, 1944; Dobzhansky, 1951). The third chromosomal rearrangements have been associated with many adaptive effects and used extensively in experimental manipulation without conclusive results. It is of some interest that a similar relationship between heterozygosity in central and peripheral populations was detected by Levin (1978) in flowering plants. Here, structural heterozygosity was found to be more common in ecologically and/or geographically peripheral populations which are subjected to major fluctuations in population size and breeding structure. That is, the reverse situation to that found in *Drosophila willistoni*. The observation that chromosomal variability changed between central and peripheral populations was associated with the occupation of a less ideal habitat in marginal populations. The direction that this polymorphic variation had taken in flowering plants and *Drosophila* was dependent on the interaction of the genetic system of the organism with the particular environmental conditions to which it was exposed. There is no reason why the observed differences should not be alternative strategies to a similar goal.

The consequence of the remarkable discrepancy in variability in inversion polymorphism in central and peripheral populations, was a series of investigations into the degree of genic variation in the same species. The results of these intensive analyses by Prakash (1973) on *Drosophila robusta*, Ayala *et al.* (1972) on *D. willistoni*, Saura *et al.* (1973) on *D. subobscura*, and Lewontin and Hubby (1966) and Prakash (1977) on *D. pseudoobscura* and *D. persimilis*, indicated that no change in allozyme polymorphism or heterozygosity could be associated with either the marginal or central distribution of populations. Brussard (1984) suggested that such a result was not surprising, since the electrophoretically detectable variants were probably neutral, and that migration could adequately account for the lack of differentiation in allele frequencies and the absence of the central/peripheral decline in heterozygosity. Thus, we might conclude that chromosomal inversions are adaptive in *Drosophila*, whereas electrophoretic differences are not. The implications of these findings with regard to the status of central and peripheral populations of allopatric species, and the changes associated with founding populations and population isolates, will become more apparent in the chapters which follow.

3.4 Stages of genic differentiation during allopatric speciation

The classical theory of genic allopatry supposes that following the isolation of two portions of the distribution of an ancestral species from gene flow, the passage of time will lead to the gradual accumulation of genic and morphological change. Consequently, if the two allopatric and differentiated populations ever come into contact again due to range expansions, and if sufficient reproductive differences have accumulated between the forms, they will be functioning as reproductively incompatible and independent species.

Despite the formidable amount of morphological data which have been used to support this concept (see Mayr, 1963; White, 1978a), a greater appreciation of the levels of genomic reorganization associated with speciation has been realized in more recent investigations. Indeed, Lewontin (1974), in his landmark work on the genic basis of evolutionary change, argued that virtually nothing is known about the genetic changes that occur during species formation. In the early 1970s, both Lewontin and Ayala attempted to conceptualize a framework for the staging of species differentiation. This may not have been absolutely successful in terms of the consistency of interpretation, but there is no doubt that the action was necessary. A consideration of the nature of genetic change associated with speciation is particularly germane to our understanding of the process of speciation.

3.4.1 Lewontin's approach (1974)

Lewontin (1974) attempted to define the processes of speciation in geographically isolated populations where there was a severe restriction on genetic exchange. He recognized three stages:

Stage 1: The gradual accumulation, after a period of geographic isolation, of genetic differences which were great enough to restrict gene exchange if the populations ever made contact. This was accompanied by niche divergence, a character which may be linked to reproductive isolation.

What kind and what degree of genic differentiation was required for reproductive isolation? Lewontin argued that primary reproductive isolation without gene flow could result from chance changes in a few loci and nothing more. That is, there was no genetic revolution in terms of multiple genic changes, but rather the gradual acquisition of a small number of fortuitous and significant changes.

Stage 2: When populations made secondary contact, two forms of interaction were possible. First, if sufficient reproductive isolation had developed while the populations were allopatric, natural selection would reinforce the reproductive barriers. That is, physiological differences which had reached fixation could reduce the fertility or viability of hybrids. Concomitantly, there would be fewer matings and selection for those characters that reduce the likelihood of mating. Second, selection would also occur for the ecological divergence of the potential species, resulting in a reduction of the degree of niche overlap, thus ensuring the stable coexistence of parental and daughter species.

What additional genic divergence is required to produce ecologically differentiated and stable members of the species? Lewontin considered that in stage 2 as much as 10% of the genome may have become differentiated (primarily in allelic frequencies), and that complete divergence involving the fixation of alleles would be rare.

Stage 3: In this stage the new species which had evolved continued to differentiate, but at this point independent divergence rather than interactive divergence occurred. Subsequently, genetic divergence continued and would be an indication of the time since cladogenesis and the rate of genic mutation.

The question of how much genic similarity existed between species which had recently diverged was raised by Lewontin. He argued that sibling species could differ from each other by from 10 to 50% of their genome, but that such a process of phyletic evolution was open-ended, since it was simply a reflection of the amount of time since initial divergence.

White (1978a) emphasized that Lewontin's third stage was not really a part of the speciation process because it was phyletic evolution and occurred subsequent to speciation. White also argued that this was the stage when major genic differences were established between species.

3.4.2 Ayala's view (1974, 1975)

Ayala *et al.* (1974) and Ayala (1975) produced a two-stage model for genic differentiation and speciation.

Stage 1: Allopatric populations of the same species differentiated genetically to the point where they would be likely to evolve into new species if they came into contact. Differentiation was promoted by genetic drift, and natural selection enforced adaptation to the local environment.

Stage 2: Populations which had become genetically differentiated while in allopatry made secondary contact and hybridized. If interpopulation hybrids had low fitness due to the disruption of coadapted gene complexes, selection would favour the further development of complete reproductive isolation.

Ayala *et al.* (1974) and Ayala (1975) used the *Drosophila willistoni* species group as a basic model, comparing genetic diversity at the stages of taxonomic differentiation from geographic populations of the same taxon to full species. This comparative approach was then extended to other *Drosophila* species, other invertebrates, fish, salamanders, lizards, mammals and plants.

The five increasingly divergent levels of cladogenesis recognized were:

Stage 1: a. Between geographic populations of the same taxon: There is no reproductive isolation and very little genic differentiation; average $I = 0.970$, $D = 0.031$. However, it is noteworthy that certain populations have substantial differences at some loci.
b. Between subspecies: These are allopatric populations in the first stage of the speciation process. They have partial hybrid sterility and thus a degree of reproductive isolation. Genic differentiation is quite substantial (23 allelic substitutions/100 loci). Average $I = 0.795$, $D = 0.230$, i.e. D is about ten times larger than between populations. Differentiation has been restricted to a few loci, otherwise they are identical electrophoretically.

Stage 2: c. Between semispecies: These are in the second stage of speciation, they are often sympatric and sexual isolation is nearly complete. Here there are a greater proportion of loci which have differentiated; average $I = 0.798$, $D = 0.226$.

d. Between sibling species: These are morphologically similar, but not identical species, and they are reproductively isolated. These species may have 58 allelic substitutions for every 100 loci. At most loci they may have either identical, or completely different, genetic constitutions (average $I = 0.563$, $D = 0.581$).

e. Morphologically distinguishable species of the same group. Easily distinguished by morphological characters. These are very different organisms with one electrophoretically detectable substitution per locus (average $I = 0.352$, $D = 1.056$).

Here, too, Ayala has included substages of genic differentiation (d and e), which could be interpreted as phyletic evolution and subsequent to the processes of speciation.

3.4.3 The dichotomy

Both Lewontin and Ayala's analyses considered genic evolution in the *Drosophila willistoni* and *D. paulistorm* species complexes using protein electrophoresis. In addition, Lewontin had analysed electrophoretic divergence in *Mus musculus* and *Drosophila pseudoobscura*, and in his later paper, Ayala (1975) considered a range of animal and plant species. Both authors reached relatively different conclusions as to what was actually happening during the speciation of these allopatric populations. The dichotomy of views can be summarized as follows. First, Lewontin argued that only minor genic differentiation occurred at stage 1 when reproductive isolation was attained, and most genic differentiation occurred subsequent to this in stages 2 and 3. Second, Ayala argued that a high level of genic differentiation occurred at stage 1, and that there was only a minor amount of genic differentiation during stage 2 when reproductive isolation was attained.

Conceptually, most of the difficulties stem from the question of why genic differentiation should occur at any set time during the speciation process. Major or minor mutations affecting reproduction could be established at any time during the isolation of species, or may not occur at all, leaving isolated species reproductively compatible. Formalized models for the staging of speciation have problems in this area.

It is noteworthy that a substantial proportion of the structural genes were changed in Ayala's first stage of speciation (10 to 25 allelic substitutions), whereas there were fewer changes after reproductive isolation had been attained. Since this trend was present in an array of organisms, this led Ayala to conclude that

the genic changes of the structural genes were not responsible for attaining reproductive isolation, but that other forms of structural genes, or probably regulatory
genes, were involved in that process. Indeed, Ayala even further divorced the role
of structural gene differentiation in attaining reproductive isolation when he cited
the case of several plant and animal species which had diverged chromosomally:

> In the instances of saltational speciation, reproductive isolation was initiated by
> chromosomal rearrangements and/or changes in mating systems; further develop-
> ment of reproductive isolation barriers had been accomplished without many allelic
> changes
>
> (Ayala, 1975, p. 61).

It would appear that many of the problems associated with models of this type
may be due to the organisms which were analysed. In their initial study, Ayala
et al. (1974) had concentrated on the *Drosophila willistoni* complex and it was from
this analysis that the basic staging model was derived. There are two problems
associated with this choice of organism. First, genic variation is greater in invertebrates than it is in plants or vertebrates. Second, within invertebrates, *Drosophila*
exhibits significantly higher genic variation than the rest of the invertebrates
(Nevo, 1978). These factors could in part account for the large amount of genic
variation which Ayala encountered in the earlier stages of speciation.

The high level of genic variation in stage 1 of Ayala's model could also be
accounted for by the form of classification which he used. By including organisms
which had diverged to the subspecies level in stage 1 rather than stage 2 (23
allelic substitutions and a significant degree of reproductive isolation), Ayala had
inadvertently shifted the area of maximum genic differentiation from his stage 2
to stage 1. This systematic problem might have been reflected in the fact that
subsequent comparisons between subspecies and semispecies (a difficult class to
accept at the best of times), revealed few additional genetic changes and only
subtly more reproductive isolation, yet the semispecies were classified as stage 2
animals. Finally, some of the forms of *Drosophila* which had been recognized as
subspecies in Ayala's staging sequence had been separated for many hundreds of
thousands of years, and were genically highly differentiated. Indeed, it is probable
that some of these would be regarded as allopatric biological species today (White,
1978a). Their inclusion in stage 1 would artificially exacerbate the level of genic
differentiation in the earliest stages of speciation.

3.4.4. Species hybrids, genic differentiation and reproductive isolation

Hybridization between species of *Drosophila* in nature is a rare event; only eight cases of interspecific hybridization have been recorded (Bock, 1984). However, many of these species can hybridize in laboratory experiments, due to the disturbance of pre-mating isolating mechanisms which prevent hybridization in the wild. Bock (1984) catalogued 266 hybridizations between 191 species of Drosophilidae which had been published up until 1984. He found that a great many *Drosophila* species were uncrossable, and interspecific hybrids could only be produced among the most closely related forms. The fertility/viability effects on the hybrids ranged from the production of a few viable embryos that died as larvae, with the cross achievable in only one direction, to the formation of normal offspring in both reciprocals of a mating. In this last category, 35 different species combinations produced normal F1 progeny. Bock (1984) suggested that this was not a demonstration of conspecificity of the parental species, since there was no evidence that these forms interbreed in nature. Rather, it indicated that major reproductive mutations affecting fertility had not reached fixation in those species involved and that pre-mating isolating mechanisms were a significant barrier to gene flow.

Coyne and Orr (1989a, b) took a slightly different approach to estimating the course of speciation and the significance of pre- and post-zygotic isolation. They used published data on 119 pairs of closely related *Drosophila* species which had known genetic distances, mating discrimination data, hybrid sterility and inviability data, and a known geographic range for the species. Their study suggested that mating discrimination and the sterility or inviability of hybrids increased gradually with time and that hybrid sterility and inviability evolved at similar rates in allopatric species. However, in sympatric species pre-mating isolating mechanisms appeared before severe sterility or inviability. Coyne and Orr (1989a) reasoned that this result reflected the increased level of pre-mating isolation produced as a form of reinforcement when allopatric species became sympatric.

Coyne and Orr made estimates on the genetic distance which was required to attain species status, i.e. with total reproductive isolation. The total pooled data produced a D of 0.53. Whereas allopatric populations had speciated when a D was 0.66, sympatric species with a D of 0.31 were completely isolated from each other.

Bock's (1984) hybridization surveys found that the most common hybrids produced between *Drosophila* resulted in sterile males and fertile females. This preponderance of male-based sterility was regarded as an example of Haldane's (1922) rule, which suggested that a high incidence of male sterility would occur in hybrids because of the great impact of the X chromosome on fertility (also see Section 5.1.1.3).

One of the most interesting findings that Coyne and Orr (1989a) made, was that a much greater genetic distance existed between species which produced sterile or inviable hybrids of both sexes, compared to those which produced sterile or inviable males only. That is, hybridizations showing sterility or inviability of both sexes are much older than those which obeyed Haldane's rule. They reasoned that the pathway to total sterility or inviability went through two phases. First, an initial accumulation of alleles causing post-zygotic isolation in male hybrids. Second, an accumulation of alleles causing isolation in female hybrids. The assumption necessary for such a model was that the genetic mechanisms causing reproductive isolation between species, operate quite differently from other forms of genetic differentiation, where additive substitutions spread throughout the genome.

Most analyses on evolutionary divergence between species have concentrated on differences in allele frequencies of single copy genes. However, it is likely that such structural loci are irrelevant to the process of speciation, although they provide excellent markers to the degree of genic differentiation within and between species. In eukaryotes, most of the gene systems that are responsible for major reproductive functions consist of multigene families which act in concert to produce an integrated end-product in the final phenotype. Presumably, the mutation of any loci in such a complex can impinge on the success of the system.

Vigneault and Zouros (1986) and Zouros et al. (1988) examined male hybrid sterility in crosses between Drosophila mojavensis and D. arizonensis. The male sterility that they examined was complicated by the fact that it was often non-reciprocal. Vigneault and Zouros (1986) were able to show that sperm immobility resulted from interactions between genes on the Y chromosome and at least two pairs of autosomes in hybrids which had the D. arizonensis Y chromosome. Zouros et al. (1988) extended this finding and demonstrated that all chromosomes apart from pair 6 (which was not tested), carried at least one gene that could interfere with normal spermatogenesis. The demonstration that hybrid sperm motility was a polygenic character belies the complexity of the situation. Several interspecific combinations of autosomes with one or other sex chromosome (X or Y) resulted in abnormal sperm. Moreover, a dominance relationship existed between heterospecific homologous autosomes in their interactions with the sex chromosomes, and the direction of the dominance depended on whether the sex chromosome was the X or Y (Fig. 3.2).

Dobzhansky (1974) analysed asymmetrical male sterility in hybrids between the subspecies Drosophila pseudoobscura pseudoobscura and D. p. bogotana and found evidence for incompatility between X-linked and autosomal genes producing male sterility. Vigneault and Zouros (1986) found four additional examples of the X

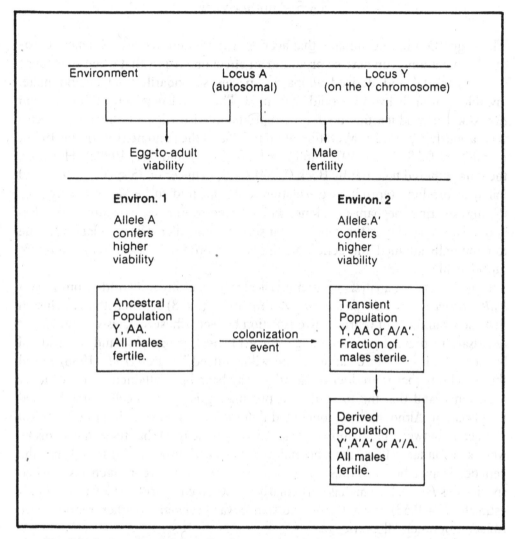

Fig. 3.2. A general model depicting the sequence of events which are necessary for the production of a new population, which when crossed with the original population, exhibits asymmetrical male hybrid sterility. (Redrawn from Zouros, 1986.)

and Y chromosomes possessing genes which interact with one or more autosomes producing asymmetrical male hybrid sterility. All studies indicate that polygenic complexes are responsible for some degree of fertility control in these *Drosophila*.

3.4.5 Another view

The suggestion has been made that accrued genetic differences, or major reproductive mutations, can have so profound an effect on fertility that when specimens
of the parental and derived allopatric species secondarily contact and mate,
inviable or infertile hybrids would be formed. The degree of genetic differentiation
necessary to produce this degree of reproductive isolation in allopatric species
was variously reported as alteration of up to 10% of the genome (Lewontin, 1974),
$D = 0.23$ to 0.58 (Ayala, 1975) and $D = 0.53$ (Coyne and Orr, 1989a). However,
the data gathered together by Bock (1984) show a cluster of 35 species which had
failed to produce reproductive isolation and appeared to be separated by pre-
mating isolating mechanisms alone, which, once overcome, permitted hybridization. The generality of the concept that speciation in allopatric populations is due
to temporally attained and genetically induced reproductive isolation is therefore
questionable.

In their detailed analysis of asymmetrical male sterility in hybrids from *Drosophila arizonensis* and *D. mojavensis*, Zouros *et al.* (1988) reached the conclusion
that the attainment of reproductive isolation between these species was no longer
necessary to prevent gene exchange, for ecological, geographic and behavioural
barriers eliminate hybridization in the wild. Indeed, Zouros *et al.* (1988) raised
the possibility that reproductive isolation may have been attained as a secondary
occurrence and that the formation of pre-mating isolating mechanisms had led
to speciation. Although they concluded that the degree of variation in ethological
characters between populations was so great as to suggest that these mechanisms
were a secondary effect, the uncertainty remains. Zouros (1974) found that the
genetic distance between these recently differentiated, but reproductively isolated
species was $D = 0.21$, which is particularly low when compared to Coyne and Orr's
estimate ($D = 0.53$) of what genetic distance was necessary to attain reproductive
isolation between allopatric species.

In some complexes, speciation in allopatry appears to have occurred without
reproductive isolation being established, despite profound genetic differentiation
of the species involved. The Australian bush rats of the genus *Rattus* are an
evolutionary exciting array of genetically and chromosomally diverse species.
Three of the species, *Rattus lutreolus*, *R. tunneyi* and *R. fuscipes*, are allopatric in
their distribution and chromosomally most similar, with *R. lutreolus* and *R. tunneyi*
($2n = 42$) differing by one pericentric inversion and *R. fuscipes* ($2n = 38$) by two
fusions (Baverstock *et al.*, 1983a). The species show profound genetic differences
between one another. Thus, Baverstock *et al.* (1986) in an electrophoretic analysis
of 55 loci found that *R. fuscipes* and *R. tunneyi* had fixed differences at 25 to
30% of their loci ($D = 0.37$) and *R. lutreolus* and *R. tunneyi* differed similarly at

approximately 20% of their loci ($D = 0.26$). Each of the species is readily distinguished by external morphology.

While all three of the species are broadly allopatric, *R. lutreolus* and *R. tunneyi* are sympatric in an area of overlap. The species do not form interspecific hybrids in the wild, presumably due to pre-mating isolating mechanisms. Mahony, Baverstock and Smith (pers. comm.) have hybridized each of these species pairs in the laboratory. Despite the genetic differences, F1 hybrids are fertile and display increased vigour, and backcrosses to the parental strains are fertile and viable.

Despite the vast amount of data which has been accrued delineating the degree of genetic differentiation in allopatric species, and demonstrating that reproductive isolation can be established by mutation in the polygenic controls on reproductive characters, there are still cases such as this which confound the very basis of allopatric speciation. These suggest that allopatric populations can diversify on most levels resulting in behaviourally, morphologically and genetically distinct species, but which have differentiated to such a small extent at the reproductive level that hybridization can occur without constraint, when pre-mating isolating mechanisms are disrupted.

3.4.6 Rapid reproductive isolation in some flowering plants

Studies on the evolution of mating systems in heterostylous plants by Barrett (1989) provide an interesting contrast to the development of reproductive isolation in animals, for they demonstrate the relatively simple changes which can lead to reproductive isolation and speciation. In most plant species reproductive isolation is associated with the divergence of allopatric populations. However, changes in genetic systems associated with polyploidy, can result in instantaneous reproductive isolation and sympatric speciation.

Barrett found that in small populations of herbaceous flowering plants, shifts in the mating system from outcrossing to selfing result in species formation by producing reproductive isolation. For example, in the *Turnera ulmifolia* complex, recombination in the distyly supergene in association with polyploidy, leads to a quantum change in mating system with reproductive isolation arising in a single step (Fig. 3.3). On the other hand, in the genus *Eichhornia*, a breakdown of heterostyly involving several changes in the mating system from outcrossing to selfing have evolved. Changes of this type can occur through simply inherited modifications of floral tracts. Moreover, the evolution of self-fertilization enhanced the likelihood of character divergence and the accumulation of postzygotic isolating mechanisms. Genetic constraints associated with floral development may play an important role in guiding the specific reproductive modifications

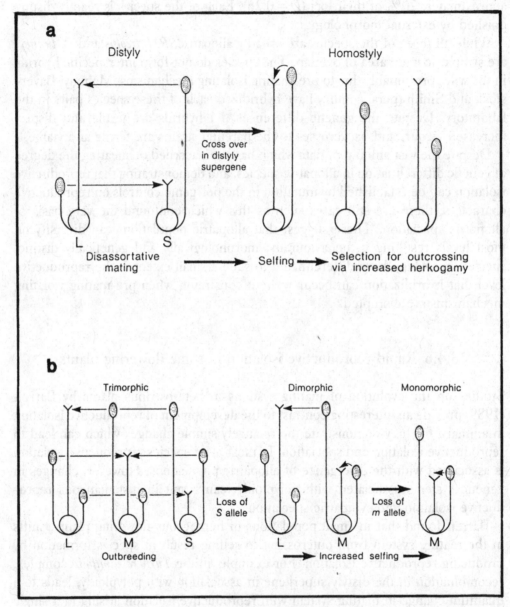

Fig. 3.3. a. A model depicting the evolutionary relationships among distylous and homo-stylous forms within the *Turnera ulmifolia* species complex.
b. A model depicting the evolutionary breakdown of tristyly to semihomostyly in the *Eich-hornia paniculata* complex. (Redrawn from Barrett, 1989.)

that can occur (Barrett, 1989). In *Eichhornia*, Barrett found that the derived homostyles are morphologically indistinguishable from the outcrossing hetero-styles, their progenitors in terms of their vegetative growth, but they cannot be crossed with them due to floral traits which impair reproductive functions. The evolution of self-fertilization in plants is frequently associated with the coloniz-ation of geographically or ecologically marginal environments. Population bottle-necks arising through founder events have been implicated in establishing self-fertilizing variants.

3.5 Concluding remarks

Electrophoresis has been used extensively as a means of obtaining a genetical insight into the nature and rate of the evolution of specific loci in speciating complexes. Needless to say, the molecular analysis of DNA sequences at particu-lar loci, or areas of the genome, can provide far more information than electroph-oresis on the degree of sequence variation between individuals, populations or species. However, the significance of this variation is often not apparent in terms of its evolutionary impact on the genome, whereas the impact of variation at some loci is immediately apparent. Thus, electrophoresis can provide a valuable perspective on our understanding of genome evolution and of levels of genic differentiation associated with speciation, although this can be readily augmented by additional perspectives by using particular molecular techniques such as se-quencing of specific loci, mtDNA analysis, or immunogenetic techniques such as microcomplement fixation.

Perhaps the most revealing contribution to our understanding of allopatric speciation processes made by this form of analysis is in the relationship between genic change and the evolution of reproductive isolation. Some of the salient features are:

1 In the majority of cases a direct correlation exists between the degree of electrophoretic divergence and the degree of population divergence.
2 Reproductive isolation between diverging populations is correlated with the accumulation of genetic differences. However, while some of these differences have a direct effect on reproductive physiology, producing hybrid infertility or inviability (*Drosophila*), in other cases speciation appears to have occurred by the accumulation of pre-mating isolating mechanisms alone. For example, in some *Drosophila* and in Australian *Rattus*, genically, morphologically and behaviourally divergent species can hybridize in the laboratory, where pre-mating isolating mechanisms

are rendered ineffective. Thus, physiological reproductive isolation is not necessarily associated with genetic differentiation (see also Chapter 7).

3 In most interspecific hybrids between *Drosophila* species, the degree of reproductive isolation was positively correlated with the degree of genetic divergence (Coyne and Orr, 1989a, b). This appeared to represent a stepwise sequence of events where F1 hybrids between genetically divergent allopatric species first exhibited male sterility and reduced female fertility. Second, allopatric species which had greater genetic distances established than these, tended to produce totally infertile or inviable hybrids.

4 Difficulties are associated with such comparisons. Neither Bock (1984) nor Coyne and Orr (1989a, b) made a distinction between chromosomally different species pairs and those which were chromosomally indistinguishable. Thus, the fertility effects attributed to genic differentiation could not be isolated from those due to chromosomal rearrangements. In this regard, the male sterility and reduced female fertility effects associated with genic differences are also known to be caused by X-autosome translocations in *Drosophila* (Lindsley and Tokuyasu, 1980), and are related to X-autosome associations in many other species (see Section 5.1.1.3).

5 The stepwise models proposed by Lewontin (1974) and Ayala (1975) associating genic differentiation and reproductive isolation will always be hampered by the uncertainty of whether changes to major reproductive loci are responsible for speciation, or whether this could be attained without any form of reproductive mutation. Concomitantly, the estimation of particular genetic distances at which reproductive isolation might be attained are 'rule of thumb' averages with quite considerable ranges. After all, reproductive isolation is not a planned event and in allopatric situations it is dependent on the vagaries of mutation, genetic drift and the loci involved.

6 It can be questioned whether the degree of evolution of single copy genes is really relevant to the speciation process. Since most gene systems which are responsible for major reproductive functions in eukaryotes are multigene families which act in concert to produce an integrated end-product, can the study of structural loci provide any more information than the degree of genic differentiation?

It would appear that the genetic basis for the most widely accepted mode of speciation, allopatric speciation, is itself open to question. It is clear that certain species have been isolated for prolonged periods and have diverged to such an extent that differences are accumulated that may prevent the production of fertile and viable hybrids if mating occurs between them. However, the precise means by which this divergence occurs and the relevance of genetic differentiation to that process, is open to debate.

4

Genetic revolution or gradual reform? Expectations of the founder effect

We are not arguing that peripheral populations are unimportant in speciation for they are, but through mechanisms other than genetic revolution. In cases involving founder events from highly inbred peripheral populations, the primary role of the founding event is to establish a geographically isolated population, and there is nothing about the founding event *per se* that induces speciation.

(Carson and Templeton, 1984, p. 120)

In the previous chapter, emphasis was placed on genic differentiation within and between established and evolving allopatric species and its relationship to the attainment of reproductive isolation. In the present chapter, a single aspect of speciation in allopatric populations is considered, that is the means by which genetic diversity is established in an isolated founding population on the periphery of a species distribution. This is a contrasting situation to the gradual accumulation of differences in established allopatric species, and is regarded as being a most important aspect in the colonization of new habitat types and in the fixation of evolutionary novelty.

4.1 The shifting balance theory (Wright, 1932, 1982a, b)

In 1932 Sewall Wright produced one of the most incisive and long-lived models for evolutionary change, called the shifting balance theory. This became the basis for our understanding of adaptive processes and of founder-induced speciation. Wright perceived evolution to be based on the action of selection on 'interaction systems' rather than on particular alleles. That is, a suite of genes which controlled the expression of a character, was modified by both pleiotropic effects and by non-additive intergenic interactions. Thus, the value of any gene was dependent on the array of other genes with which it was associated, and the fitness of that interaction would vary with the particular combination of genes present. Optimal gene combinations would be selective peaks. Moreover, selection operated on two levels: genic and genotypic. Genic selection occurred among individuals for alleles at particular loci, which produced the most favourable reaction with other loci and therefore bound a population to a selective peak. Genotypic selection operated on differentiated local populations which had diverse interaction systems and had reached other selective peaks (Wright, 1982a).

To appreciate the action of these two levels of selection, Wright (1931, 1932) placed local populations of a species in multidimensional space. That is, by including a dimension for each allele other than the leading one at each locus, defining coordinates according to the allelic frequencies, and adding a dimension for the selective value for each population, Wright was able to provide a multidimensional surface of selective values, the topography of which had numerous selective peaks and saddles (Fig. 4.1).

Wright argued that populations tend to occupy a peak, although the peak itself could change as conditions changed. Thus, the populations must themselves change to maintain their position at a selective peak, or at least in its vicinity. The shifting balance process consists of a shift across a saddle from the control by a lower selective peak to control by a higher one. Species which had an unsuitable population structure could not speciate by changes of this type, although they could change in other ways. For example, those species with high population density or wide dispersion of offspring, could not move from the adaptive peak they occupy because of selection, and were in near stasis in the vicinity of that peak. Equally, a wide-ranging species which was sparsely distributed locally and had a more restricted dispersion could, by limited operation of the shifting balance, gradually improve its adaptive position.

The shifting balance theory could lead to speciation, in those species which were extensively distributed, but in which local populations or demes existed as small colonies frequently subjected to extinction and re-founding by stray individuals. Major adaptive changes might be facilitated by numerous minor alterations, or by major mutations which had been kept at a low frequency by deleterious side-effects. However, the determining factor for rapid change required the presence of one or more vacant niches, whether this resulted from the entry of a species into new territory, the elimination of other species from that niche, or the fixation of an adaptation which modified the lifestyle of that species. In such a situation, the following types of change could occur (Wright, 1982a, b):

1 A major mutation could occupy a new niche despite deleterious side-effects, if there was an absence of competition.

2 If the shifting balance theory acted in concert with a major mutation, the fixation of this change would be facilitated. That is, nearly neutral modifiers might reduce the deleterious side-effects of the major mutation. Subsequent fixation of such a gene/modifier complex could occur if the modifiers reached a sufficiently high frequency to cross the saddle that pulls the major mutation to a higher peak.

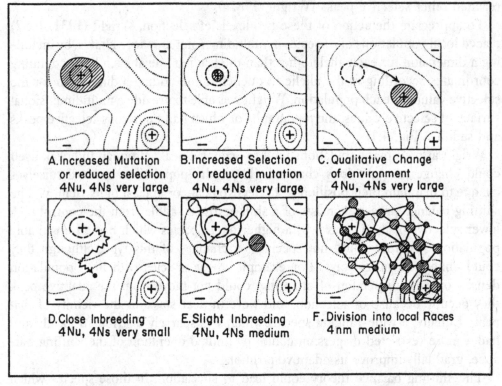

Fig. 4.1. The six regimes illustrated (A to F) represent a portion of the multidimensional array of genotypes of a population with fitness contours. The initial field occupied is designated by broken lines, whereas the derived field is crosshatched. The course of change is indicated by arrows. N (total population); n (local population); u (mutation); s (selection); m (migration). (Redrawn from Wright, 1982b.)

A. Recurrent mutation is balanced by ineffectual selection.

B. Mass selection in a panmictic population constrained by continuing similar conditions.

C. When conditions change the species itself changes to a new peak.

D. Here the population has become so small that accidents of sampling overwhelm selection.

E. The effects of sampling accidents and selection on loci in a small isolated population are small, so that the adaptive fitness of the population wanders off a peak and may eventually move to a new peak.

F. A species which is subdivided into demes where, because of small population size and isolation accidents of sampling, the weak selective pressures are overwhelmed.

3 A change of this type would result in incipient reproductive isolation followed by full speciation under selection against hybridization.
4 A population structure which favoured the shifting balance process, would also favour the fixation of chromosomal rearrangements which lead to speciation because of the aneuploidy of heterozygotes (see Chapter 5).
5 Multiple founding events, particularly by single fertilized females, give an appreciable opportunity for the fixation of negatively heterotic chromosomal rearrangements which may lead to incipient speciation.

The introduction of the concept of peak shifts, and the view that populations which were subdivided into small semi-isolated demes would be most suitable for establishing new species, was not lost on the scientific community. A spate of models defining possible mechanisms for the process of speciation in founding populations eventuated.

4.2 The founder effect (Mayr, 1954, 1982b)

Mayr (1954) proposed a form of allopatric speciation which was based on what he termed the founder principle. This was initially proferred as a very rapid means of genic speciation, but in its more recent guise as 'peripatric speciation' it may be more accurately termed as a mechanism for chromosomal speciation (Mayr, 1982b). The most significant difference between this concept, in its original form, and that of classical genic allopatry, lies in the fact that the derived population is very small and is either a relic, or a founder population. In this respect, Mayr borrowed from Wright's most likely scenario for a speciation event. However, where Wright emphasized that the action of random drift was instrumental in providing the base material on which natural selection could operate, Mayr (1963) argued against the role of drift and reasoned that selection was the primary mechanism in founding populations. Mayr (1954) invoked the founder effect to account for those many situations where he observed rapid speciation in peripheral or island populations. In its simplest sense, a genetic revolution was necessary to break the nexus of the stabilizing effect of gene flow. Such a revolution could occur, if a stable outcrossing organism with a high level of heterozygosity and a homeostatic buffering mechanism was forced by contraction in population size into an inbreeding situation. Homeostatic mechanisms would be disrupted, heterozygosity reduced and unusual homozygotes selected for. Rapid genetic change could therefore be achieved in a population isolate by extreme selective pressures induced by substantial fluctuations in population size. These 'bottle-

necks' forced 'one well-integrated and stable condition through a highly unstable
period to another period of balanced integration' (Mayr, 1963, p. 538). The
product of these interactions was the rapid acquisition of isolating mechanisms,
profound morphological alteration, ecological shifts and speciation of the founding
population.

Most island populations or peripheral isolates which were established by founders, became extinct because of the high levels of inbreeding which increased
homozygosity and thus led to inbreeding depression. In a founding colony, Mayr
(1963) considered the most important event to be the sudden conversion of a
population from an open panmictic situation to a small inbred closed population.
He suggested that the suddenness of this shift was decisive. Thus, 'the mere
change of the genetic environment may change the selective value of a gene very
considerably' (1963, p. 533), (Fig. 4.2).

The salient features which Mayr (1982b) argued would be present in a founder
population include:

1 The loss of genetic variability.
2 A greatly increased level of homozygosity, which would affect the selective value of many genes and the internal balance of the genotype.
3 The possible conditions for an occasional genetic revolution, although
   a genetic revolution was a rare occurrence and was not responsible for
   all instances of speciation.
4 The capacity of the population to pass through heterozygosity rapidly,

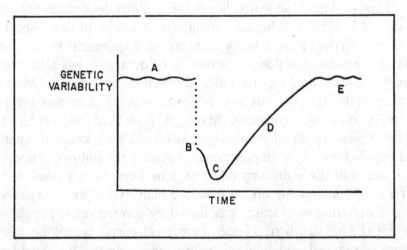

Fig. 4.2. The loss and gradual recovery of genetic variation within a founder population:
A, parental population; B, founders; B to C, genetic revolution; D, recovery of population
in suitable niche; E, new level of variability. (Redrawn from Mayr, 1954.)

even when heterozygotes had reduced fitness, due to the presence of negatively heterotic chromosomal rearrangements.

Mayr suggested that a genetically unbalanced population of this type would be ideally situated to shift into a new niche, since genetic reorganization had the potential to 'loosen up' genetic homeostasis and thus facilitate morphological change.

Mayr (1982b) renamed the founder effect as the peripatric mode of speciation. Chromosome change was thought to be the most likely mechanism responsible for peripatric speciation, although Mayr (1982b) also raised the possibility that transposable elements and regulatory genes may also play a role. Nevertheless, in an earlier publication, Mayr (1963) emphasized that the reduction in the level of genetic variation in a population isolate was a most important characteristic of the founder effect, for the selective forces unleashed by increased homozygosity could affect the whole genome and thus provide for a genetic revolution and subsequent speciation. This is precisely the area which received criticism from other workers; many of these criticisms are trenchant and worth considering:

1 In one of the more circular arguments, Templeton (1982) reasoned that since founders came from peripheral demes, it was difficult for Mayr's model of a genetic revolution with increased homozygosity to operate because many peripheral demes were already characterized by inbreeding and increased homozygosity (but see Section 3.3).

2 Lewontin (1974) argued that while a founder event could affect the frequency of a particular allele, it lacked the capacity to modify the level of genotypic markers such as heterozygosity.

3 Carson and Templeton (1984) pointed to the fact that a population's ability to respond to selection was directly proportional to the amount of genetic variation present. However, in Mayr's model, genetic variation demanded a response when genetic variation was at a minimum and response was thus impossible.

4 Barton and Charlesworth (1984) reasoned that the loss of variability was itself insignificant. That is, a major loss of variability could only occur in populations so small that they were likely to become extinct, and where genetic divergence was impeded by the reduction in variability. They also argued that it was doubtful whether increased levels of homozygosity could increase the chance of a peak shift (a change from one adaptive peak to the next), and that reproductive isolation was much more likely to result from a series of small steps or peak shifts, than a single genetic revolution.

4.3 The flush–crash–founder cycle (Carson, 1975, 1982a)

Carson (1975) had an essentially bidimensional view of the genome. That is, the genome was comprised of an 'open' system of polymorphic loci that recombined readily, without having any deleterious effects on viability. A second component, the 'closed' system, consisted of blocks of genes forming co-adapted, internally balanced gene complexes. These may have been stabilized by the presence of inversions which minimized recombination. Crossing-over within these 'closed' systems resulted in greatly reduced viability. Carson argued that the contents of the 'closed' systems varied between species, but that they were necessarily uniform within species, and that the 'closed' systems were under direct manipulation in founding populations. It is noteworthy that Templeton (1980) equated the 'open' systems with those regions containing the isozyme loci, whereas the 'closed' systems contained the regulatory genes controlling fundamental developmental and physiological processes.

The flush–crash–founder cycle, when viewed as a model for speciation, included many similarities to the founder events described by Mayr (1954). However, it differs in that the major effect is not the fixation or loss of alleles through drift (although this may occur), but a disruption of the closed system. Thus:

1 Following the relaxation of selection, an outbreeding population of an ancestral species underwent a massive flush in numbers.
2 When selection was re-imposed, the population size crashed dramatically, leaving a founder individual in an uninhabited area.
3 Since natural selection had been relaxed during such a cycle, the individual which survived may have been aberrant, with a disorganized co-adapted balance of the 'closed' system.
4 This individual may have successfully founded a small inbred population. The action of selection on the genetic system of this aberrant population would result in a new co-adapted 'closed' system. Carson reasoned that if several of these flush–crash–founder cycles occurred, speciation could also occur (see Fig. 4.3).

White (1978a) regarded this as a particularly speculative model. He pointed out that there was no evidence for two kinds of adaptive genetic systems, nor did he support the view that population 'flushes' played a role in speciation.

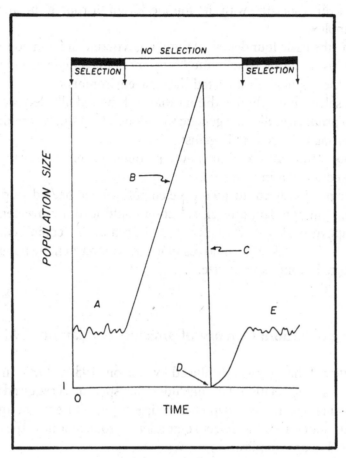

Fig. 4.3. A diagramatic representation of the flush-crash-founder cycle: A, a population under natural selection; B, a population flush; C, a population crash; D, a founder individual; E, a new population which develops and is under natural selection. (Redrawn from Carson, 1975.)

4.4 The founder–flush model (Carson, 1968; Carson and Templeton, 1984)

This model differs from the flush–crash–founder cycle in that the founder event is itself primary, and population changes which may lead to speciation occur subsequent to this.

Carson proposed that drastic events were necessary to modify the closed system, and suggested the following characteristics for the founder–flush model (from Carson and Templeton, 1984):

1 Genetic drift associated with the founder event disrupted the ancestral gene complex.
2 Following the basic founder event there was a period of rapid population growth (flush), during which little of the ancestral genetic variation was lost and the population saturated the new environment.
3 Relaxed selective conditions during the flush period allowed recombination to produce additional genetic variation. This extra recombination would normally be selected against.
4 As the population saturated the new environment, selection reappeared and caused a crash in numbers.
5 Since the previously co-adapted gene complexes had been disrupted by the change in population level, selection could lead a movement to a new adaptive peak, where a new set of co-adapted complexes were produced. This peak shift could result in speciation when the population again started to increase in size.

4.5 The organizational theory of speciation (Carson, 1982a, 1985)

The organizational theory was introduced by Carson (1982a, 1985) in an attempt to unify his views on speciation. He concluded that speciation required the disruption of the parental species' co-adapted genetic system, before the reorganizational process could, under a strong selective gradient, produce a new species with its new co-adapted gene complex.

To achieve this disorganization of the parental genome, significant stochastic forces were required to break the polygenic balances. Carson (1985) considered that chief among these forces was the founder effect, wherein an allopatric daughter population was established in allopatry by one or a few migrants. The major effect was not the fixation or loss of alleles through drift, but a disruption of the polygenic balances through gametic disequilibrium. Carson (1982a) believed that the founder event itself could powerfully disorganize the genome, by upsetting the selection processes which impinge on multilocus systems controlling the integrated developmental, physiological and behavioural traits. Other forces which could disorganize this 'closed' system included:

1 A flush–crash–founder cycle which operated as an adjunct to the founder event.
2 Reduction in population size to a vestige of what it was, could have brought into play those forces responsible for the founder–flush model, where low population numbers destabilized the selective regime. The

genetic composition of the population could have wandered off its pre-existing adaptive peak to a new adaptive peak.

3 The act of hybridization could have broken up the co-adapted gene complexes and released the population from stabilizing selection.

4 Carson argued that Wright's (1932) shifting balance theory was brought into force when a parental population was broken into semi-isolated components. Any one of the isolates could have had its polygenic balance disorganized and provided the opportunity for a shift in adaptive peak.

All of the above-mentioned disorganizational characteristics set the stage for the reorganization of the genome when natural selection was reimposed. Carson believed that the selection pressures that operated during reorganization were directional rather than stochastic, and that the two key areas of operation were either adaptations leading to altered sexual reproduction, or were environmental adaptations.

Carson's model required powerful stochastic changes to disorganize the co-adapted genetic systems. Those mechanisms which he cited are formidable, and involve considerable attenuation of the 'closed' component of the genome. Modification of this system would result in profound effects on viability, development and physiology. This cannot be interpreted as a gradualistic model for speciation, for it involves basically revolutionary concepts. However, there is no more evidence to support this theory, than there is to support the subsidiary concepts involved in it.

4.6 Genetic transilience (Templeton, 1980, 1982)

This model suffers from an unfortunate title and, as Mayr (1982b) and Barton and Charlesworth (1984) point out, it was derived from a term proposed by Galton (1894) which originally described a form of blending inheritance, and is thus quite inappropriate in discussions on genetic revolution. However, while suggesting that peripheral populations are most significant in speciation, Templeton's model proposed that genic changes are gradual and not revolutionary and the following conditions were specified: 'the large panmictic ancestral population represents an optimal structure for yielding genetic transiliences because there is a large potential for inducing a selective bottleneck by using the genetic variability carried by individual founders' (Templeton, 1980, p. 1018). More recently, he considered that this ancestral population should be outcrossed and polymorphic for co-adapted gene complexes which were centred on major loci (Carson and Templeton 1984).

The characteristics of Templeton's (1980) model include:

1 Small founding populations were characterized by inbreeding and gametic disequilibrium.
2 Alleles could be selected for on their homozygous fitness effects, and for their effects on a more stable genetic background.
3 Selection was directed to those multilocus systems controlling integrated developmental, physiological and behavioural characteristics (major loci), which had strong epistatic interactions with other loci (modifier loci).
4 Thus, the fixation of an allele at one critical regulatory locus could have sequential fitness effects at other loci, resulting in a cascading fitness effect. That is, in response to drift and selection, a shift was made to a new adaptive peak (a genetic transilience).
5 Peak shifts were most likely to occur when the founder event caused a rapid accumulation of inbreeding, without a severe reduction in genetic variability.

Templeton (1980) also included the following provisos:

1 A substantial proportion of the genetic variability in a founding population was carried in the form of individual heterozygotes.
2 Mutation was not an important source of variability.
3 When examining the genetic consequences of the founder effect, average observed heterozygosity was a more important parameter than the level of polymorphism.
4 Rare alleles were lost, and resulted in a reduction in allele number, but overall genetic variability, as measured by the percentage of polymorphic loci, or individual heterozygosity, could remain quite high, or be minimally increased, by a founder event.

Templeton therefore envisaged that reproductive isolation in a founding population was more likely if changes occurred at a few major loci followed by co-adaptation at modifier loci, rather than by a single genetic revolution. He also emphasized that genetic bottlenecks would increase the level of homozygosity. Consequently, 'alleles will be selected more for their homozygous effects than they previously were' (p. 1015).

4.7 The dilemma: is it shifting its balance or foundering?

The involvement of founding populations in the process of speciation is a problematical area and remains as one of the more confused and most changed aspects of speciation research. Attitudes range from the denial of a role for the founding population in cladogenic processes (Barton and Charlesworth, 1984), to its pre-eminence as a means of colonizing speciation (Templeton, 1980). Nevertheless, the founder effect should not be seen as a replacement, or alternative, to classical allopatric speciation, but as an adjunct to it. As such, the changes which may occur in founding populations can provide a means for rapid speciation, in particular situations.

Our dilemma is more perverse than this for the concept of a founder effect has become a much changed beast since it was introduced by Mayr (1954). Indeed, the version advocated by Templeton (1980) shared little in common with the original Mayrian view, other than that it described hypothetical events in founding populations. In its basic format, Mayr's concept remains unaltered; a population isolate from an outbreeding parental population is forced to become a small inbred colony. Selective pressures destroy the homeostasis of a genetic system which has been adapted to a panmictic population, thus reducing genetic variability, increasing homozygosity and forcing change from the previously existing stable condition to a new stable condition. The concept remains the same, but in his 1982b publication Mayr suggested that the most likely means of mediating this process was chromosomal mutation. This possibility did not exist in 1954, 1963 or 1970 (and see introductory quotation to Chapter 9). The founder effect is now peripatric speciation. This is most similar to Grant's quantum speciation (1963), a model which was put forward as a chromosomal extension of founder-mediated speciation and which in part had been derived from Wright's (1932) shifting balance theory (see Section 9.2.2 for detail).

The likelihood that a genetic revolution could occur in a founding population was criticized by Templeton (1980), who argued that genetic variability would remain at a high, or increased level, and that peak shifts occur when major regulatory genes are modified by the fixation of alleles at critical regulatory loci. Carson and Templeton (1984) regarded their founder–flush and genetic transilience models as being analogous in respect to the retention of high levels of variability and neither regarded founder events as a genetic revolution. Indeed, the introductory quotation to this chapter leaves little doubt as to what Carson and Templeton (1984) meant: they saw founder events as a means of establishing new isolated populations, not for establishing evolutionary novelty which could induce speciation.

Contrast this view with the organizational theory of speciation (Carson, 1982a,

1985). Here the conditions of the founding population were so extreme as to disrupt the polygenic balances through gametic disequilibrium, a concept not unlike the Mayrian destruction of homeostatic mechanisms. Carson (1985) tacitly accepted that a fixation, or loss, of alleles through gene frequency drift could occur, which would appear on face value to agree with the Mayrian founder effect and to contradict the views expressed in Carson and Templeton (1984). Indeed, the mechanisms Carson opted for in his organizational theory of speciation, which were powerful enough to break up the closed system of the genome, included the founder effect, hybridization, the flush–crash–founder cycle and a founder–flush situation. Thus, while the organizational model is unificatory, it also incorporates basically incompatible mechanisms and has many of the attributes of a genetic revolution, a concept which Carson previously opposed (Carson and Templeton, 1984).

Carson (1990) presented a challenging reappraisal of the impact of population bottlenecks on the genetic variability in founding populations. He argued that most models predicted a decrease in the level of genetic variability in populations which were reduced to a very low level at different stages of their evolution. However, experimental data derived from the analysis of chromosomal polymorphisms in a bottlenecked population of *Drosophila silvestris* (Carson and Wisotzkey, 1989), and morphological analysis of characters which had been through a population bottleneck in *Musca domestica*, showed an increased level of genetic variation (Bryant and Meffert, 1988).

It is possible that an isolated population bottleneck could provide significant shifts in the organization of a gene pool of a species, resulting in an increase in the degree of genetic variability released (Carson, 1990). Such an increase in variability could result from the conversion of balanced epistatic variance to additive variance that then became available to selection. These effects may be greatest on the inheritance of quantitative characters, releasing new variance through the disruption of covariance matrices that underlie and interrelate quantitative traits. Thus, character change in adaptation and speciation may in some instances be promoted by founder events. Indeed, Carson raised the possibility that stress on the genetic system produced by bottlenecking may activate transposable elements producing mutation, hybrid dysgenesis and crossing-over. Such an hypothesis would appear to be at odds with Carson's earlier views and raise the possibility for chromosomal mutation and a genetic revolution (see Sections 4.3 to 4.5 and 10.2.2).

The legacy drawn from Wright's (1931, 1932, 1982a, b) view of rapid speciation in founding populations, not only provided the means for speciating genically or chromosomally, but introduced the concept of peak shifts, major and minor genes, integrated complexes of genes and intergene interactions. While Wright's shifting balance theory has been heavily utilized by all of the authors whose models have been discussed in this chapter, this is definitely not the case when the validity of mechanisms for chromosomal speciation are considered. Indeed, while Grant (1963) provided a role for chromosomal speciation in founding populations and Mayr (1982b) accepted these as a most reasonable mechanism for inducing rapid speciation, neither Carson (1982a) nor Templeton (1982) regarded them as suitable vehicles for initiating speciation. Indeed, this might account for their view on the role of founding populations (Carson and Templeton, 1984). Wright (1982a, p. 10) made a salutary comment in regard to the role of founding populations:

> Finally, the same population structure is favourable both for adaptive character change due to peak-shifts and to incipient speciation from local fixation of a chromosome rearrangement (Wright, 1940, 1941). This population structure is one in which the population is broken up into numerous small colonies, frequently subject to extinction and re-founding by stray individuals (perhaps a single fertilized female) from the more flourishing colonies. This situation is obviously very favourable for a peak shift. If one of the founders happens to be heterozygous for a rearrangement that has been kept at low frequency by selection against the heterozygotes because of high aneuploidy, there is an appreciable chance that the arrangement may drift past the barrier and become homozygous. Whether the peak-shift or the fixation of the rearrangement occurs first, occurrence in a colony gives a favourable start for a new species.

The point appears to be that the different founder effect models considered in this chapter are constrained by the particular views of the authors involved. In this respect, those who do not believe that reproductive isolation can possibly be attained by negatively heterotic rearrangements, appear to have defined models which may have population parameters incompatible with the fixation of such arrangements, or simply fail to consider their fixation. On the other hand, models may have been produced which provide the maximum likelihood for reproductive isolation being attained by major genic changes. That is, the role of founding populations as a novel means of establishing differentiation is based on a predetermined position rather than on observation.

Indeed, this view is supported when recent attempts at computer simulation of the shifting balance theory are considered. Lande (1985) had little difficulty in modelling a scenario where negatively heterotic chromosome changes could reach fixation in founding populations by peak shifts (see Section 6.5 for detail). Rouhani and Barton (1987) provided a case for the fixation of neutral characters in a similar situation. However, Charlesworth and Rouhani (1988) argue that such results are not reasonable and that the probability of a peak shift that induces a significant degree of reproductive isolation is low. This tends to confirm the view that models are only as good as the assumptions made in their construction.

4.8 Genetic variability and the founder effect

Substantial criticisms of Mayr's founder effect/peripatric speciation, by Templeton (1980), Charlesworth *et al.* (1982), Barton and Charlesworth (1984) and Carson and Templeton (1984), have been made on the grounds that there is no evidence for increased homozygosity and loss of genic variation in those population analyses which have been undertaken using protein electrophoresis. Much of the criticism has been based on the analysis of certain lineages of Hawaiian *Drosophila*, where there is an absence of any correlation between changes in the proportion of polymorphic loci and reduction in heterozygosity associated with past founder events (Sene and Carson, 1977; Craddock and Johnson, 1979; Carson and Templeton, 1984). These workers largely agree that data obtained from Hawaiian *Drosophila* provide direct evidence against the founder effect as proposed by Mayr. Carson and Templeton (1984) have suggested that the data give tacit support for the founder–flush and genetic transilience models. The question is, of course, is this a realistic interpretation of the situation, and does the evidence provide support for any of the models which have been advocated?

The Hawaiian *Drosophila* are a highly complex group of species with lineages of diverse origin. Some of these lineages have been used by Carson as examples of classic founding situations, a view given substantial support by Mayr (1982b). While the electrophoretic data lend no support to the Mayrian view of a founder effect, more recent analyses comparing electrophoretic genetic distances and the evolution of mtDNA in the same lineage provide a relatively different picture.

A comparison between the *alpha* and *beta* lineages of the *planitibia* subgroup of Hawaiian *Drosophila* was made by Templeton (1987), using genic distances on isozyme data derived from Johnson *et al.* (1975) and mtDNA distances obtained from DeSalle (1984). These lineages were chosen because the *alpha* lineage (*D. cyrtoloma*, *D. melanocephala* and *D. neoperkinsi*), was thought to have speciated by the geographic subdivision of a once continuous population when the Hawaiian islands of Maui, Molokai and Lanai were isolated by a change in sea level. The *beta* lineage (*D. silvestris*, *D. differens*, *D. planitibia* and *D. hemipeza*), on the other hand, is found on islands which were never connected, because of the presence of deep ocean channels which separate them. These populations are thought to have been derived from founder individuals (Fig. 4.4).

Comparisons of the genic distances showed that interlineage distances are greater than intralineage distances, and that both lineages have the same rate of evolution. In contrast, the mtDNA distances showed that the interlineage distance is only slightly larger than the intralineage distance for the *beta* group, thus suggesting a much faster rate of evolution in the mtDNA genome in the *beta* animals than the *alpha* animals. Templeton (1987) interpreted these data as indi-

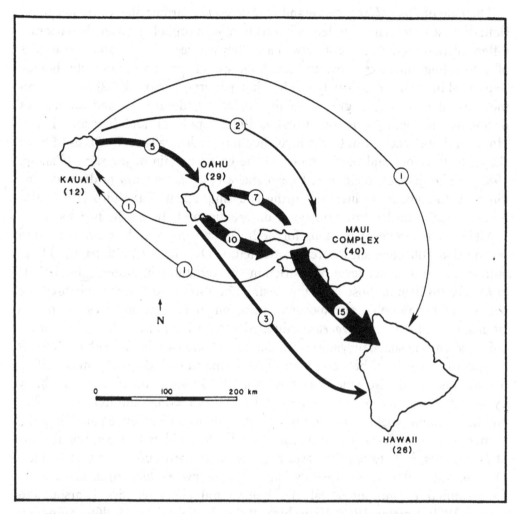

Fig. 4.4. The founder events involved in the radiation of the picture-winged *Drosophila* in the Hawaiian islands. The number of species is in parentheses, whereas the width of the arrows indicates the proportion of those originating from founders. (Redrawn from Carson, 1983; Carr *et al.*, 1989.)

cating that founder events accelerate the rate of mtDNA evolution, when compared to the nuclear DNA component.

Carson and Templeton (1984) emphasized 'that both founder–flush and genetic transilience theories predict that different loci or types of genetic variation will behave differently during founder and subsequent events' (p. 125). Thus, 'the support for founder-induced speciation from natural examples should not be based upon only one type of genetic variation, but rather, on the joint pattern defined by different types of genetic variation' (p. 125).

The case of *Drosophila silvestris* and *D. heteroneura* further illustrates this point. Both of these species have undergone considerable morphological and behavioural differentiation on the island of Hawaii and they are regarded as typical examples of a founding situation. Nevertheless, both species are chromosomally homosequential in terms of their polytene banding pattern (Carson, 1982b, 1983), and they also show a high degree of genic similarity without any fixed differences distinguishing them (Sene and Carson, 1977; Craddock and Johnson, 1979). However, both species can be distinguished morphologically on the basis of head shape, body colour and pigmentation of the costal margin of the wing (Carson, 1982b). The distinctive elongated head shape of *D. heteroneura* (a derived condition) is thought to be involved in the courtship ritual. Kaneshiro (1983) also found a significant level of ethological differentiation between the two species.

MtDNA restriction site maps for both *D. heteroneura* and *D. silvestris* from seven major collecting sites were constructed by DeSalle *et al.* (1986a, b). These authors found a higher level of restriction site variability in eastern populations of *D. silvestris* than in those from the west. They also found that this mirrored the amount of polymorphic chromosomal variation in eastern and western populations. In addition, *D. heteroneura* exhibited a lower level of restriction site variability and chromosomal polymorphism than did *D. silvestris*, but as with *D. silvestris* a concordance existed between both of these metrics. Perhaps the most telling feature was that the isozyme data showed no difference in the level of variability between the two species (Sene and Carson, 1977; Craddock and Johnson, 1979). Neither chromosomal nor electrophoretic data showed divergence paralleling the numerous fixed differences present in the mtDNA, either between populations of *D. silvestris*, or between the eastern and western distributions of that species. DeSalle *et al.* (1986b) subsequently found that the two *D. silvestris* lineages could be identified by morphological and behavioural characteristics (Carson and Bryant, 1979; Carson, 1983; Kaneshiro, 1983). DeSalle *et al.* (1986b) were then able to construct a phylogeny for the *D. heteroneura/D. silvestris* complex using mtDNA restriction site analysis.

Much emphasis has been placed on the apparent failure of electrophoretic data to produce results which are equitable with certain founder effect models. Not surprisingly, both Carson (1982a) and Mayr (1982b) took the view that the value of the electrophoresis of soluble proteins was itself questionable. Mayr (1982b, p. 8) made the following unambiguous statement: 'Nevertheless, this method [electrophoresis] supplied the extremely important evidence that in speciation there is no major involvement, if any at all, of the classical enzyme genes. There are multiple proofs of this conclusion, perhaps the most convincing being that the rate of isozyme replacement is no more rapid in actively speciating phyletic lines than in conservative ones'.

In reality, the isozyme data could be interpreted as suggesting that no founder effects or population bottlenecks had occurred, or, alternatively, that founder events had occurred, but that these had been followed by substantial population increases (Templeton, 1987). Any detectable isozyme-related founder effect had been masked by the dynamics of the speciating populations. Clearly, the mtDNA data indicate that the latter explanation may be the most acceptable and that the founding populations had undergone considerable genomic reorganization during speciation. Whether this may also be interpreted as a genic revolution is open to debate. Nevertheless, the data obtained from the mitochondrial genome also suggest that some founding populations have a higher rate of evolution, and this appears to lend some support to Mayr's concept.

If Templeton is correct, it would appear that we are no longer in a position to interpret what has happened in the evolution of a founding population from electrophoretic data. Predictions on the increase or decrease in heterozygosity, or changes in the proportion of polymorphic loci are blinded by the sequence of changes present in founding populations. That is, the analysis of species which are thought to have been the product of a founder event does not provide information on changes associated with the founder effect, but with post-founder events.

In conclusion, a consideration of the different founder effect concepts suggests that differentiation in founding populations is a means for rapidly establishing cladogenic differences. It would appear that severe selective regimes associated with the bottlenecking of founding populations could lead to the fixation of rare mutants as homozygotes. Indeed, the dynamics of founding populations provide the only realistic opportunity for the fixation of profoundly deleterious rearrangements which characterize many derived species. The models generally opt for either major genes (regulatory genes, phenodeviants), complexes of genes (interaction systems, the closed system, major modifier complexes), and/or negatively heterotic chromosomal rearrangements as being the chief agents in attaining the degree of reproductive isolation necessary for speciation. In all cases, the fixation of one or more of these agents in a population isolate is necessary to make this transition. Species most suited to this mode of speciation are low vagility forms characterized by a highly subdivided population structure of small isolated demes, which experience numerous founding and extinction events. The fixation of rare mutant genes or chromosome changes which have significant and sometimes deleterious side-effects may be the driving force toward speciation in such populations.

5
Chromosomal rearrangements as post-mating isolating mechanisms

Over 90 per cent (and perhaps 98 per cent) of all speciation events are accompanied by karyotypic changes and in the majority of these cases, the structural chromosomal rearrangements have played a primary role in initiating divergence.

(White, 1978a, p. 324)

and

I contest this view and consider that in most cases their fixation is likely to be merely an incidental accompaniment of small population effects and forced selection for reorganization as the species is formed.

(Carson, 1982a, p. 425)

The view that the fixation of chromosomal rearrangements can act as a barrier to gene flow has as its very basis the assumption that chromosomally differentiated forms have the potential to produce hybrids when the parental and daughter populations come into contact. It is at the time of hybridization that chromosomal rearrangements may act as post-mating isolating mechanisms by preventing hybrid formation or conveying deleterious fertility or viability effects to the hybrid individual. That is, there is no requirement that parent/daughter populations have acquired genetic or morphological differences, pre-mating isolating mechanisms, or major gene differences affecting reproductive capacity. Chromosomally divergent populations may be identical in all respects, but for the structural rearrangement which distinguishes them. Thus, there is no need for prolonged isolation to allow for the fixation of numerous morphological and genetic differences. Rather, the duration of isolation required is the time necessary for the negatively heterotic change, or changes, to reach fixation in the daughter population. It is possible that this could be a very rapid process in an isolated and colonizing population where meiotic drive occurred. It would be less rapid if these criteria were not present, but probably no longer than the time taken for a single gene difference to reach fixation within a population.

5.1 Types of structural rearrangements implicated in reproductive isolation

If negatively heterotic rearrangements are to reach fixation in isolated populations, a juxtaposition must occur between the impact of the rearrangement on fertility or viability in the structural hybrid and the ultimate survival of that organism. It would appear unlikely that a severely deleterious rearrangement which produced high levels of infertility, or which greatly affected viability, would be able to reach fixation in a population isolate because of the effect on the carrier. Hybrids which are rendered totally inviable or infertile by the rearrangement which they carry in heterozygous form cannot contribute.

White (1978a) recognized these difficulties and suggested that in those cases where species were distinguished by multiple fixed rearrangements it was probable that each rearrangement would have had a minor effect on fertility and would have been fixed independently and sequentially. Such rearrangements would be able to act as profound isolating mechanisms, since their independent fertility effects could act in concert in an interspecific or interracial hybrid.

However, such an explanation cannot account for the fixation of profoundly deleterious rearrangements which occur as single or multiple fixed differences between species, and which are often encountered. Rearrangements such as sex-linked translocations, or tandem fusions, which lead to the sterility of one sex, or at least 50% infertility in a carrier, are commonly encountered. Equally, multiple rearrangements which have less damaging fertility effects, but which are thought to have reached fixation simultaneously have done so despite their additive fertility effects.

The most reasonable explanation for the fixation of these particularly deleterious chromosomal rearrangements must be sought in the dynamics of the populations in which they reached fixation (see Chapter 6), and the likelihood that such rearrangements were driven to fixation by meiotic drive.

5.1.1 Potentially negatively heterotic changes

Of the spectrum of chromosomal changes that can occur, it is only those few rearrangements which are negatively heterotic, or are potentially negatively heterotic, that can possibly play a role in cladogenic processes. These include those rearrangements which malsegregate at meiosis in structural hybrids, producing aneuploidy, duplications and deficiencies, or which disrupt meiosis entirely. Rearrangements that have the capacity to be powerfully negatively herotic are tandem fusions, reciprocal translocations, centric fusion or fission, multiple centric fusions, multiple centric fusions which share brachial homologies, X-autosome translocations and pericentric and paracentric inversions (see White, 1973). However, if normal segregation of the meiotic products occurs in heterozygotes for these changes they will not form a post-mating isolating mechanism, but will occur as chromosomal polymorphisms. It is not at all certain that these disruptive changes will always be involved in speciation, although they clearly have the potential to be.

5.1.1.1 Tandem fusions

Tandem fusions may be formed by the direct fusion (without chromosomal loss), of telomeric–centromeric and telomeric–telomeric chromosomal regions. They are primarily observed in lineages of species with acrocentric karyotypes and have not been encountered as balanced polymorphisms. When two acrocentric chromosomes fuse tandemly producing a large acrocentric with two centromeres, one of the centromeres can be deactivated and may remain as a C-positive heterochromatic region. This region may diminish with time and eventually disappear. Tandem fusions occur between euchromatic chromosome regions and do not generally involve heterochromatic arms (Elder and Hsu, 1988).

Heterozygosity for tandem fusions in uniarmed chromosomes can have a considerable impact on fertility resulting in an automatic reduction in fertility by 50%. Such a high level of infertility can lead to incipient speciation (Fig. 5.1) (White, 1973; Wright, 1982b). Since the introduction of G and C-banding techniques, it has become apparent that powerfully negatively herotic changes, such as tandem fusions, have played a significant role in chromosomal speciation. The difficulty of determining homologies had in the past led to an underestimation of their significance. However, numerous species are now known to be distinguished by single tandem fusions: *Uromys*, Baverstock *et al.* (1983a); Cricetine rodents, Engstrom and Bickham (1982) and Baker *et al.* (1983a); *Gehyra australis*, King (1982); and *Gehyra purpurascens*, Moritz (1984). In other cases multiple tandem fusions have prevailed. In the muntjac deer some 17 tandem fusions and three Robertsonian fusions have been established in the lineage leading to *Muntiacus muntjak vaginalis* (Liming *et al.*, 1980). In the cotton rats of the genus *Sigmodon*, nine tandem fusions separate *S. mascotensis* from *S. hispidus* and ten (nine of which are the same) separate *S. arizonae* from *S. hispidus*. Six Robertsonian fusions are also present in this dichotomy (Elder and Hsu, 1988).

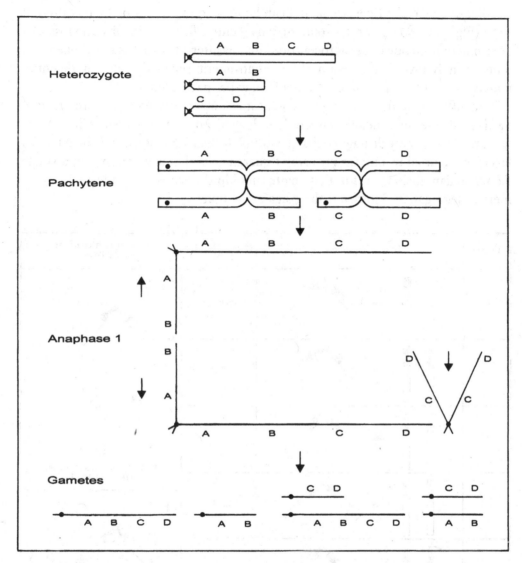

Fig. 5.1. Meiotic segregation in an individual heterozygous for a tandem fusion involving acrocentric chromosomes. Here, 50% of all gametes carry duplications or deficiences. (Redrawn from White, 1973.)

5.1.1.2 Robertsonian fusions and reciprocal translocations

Robertsonian fusions, centric fusions or chromosome fusions form the most common type of structural rearrangement encountered between species or chromosome races. They predominate in species with an acrocentric or telocentric karyomorph and frequently form balanced chromosomal polymorphisms. They always involve the fusion of two uniarmed chromosomes and result in the production of a metacentric element and can be formed in a series of subtly different ways (Fig. 5.2). Some chromosome fusions (centric fusions), involve the loss of a centromere from one of the acrocentric elements after a two-break rearrangement, others simply involve the fusion of paracentromeric regions producing dicentric metacentrics. One of these centromeres may be inactivated.

The difficulty in determining which chromosomal rearrangements are going to produce deleterious meiotic effects has dogged population cytogenetics. Apart from those markedly damaging changes which induce lethality, or produce known levels of aneuploidy, one simply cannot guarantee with any certainty that a single Robertsonian fusion, or for that matter multiple fusions, will adversely affect meiotic segregation. However, the rules are simple:

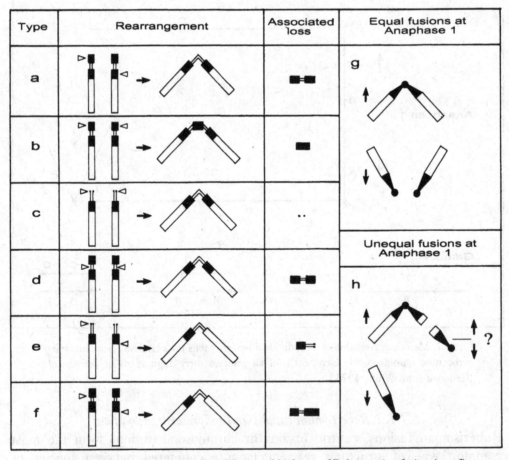

Fig. 5.2. This figure summarizes six of the possible forms of Robertsonian fusion (a to f). C-positive paracentromeric heterochromatin is shown in black. The possible segregation difficulties at anaphase associated with equal and unequally sized fusion chromosomes are shown in g and h. (Redrawn in part from John and Freeman, 1975.)

1 The effect on fertility of heterozgyosity for any single chromosome fusion will be proportional to the capacity of the meiotic system of that organism to segregate the meiotic products in a balanced fashion.
2 Not all chromosome fusions are going to be negatively heterotic, just as they will not all be positively heterotic or neutral.

Consequently, the segregation patterns of heterozygous rearrangements at metaphase and anaphase 1 in meiosis will determine whether such changes will enter the population as a balanced or transient polymorphism, or as a negatively heterotic change which can take part in cladogenesis.

There are, of course, a number of aspects of chromosome structure which may enhance the deleterious fertility effects of chromosome fusions. These are:

1 When single chromosome fusions are present, it is probable that substantial differences in the arm lengths of the chromosomes involved would increase the incidence of malsegregation due to orientation problems in meiosis (see Fig. 5.2 and also White, 1973). That is, it is possible that equally armed chromosomes may segregate more easily because of simple mechanical advantages, although this does not mean to say that equal armed fusions cannot be negatively heterotic.
2 When a series of chromosome fusions which share monobranchial homologies have been established, malsegregation may be enhanced by a larger number of chromosomes being involved in the multivalent, thus increasing the chance of segregation-induced errors (Fig. 7.3). Similarly, the involvement of chromosomes with disproportionate chromosome arm lengths may induce malorientation and malsegregation when complex chain multiples are forced through meiosis.
3 Pairing failure in the paracentromeric regions of trivalents or multivalents may lead to non-homologous pairing, or association with unpaired chromosomes such as the sex chromosomes, thus involving the X-inactivation process and its profound fertility effects.

In the case of reciprocal translocations, heterozygosity may result in the production of multivalents (chain or ring) at metaphase, which segregate to produce 50% euploid and 50% aneuploid gametes. Such a high level of aneuploidy invites the production of inviable or lethal gametes (White, 1973). However, there are also numerous cases where apparently detrimental reciprocal translocations which form complex multivalents have no problems with meiotic segregation (see Fig. 5.3). The evening primrose (*Oenothera*), which forms rings at metaphase which orientate in a zig-zag pattern at anaphase, is a well-established example (Cleland 1962) (Fig. 5.4). Segregational problems and effects on fertility and viability in

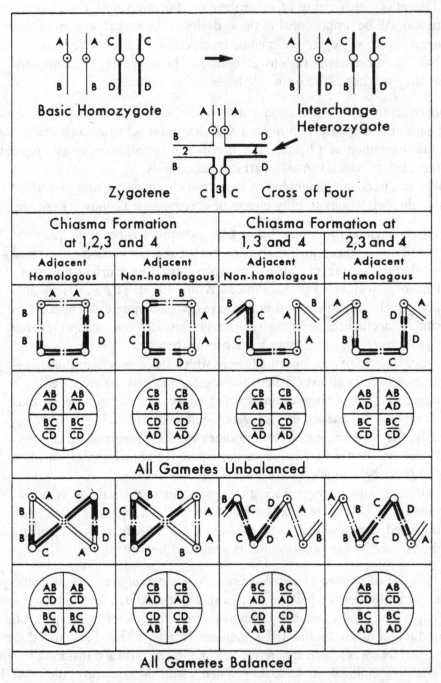

Fig. 5.3. Structural heterozygosity for a single reciprocal translocation can lead to the formation of both balanced and unbalanced gametes, depending on the form and orientation of multivalents at meiosis. (Redrawn from John, 1976.)

reciprocal translocations may be enhanced by pairing failure and the involvement of sex chromosome associations.

5.1.1.3 X-chromosome effects – a special case?

The association of sex chromosomes and sterility was first made by Haldane (1922) who commented 'When in the F1 offspring of two different animal races one sex is absent, rare or sterile: that sex is the heterozygous (heterogametic) sex' (p. 101). Coyne and Orr (1989b) argued that this correlation should not be connected with sex, but with sex chromosomes, for the genes having the greatest effect on hybrid sterility and inviability are X-linked. That is, post-mating reproductive isolation is caused largely by genetic changes on the X chromosome.

Of the numerous hypotheses that Coyne and Orr considered to account for the Haldane effect they observed in hybrid *Drosophila*, they opted for the difference in evolutionary rate between autosomes and sex chromosomes. That is, in all cases except for selection affecting the homogametic sex alone, increased recessivity resulted in further evolution of the sex chromosomes. Thus, alleles that ultimately cause sterility, or inviability, tend to accumulate on the sex chromosomes if they are recessive and have favourable effects on the normal genetic background. Second, they also suggest that both male and female sterility/inviability alleles accumulate faster on the X chromosomes than the autosomes. Third, more alleles causing male than female hybrid sterility and inviability will accumulate per unit of time and these genes will cluster on the X faster than on the autosomes. The critical assumption is that sterility/inviability genes are different to those causing morphological or behavioural differences between species. This approach contradicts the modern synthesis 'that reproductive isolation arises simply as a pleiotropic byproduct of the divergence of "ordinary" genes via natural selection or genetic drift' (Coyne and Orr, 1989b, p. 197). However, it provides a convenient and plausible pleiotropic argument for the profound bias toward sterility in male hybrids compared to females.

This support and explanation for the Haldane effect was based on extensive surveys of interspecific hybridization events between *Drosophila* species accumulated by Bock (1984), and Coyne and Orr (1989a). Unfortunately, although both studies indicated that a Haldane effect occurred, neither was sufficiently discriminatory in its handling of the data to prevent the conclusion being reached that the differences could be due to other factors. In particular, Coyne and Orr made no attempt to distinguish between those hybridizing species pairs which had structurally reorganized karyotypes and those which did not. Thus, all fertility effects in F1 hybrids due to structural hybridity were ignored.

The fact of the matter is that absolute male sterility associated with female fertility is a hallmark of 75% of all translocations of the X chromosome and autosomes in *Drosophila* (Lindsley and Tokuyasu, 1980). This sterility is chromosomal and not genic. Such a conclusion is supported by the following evidence from *Drosophila*:

1 Other chromosomal rearrangements not involving the X and autosomes are male fertile.
2 All translocations with new terminal break-points in the X and autosomes, or associated heterochromatin, are fertile, whereas others are sterile.
3 The male sterility of X-autosome translocations is dominant. Inactivation of genes in the vicinity of a break-point is expected to be recessive. However, it is not possible to 'rescue' fertility by adding a duplication that covers the X-chromosome break-point.
4 Male sterile translocations share a commonality in effect on spermiogenesis (incomplete elongation of the sperm nucleus).

It would be unreasonable to suggest that genes inactivated at so many different break-points all had the same effect (Lindsley and Tokuyasu, 1980).

Lifschytz and Lindsley (1972) argued that reciprocal X-autosome translocations interfered with the asynchronous inactivation of the X chromosome and the autosomes. It is worth noting that X-autosome pachytene associations are a common occurrence in meiosis of structural hybrids between vertebrate species. Here, unpaired synaptonemal complexes, or chromosome arms in heterozygous multivalents, mispair or associate with sex chromosomes (see Section 7.3). The fertility effects of such associations produce male sterility and impaired fertility in females, the same effect that Haldane detected. Thus, although different forms of sex-determining mechanisms are present in *Drosophila* (the genic balance system) compared to vertebrates (the dominant Y system), the potential for sex chromosome/autosome interaction due to structural rearrangements cannot be ignored in either group.

There are numerous chromosomally derived species, in many bisexual vertebrate and invertebrate lineages, in which heteromorphic sex chromosomes are present, and which are distinguished by either structurally reorganized sex chromosomes or X-chromosome/autosome rearrangements. X–autosome rearrangements are a powerful reproductive isolating mechanism due to their direct impact on sex-determining mechanisms or on the X-inactivation process (see Fig. 7.5).

5.1.1.4 Inversions

Pericentric and paracentric inversions form a recognizable second class of rearrangement. Both forms of inversion have the propensity to produce lethal or

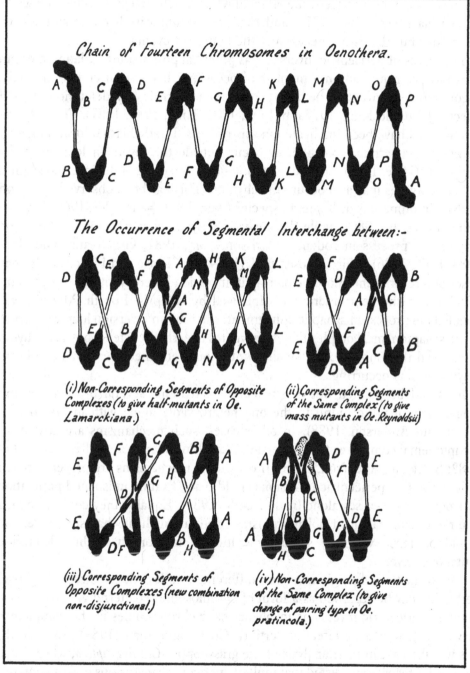

Fig. 5.4. Complex heterozygotes in *Oenothera* showing normal and anomalous pairing in a chain and rings of different sizes. The first two rings, (i) and (ii), yield mutant gametes due to crossing-over in differential segments producing an interchange. Type (iii) fails to produce viable mutants, whereas type (iv) does. (From Darlington, 1931.)

deleterious meiotic products, the effect of which can be maximized by the location of chiasmata (see White, 1973, and Fig. 5.5) resulting in dicentric and acentric fragments, but this is not necessarily the course of events.

The status of pericentric inversions as potential post-mating isolating mechanisms has been queried on a number of occasions, because of certain anomalies encountered in inversion heterozygotes which override any deleterious meiotic effects (John and King, 1977a, b; John, 1981; King, 1984, 1987). Four separate mechanisms have been implicated in this process. First, chiasma location may be changed to a terminal position, or occur outside the inversion loop formed at pachytene (Hale, 1986). Second, non-homologous pairing of the inverted segment in heterozygotes may also occur at pachytene. This has been observed by meiotic analysis in numerous orthopteran species (Nur, 1968; Schroeder, 1968; John and King, 1977a, b, 1980), while the analysis of synaptonemal complexes has confirmed this process in rodents (Davisson et al., 1981; Greenbaum and Reed, 1984; Hale, 1986). Third, Moses et al. (1984) and Tease and Fisher (1988) have described reverse chromosome pairing in inversion heterozygotes, i.e. association of telomeric and centromeric ends followed by synapsis. Fourth, Moses (1977) detailed the process of synaptic adjustment, whereby an inversion loop was formed in the synaptonemal complex of heterozygotes in early pachytene, but as pachytene progressed the loops became smaller until none remained, i.e. non-homologous resynapsis was occurring.

It should be pointed out that in certain situations several of these synaptic mechanisms may be found in the one heterozygote along with inversion loops (Stack and Anderson, 1988). In other cases, such mechanisms are absent and synaptonemal complexes form inversion loops (Poorman et al., 1981; Moses et al., 1982; Stack and Anderson, 1988). If overriding mechanisms are present, there is little doubt that pericentric inversions could reach fixation in isolated populations by stochastic processes alone (sensu Lande, 1984). Equally, they could remain as neutral and/or balanced polymorphisms. Without such mechanisms, inversions would be faced with the same difficulties of fixation that other deleterious mutations have.

The formation of single inversion differences, unless demonstrably negatively heterotic, may be regarded as a form of chromosome repatterning. Nevertheless, multiple pericentric inversions involving many chromosomes in the complement may have a cumulative effect on fertility. Coates and Shaw (1984) proposed that this was the case in their analysis of the grasshopper *Caledia captiva*, where inviability of F2 hybrids suggested that multiple pericentric inversions had the potential to be effective post-mating isolating mechanisms. However, in many of the species in which multiple pericentric inversions had been encountered, such as the rodents *Neotoma micropus* (Warner, 1976) and *Peromyscus* species (Greenbaum et al.,

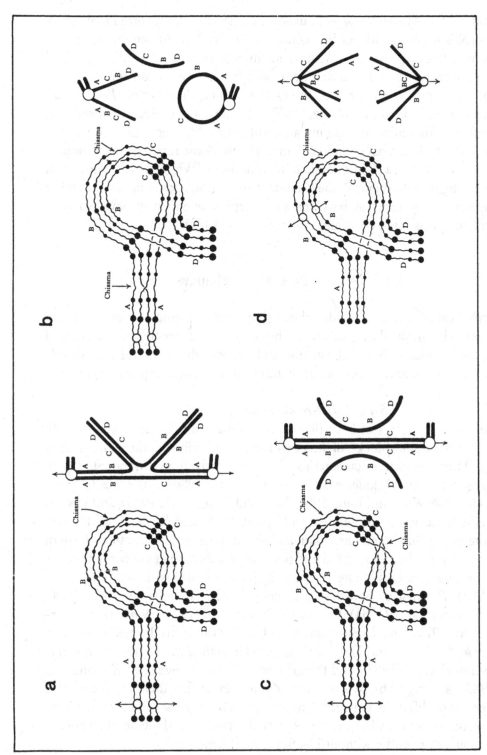

Fig. 5.5. a, b and c. Heterozygotes for paracentric inversions at pachytene (left) and at first anaphase (right), showing the differing meiotic products which vary according to the position of chiasmata within, or outside, inversion loops. d. A heterozygote for a pericentric inversion at pachytene (left) and first anaphase (right), showing the unbalanced meiotic products when cross-over occurs within the inversion loop. (Redrawn from White, 1973.)

1978), and the fish *Ilyodon furcidens* (Turner *et al.*, 1985), they involved high levels of polymorphism and could not be claimed to be involved in speciation. There are, however, lineages of species, the constituents of which are distinguished by fixed pericentric inversion differences, which may have been implicated in speciation. The phylogeny for the genera *Pan, Homo, Gorilla* and *Pongo* is an interesting example (Sites and Moritz, 1987). The impact of pericentric inversions on speciation is therefore ambiguous and will probably remain so. Paracentric inversions, which are a most common form of chromosome repatterning in organisms such as the Diptera, Chironomidae or Simulidae (White, 1973), and which are also potentially deleterious in their heterozygous state, appear to be in a similar situation; they often occur as balanced polymorphisms and may have played a minor role in speciation (see White, 1973; Zouros, 1982).

5.1.2 Neutral or adaptive changes

The third class of change includes those structural rearrangements which are thought not to be involved in speciation. The two forms of repatterning implicated are, first, the addition of heterochromatin, and, second, those neutral or positively heterotic rearrangements which occur as transient or balanced polymorphisms.

5.1.2.1 Heterochromatin addition

A vast number of species of both plants and animals are characterized by both fixed and polymorphic blocks of heterochromatin distributed throughout their genomes. These include organisms as diverse as hedgehogs (Mandahl, 1978), amphibians (King, 1990), killer whales (Árnason *et al.*, 1980), numerous Orthoptera (Webb, 1976; King and John, 1980; John and King, 1982, 1983), and flowering plants (the Liliaceae, Greilhuber and Speta, 1976, and *Scilla*, Vosa, 1973).

It is noteworthy that heterochromatin addition is a particularly common form of change in rodents, where both fixed and polymorphic differences occur. Examples showing pronounced C-banding variation include: *Thomomys*, Patton and Sherwood (1982); *Peromyscus*, Bradshaw and Hsu (1972), Greenbaum *et al.* (1978), and Robbins and Baker (1981); *Onychomys*, Baker *et al.* (1979); *Neotoma*, Mascarello and Hsu (1976); and *Uromys*, Baverstock *et al.* (1982). In the pocket gopher, *Thomomys bottae*, Patton and Yang (1977); the white-tailed rat, *Uromys caudimaculatus*, Baverstock *et al.* (1982); and the grasshopper *Atractomorpha similis* John and King (1983) (see Fig. 5.6), hybridization studies reveal that there is no evidence for reduced hybrid fertility between cytotypes possessing gross structural heterozygosity for the amount and location of heterochromatin. In the case of *A. similis*, hybridization was possible without ill-effects to at least F3.

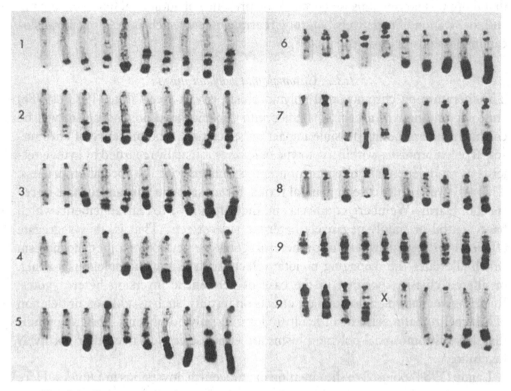

Fig. 5.6. Chromosomal polymorphism variants for each chromosome of *Atractomorpha similis*. Note the enormous differences in the position and quantity of C-positive heterochromatin in each chromosome. (From John and King, 1983.)

A number of claims have been made associating the involvement of heterochromatin addition to speciation. Yunis and Yasmineh (1971), Corneo (1976), Hatch *et al.* (1976) and Fry and Salser (1977) have all argued that satellite DNA, which is located in constitutive heterochromatin, is fundamental to the pairing process in meiosis. They suggested that fertility barriers could be formed between species or cytotypes which differ in the chromosomal distribution of heterochromatin. The supposition is that pairing in meiosis would be disrupted. John and Miklos (1979) and Miklos *et al.* (1980) dealt thoroughly with these arguments in both insects and mammals, where heterochromatin variation failed to show any demonstrable impact on pachytene pairing.

Miklos *et al.* (1980) have considered a second possible role for heterochromatin/satellite DNA addition in speciation. That is, that large amounts of heterochromatin, or satellite DNA, in the genome, allow other types of chromosome rearrangements to occur, or facilitate their fixation, and that these structural rearrangements are major factors in reproductive isolation. These authors found that there was no relationship between the presence of additional heterochromatin and the fixation of negatively heterotic rearrangements such as reciprocal translocations, fissions or fusions.

5.1.2.2 Chromosomal polymorphism

The literature on chromosomal polymorphism is vast (see White, 1973, 1978a) and only one narrow aspect of studies into chromosomal polymorphism will be considered here, i.e. its possible impact on speciation. The presence of chromosomal polymorphisms within a cytotype or species is usually regarded by cytogeneticists as an independent phenomenon not associated with the speciation process. This may well be the case in general terms, for most are in a balanced state in respect of Hardy–Weinberg equilibria and they often involve rearrangements which have neutral or mildly negatively heterotic side-effects. That is, they segregate efficiently at meiosis when fusion heterozygotes, or have some other mechanism which alleviates the damaging meiotic effects such as non-homologous pairing, or altered chiasma position in the case of pericentric inversion heterozygotes. Rearrangements which do not have effects on fertility, such as addition or deletion of heterochromatin, have no difficulty in forming polymorphisms. The end point is that these chromosomal polymorphisms are either selectively neutral or positively heterotic.

Lande (1984) pointed to the situation of paracentric inversions in Diptera. Here there was no selection against heterozygotes, for aneuploidy associated with these inversions was overcome by the absence of recombination in males. Moreover, female aneuploid gametes which were produced by recombination within a heterozygous inversion, generally segregated to the polar bodies rather than to the egg nucleus. The observation that nearly half of the these 'quasinormal heterozygotes' had a heterozygous advantage and only a few were at a homozygous advantage, explained why stable paracentric inversion polymorphisms were common in *Drosophila* and other Diptera.

In reality, all rearrangements which pass through a heterozygous condition are polymorphisms in a cytological sense. This distinction between the different forms of polymorphism is the time that they remain in this multiformed state. That is, a negatively heterotic rearrangement may exist as a transient polymorphism for a very short time before the new rearrangement reaches fixation by chance, or is driven to fixation. In contrast, some transient polymorphisms which have less deleterious side-effects, or are neutral, may remain in populations for hundreds if not thousands of years before collapsing into monomorphism. Presumably other balanced polymorphisms remain as just that until that species becomes extinct.

There has been some debate as to whether polymorphisms could actually collapse into monomorphism (see Shaw *et al.*, 1986). This is an odd argument, if one accepts the view that all chromosomal rearrangements must go through a polymorphic state (of unknown duration) before they become monomorphic. Whether they 'collapse' or otherwise is less than significant, for their transience is proved by the numerous examples of derived monomorphism. White (1973) viewed polymorphisms as being adaptive in certain conditions and suggested that they were maintained by selective gradients. Thus, a change in environmental parameters could lead to a collapse of this gradient and may result in a collapsed polymorphism. This is certainly the view advocated by King (1981, 1982). There are numerous publications describing clines in the distribution of different chromosomal polymorphisms in a range of organisms (see da Cunha and Dobzhansky, 1954; Carson, 1955; da Cunha *et al.*, 1959; White, 1973 and Section 3.3). Both environmental and geographic gradients have been defined and correlated with variation of this type. For example, Shaw *et al.* (1985) proposed that variation in the structure of the karyotype in terms of the centromere position in inversion heterozygotes, may provide an adaptive system independent of other forms of genetic variation. Indeed, they found a correlation between bivoltinism with a metacentric karyotype and univoltinism with an acrocentric karyotype in the grasshopper *Caledia captiva* (but see Section 11.3).

Chromosomal polymorphism has not generally been associated with concepts of speciation, although the triad hypothesis of Wallace (1966) is a clear exception. This is dealt with more fully in Section 9.1.1.

5.2 Chromosome change and speciation: the theoretical approach

In the previous section the difficulties associated with ascertaining a precise role for particular structural rearrangements in speciation have been discussed. Chief among these is the difficulty of ascertaining the meiotic impact of a rearrangement as a structural heterozygote. Nevertheless, a strong case exists that neither chromosomal polymorphisms for any structural rearrangement, nor fixed addition differences are involved in that process.

With this in mind, it is apparent that a number of theoretical overviews, which are widely cited in the literature, have attempted to assess the role played by chromosomal rearrangements in speciation. These have used very large databases encompassing most animals which have been studied cytogenetically in particular systematic groups. In these studies, rearrangements were grouped into crude classes for ease of handling. Conclusions were then reached as to the role of chromosome change in speciation. Unfortunately, these databases had considered rearrangements not associated with speciation.

For example, in a paper directed at investigating the relationship between karyotype alteration and species differentiation in mammals, Imai (1983) made the following assessment of the chromosome rearrangements he used: 'Visible chromosome rearrangement denotes in this paper a series of rearrangements changing chromosome number and/or arm number (i.e. Robertsonian rearrangement, pericentric inversion, or tandem growth of constitutive heterochromatin in short arms)' (p. 1154).

Further on, Imai (1983) stated: 'In a species having chromosome polymorphisms, I count each homomorphic karyotype in a chromosome polymorphism as one' (p. 1155). That is, the analysis that Imai made was, to a large extent, based on chromosomal changes that have an ambiguous role in speciation (addition of constitutive heterochromatin and chromosomal polymorphism). Interestingly enough, of the 967 mammals analysed, 293 were from the Rodéntia, a group in which these types of chromosome change predominate. Clearly, these data cannot provide a valid comment on speciation, and it is not surprising that Imai found against a role for chromosomal rearrangements in speciation.

While Imai's paper argued against chromosome change acting as a primary agent in speciation, a series of other papers which are equally as suspect, have supported the concept of chromosomal speciation. Wilson *et al.* (1975), Bush *et al.* (1977), and Larson *et al.* (1984) have proposed that a relationship exists between social structuring of animal populations, rates of speciation and the involvement of chromosomal rearrangements in those processes. The earlier publications have been widely cited in studies involving chromosome change and speciation, although on some occasions not favourably so (see Bengtsson, 1980, King, 1981; and Charlesworth *et al.*, 1982). Indeed, King (1980, 1981) pointed out that much of the material used by Wilson *et al.* (1974, 1975) when discussing reptiles and amphibians was unreliable. When considering chromosomal change, these authors included modifications to chromosome number and arm number. That is, fusions, dissociations and also pericentric inversions and chromosome additions were utilized (see Wilson *et al.*, 1974, fig. 1, p. 5056). Moreover, they made no distinction between those rearrangements which were polymorphic or fixed, thus casting doubt on any conclusions relating to chromosome change and speciation.

Bush *et al.* (1977) tested the hypothesis that small demes promoted rapid speciation by estimating the rate of chromosomal evolution in 225 genera of vertebrates. The authors used what they considered an improved means of estimating the rates of karyotypic change, by considering the range in chromosome number and range of chromosome arms per genus. They argued that this approach could estimate the minimum number of chromosomal mutations which must have occurred to produce the observed range of karyotypes. They comment: 'Fusions, fissions, and whole-arm inversions are detected, whereas all paracentric inversions, most reciprocal translocations and many pericentric inversions are not.

All of the latter are unquestionably important in chromosome evolution and can serve as sterility barriers in speciation' (p. 3942).

This improved means of estimating the rate of chromosomal evolution and speciation is still based on the erroneous assumptions found in their earlier publication (Wilson *et al.*, 1974). Heterochromatin addition, a means of chromosomal repatterning common in vertebrates, has been included, as have other rearrangements not associated with speciation. It remains unclear whether chromosomal polymorphism has been included or not. Conclusions reached by these authors on the utilization of structural rearrangements in speciation, or on the lifestyle characteristics which support such processes, must remain unsubstantiated.

Larson *et al.* (1984a), when analysing the relationship between rates of chromosome change, social behaviour and speciation, used the same types of raw data, based on the number of chromosomes and chromosome arms. The authors acknowledged: 'The structural rearrangements revealed comprise a subset of the total rearrangements taking place: Robertsonian rearrangements and whole-arm inversions are counted but paracentric inversions are not counted, nor are most interstitial additions and deletions of chromatin' (p. 216).

It would appear that tandem fusions were not counted, and that whole-arm chromosome additions were. Moreover, it is still not clear whether polymorphisms have been included or not. Once again, the database incorporated chromosome changes which are irrelevant to the speciation process, and conclusions pertaining to chromosomal involvement in speciation derived from these data are also spurious.

The conclusion one might draw from the studies outlined above, is that a far more rigorous analysis of databases is necessary if meaningful conclusions are to be reached. This may have been achieved if chromosomal data were subdivided into those rearrangements which are implicated in chromosomal speciation and those which are not. Indeed, further subdivision into individual types of chromosomal rearrangements may have been advantageous. Arbitrary subdivisions of this nature, when related to such parameters as social structuring, vagility and population size, could thus provide a more realistic appraisal of the relationship between chromosomal change and speciation than does the current approach. However, the final arbiter which decides whether a particular rearrangement can or cannot become involved in the speciation process, is the meiotic system of the individual. A complex structural rearrangement may segregate in a balanced fashion in one organism and not in another, and in the final analysis this single event decides whether a rearrangement can ever play a role in speciation.

5.3 Concluding remarks

In this chapter, the problems associated with interpreting the action of structural chromosomal rearrangements as post-mating isolating mechanisms have been outlined. It is clear that there is never any certainty of how a particular structural rearrangement will cope with gametogenesis and the mechanics of cell division. Some structural rearrangements which were in the past regarded as cases of presumptive negative heterosis, appear to be capable of forming balanced polymorphisms, despite limited fertility effects. Others which have been explained away as being insignificant, because of mechanisms which avoid malsegregation (pericentric inversions), appear to produce profound viability effects when the impact of recombination on the modified chromosome structure is brought into play. Others still, such as pairing breakdown in multivalents and the sterility produced by X-autosome association, have in the past not been given adequate consideration as effects of rearrangements on fertility or viability.

The two areas discussed in this chapter, the types of structural changes which could be considered relevant to speciation and the use of these changes in theoretical modelling, were directed at clarifying this most complex situation.

There appears to be little doubt that the various forms of translocation which fail to segregate efficiently at meiosis have the potential to act as profound post-mating isolating mechanisms by producing unbalanced and inviable zygotes or infertile hybrids. In particular, tandem fusions, reciprocal translocations, multiple independent fusions/fissions with additive fertility effects and multiple fusions that have monobrachial homologies and form chains or rings in meiosis are all implicated in this process. It is probable that some single fusions have substantial fertility effects when trivalents formed in structural heterozygotes mispair to the X chromosome, others may not. Single chromosome fusions appear to distinguish many speciating complexes and can no longer be ignored as potential isolating mechanisms. Indeed, as few as two Robertsonian fusions which share a monobrachial homology can produce complete male sterility in structural heterozygotes (see Chapters 7 and 8). The evidence suggests that any of these rearrangements which have a high level of association with sex chromosomes due to pairing failure in multivalents, will have a high level of sterility. The great variation in fertility effect between similar Robertsonian fusions may be related to this pairing breakdown and sex chromosome association. Multiple pericentric inversions may also form reproductive barriers, although the status of a single, or a few, inversions in maintaining such barriers is generally unknown. Chromosomal polymorphisms and heterochromatin addition do not appear to be implicated in speciation processes.

An extensive literature is based on sophisticated computer modelling. Here, large databases of observed chromosomal variation in plants and animals have been constructed. These have been analysed from several perspectives and have provided arguments for and against the role of chromosome change in speciation, the rate of chromosomal evolution and have examined a variety of correlates associated with chromosome change. This sophistication has been largely wasted on low grade information at the database level. This has not only been degraded because of the inclusion of chromosomal polymorphisms, or cases of heterochromatin addition, but the basic assumption is at fault, i.e. that any chromosomal rearrangement observed may have been implicated in speciation. One can only safely say that chromosome rearrangements may be associated with speciation if the segregation patterns of those specific changes are known, or if changes of a profoundly negatively heterotic nature are fixed. Theoretical studies of that type can only produce meaningful results if they consider particular types of rearrangements in isolation.

In this chapter, the potential for chromosomal speciation is made apparent by the numerous types of chromosomal rearrangements which can exhibit deleterious meiotic effects capable of acting as the most profound post-mating isolating mechanism.

6
The fixation of chromosomal rearrangements in isolated populations

Thus, we do not feel that there is convincing evidence that chromosome differentiation greatly facilitates genetic divergence or speciation.

(Futuyma and Mayer, 1980, p. 262)

and

It is no longer acceptable for the advocates of allopatric speciation to dismiss non-allopatric models of speciation on the grounds that they lack convincing evidence when the counter arguments they themselves use are tenuous and based on equivocal evidence and speculation

(Bush, 1982, p. 128)

Isolated founder populations, which are characterized by severe selection gradients, intensive inbreeding and population crashes, provide an ideal medium for the fixation of negatively heterotic chromosomal rearrangements. However, the elevated incidence of mutation and the specific types of rearrangements often encountered, suggest that other factors are involved. Four of the factors which have a direct impact on the likelihood of a particular chromosomal rearrangement being established in a plant or animal population, are innate to those organisms. These are, first, the rate at which particular chromosomal mutations occur in that species or population. Second, whether the type of rearrangement, or its position, is a randomly determined process, or whether certain changes at particular sites are in some way selected for. Third, whether the rearrangements which are established occur as individual entities and subsequent new changes are fixed sequentially, or whether these rearrangements occur as simultaneous multiple events involving two or more chromosomes, or possibly the whole genome. Fourth, whether the rearrangement is actually driven to fixation by distortion of the segregation pattern, or whether this is a purely random process.

One or all of the above factors can have a significant impact on the likelihood of a particular rearrangement, or number of rearrangements, reaching fixation in animal or plant populations. In the following discussion chromosomal amplification events, breaks and DNA insertions are broadly included as chromosomal rearrangements.

6.1 Chromosomal mutation rate

One thing in life is certain: if you don't make change you don't get change. This statement is particularly true of chromosomal speciation; if there are no chromosomal mutations, or a low level of these, there is little opportunity for chromosomal involvement in speciation. Organisms which have a stable karyotypic format which cannot readily form chromosome breaks, or particular types of chromosomal rearrangement (*sensu* King, 1981), may not have the structure or capacity to speciate chromosomally.

It is important to make a distinction between the chromosomal mutation rate and the rate of chromosomal evolution. This is necessary, for the former reflects the base level at which novel rearrangements are detected in plant or animal populations, whereas the latter includes those rearrangements which have been established as polymorphisms, or fixed differences between populations. A great many comparative studies have been made on the latter, but relatively few analyses have been made on the former.

With these complexities in mind, it is important to consider what chromosomal mutation rates are actually found in plant and animal populations. White (1978a) determined a mutation rate for novel chromosomal rearrangements in organisms as diverse as lilies, grasshoppers and men, to be in the order of 1 in 500 individuals. With this measurement White was providing a rule-of-thumb estimation. That is, not only were plants and animals lumped together, but so were an array of different forms of chromosomal mutation.

Lande (1979) considered several different forms of chromosomal rearrangement in animals and produced a range of estimates. Thus, he found that reciprocal translocations appear to occur at a spontaneous mutation rate of between 10^{-4} and 10^{-3} per gamete per generation. In producing this estimate, Lande used several *Drosophila* species where rates range from 2 in 5600 for wild populations of *D. ananassae* (Yamaguchi *et al.*, 1976) to 1 in 531 for those of *D. melanogaster* (Berg, 1941).

Spontaneous chromosome fusions appear to occur at a higher rate, and in the case of the chromosome 21 fusion producing Down's syndrome, a rate of 3×10^{-5} per gamete per generation was estimated (Hamerton, 1971). Spontaneous fusions of the smaller chromosomes in man occurred at 10^{-4} (Lande, 1979), whereas pericentric inversions in man occurred at 0.4×10^{-4} (Lande, 1979). In *Drosophila melanogaster*, Yamaguchi *et al.* (1976) found the spontaneous pericentric inversion rate to be 3×10^{-3}, although these estimates were obtained from laboratory lines which may have contained transposable elements which increased the mutation rate.

In gekkonid lizards of the *Gehyra australis*, *G. pilbara* and *G. variegata – punctata* complexes, King (1983a) encountered possible rare mutants involving pericentric inversions and chromosome fusions at rates ranging from 1 in 6 to 1 in 41 using standard unbanded chromosome morphology, i.e. an average rate of 1 in 24 for 337 specimens. Using C, G and N-banding of specimens from the *Gehyra variegata–punctata* complex Moritz (1984, unpublished) detected 51 novel rearrangements in a sample of 350 animals (1 in 7), although neither King (1983a) nor Moritz could avoid the possibility that certain of these changes (particularly the inversions), were low-frequency polymorphisms. In *Gehyra purpurascens*, Moritz (1984) found G-banded variants in the W chromosomes of 30 females analysed (1 in 5). Porter and Sites (1987) detected a high level of spontaneous chromosomal mutations during their population analysis of the iguanid lizard *Sceloporus grammicus*. Here, 5 of 31 males examined had chromosomal mutations in their germ line, i.e. a rate of 1 in 6.

Recent estimates on plant populations suggest that chromosomal mutation rates can be extraordinarily high on a genomic level and even higher if particular chromosomes are considered. Parker *et al.* (1988) made an intensive investigation of the dioecious angiosperm *Rumex acetosa*. They determined that the frequency of novel chromosomal rearrangements in the euchromatic portion of the genome was 1 in 50 and that the rearrangements involved were centric shifts, although some interchanges and deletions were recorded. These data were obtained from plants grown from wild collected seeds and also from controlled crosses.

Rumex acetosa is cytologically interesting, for the two Y chromosomes are heterochromatic and hypervariable in their morphology. This variation takes the form of centric relocation within the Y chromosomes, although interchange between Ys also occurs. Wilby and Parker (1987) analysed 280 males from 28 wild populations, and used Feulgen staining rather than chromosome banding techniques. Even so, they encountered 129 variants, and on average each variant was represented 1.45 times. Thus, in *R. acetosa* the chromosomal mutation rate is some two orders of magnitude higher in the Y chromosomes than that in the autosomes (Parker *et al.*, 1988). Moreover, the general mutation rate in the autosomes and X chromosome is itself an order of magnitude greater than that found in most animals, i.e. 1 in 50 compared to 1 in 500 (White, 1978a). If numerical changes are considered, the rate rises to 1 in 33. In the Y–Y chromosomes, centromere relocation occurs at a rate of 1 in 80 and Y–Y interchanges at a rate of 1 in 70 in controlled crosses.

Parker *et al.* (1988) also point to the high mutation rate in the polytypic *Scilla autumnalis*. Here some 5 to 10% of mature plants have unique chromosomal rearrangements most of which are centromeric shifts (Ainsworth *et al.*, 1983). Thus, the rate of chromosomal mutation in these plant species is much higher than would have been expected, and may account for the great chromosomal diversity found within and between species.

Studies of naturally occurring or experimentally induced hybrids have shown extraordinarily elevated mutation rates. Shaw *et al.* (1983) detected both increased mutation rate and numerous multiple simultaneous mutations within hybrid back-cross progeny in the grasshopper *Caledia captiva*. A similar increased mutation rate was observed in *Atractomorpha similis* hybrids by Peters (1982). Hägele (1984) found an enormous increase in the mutation rate in *Chironomus thummi thummi* × *Chironomus thummi piger* hybrids, involving breaks in salivary gland chromosomes in from 3 to 79% of nuclei, and the incidence differed between the reciprocal crosses. Naveira and Fontdevila (1985) found that chromosomal mutation frequency was 30 times greater in hybrid males than in females of *Drosophila serido* and *D. buzzatii* crosses. Indeed, mutation is known to be inordinately high when *I* and *P* factors, the transposable elements responsible for hybrid dysgenesis, are in a destabilized state in *D. melanogaster* (Bregliano and Kidwell, 1983). The foldback family of transposable elements is also known to be capable of inducing a high frequency of chromosomal rearrangements in *Drosophila* (Collins and Rubin, 1984). It is probable that the act of hybridization destabilizes the genome and results in the release of mutator activity of this type. The effects of transposable elements will be dealt with more fully in Section 10.2.

The above-mentioned data not only tell us that chromosomal mutations can occur at a far greater rate than previously expected. They also tell us that there are potential differences between species and between elements within genomes. That hybrid zones can produce such a stream of mutant chromosomes has been seized on as a potential mechanism for speciation (see Section 9.3.1).

6.2 Random or non-random chromosomal rearrangements

Numerous lineages are characterized by species or cytotypes which are distin-guished by single or multiple rearrangements of the same type. For example, in the genus *Mus* most forms differ by chromosome fusions (Jotterand-Bellomo, 1986), although many types of rearrangements such as pericentric inversions

(White and Andrew, 1960, in *Keyacris scurra*), or large numbers of chromosome additions (John and King, 1983, in *Atractomorpha similis*), can be involved.

The very nature of the random mutation concept has been questioned by this form of observation. White (1973, 1975, 1978a) coined the term *karyotypic orthoselection* to explain the tendency for the same type of rearrangement to reach fixation in different chromosomes of the same species, or complex of species. He argued against what he termed the *differential origin hypothesis*, that is, that particular rearrangements simply occurred more frequently in some groups than in others. He reasoned that a random array of rearrangements were produced and that 'orthoselection' resulted in the restricted nature of rearrangements fixed. White provided five possible explanations for karyotypic orthoselection. The first of these supposed that similar chromosomal rearrangements had similar phenotypic effects which were adaptive in the same environment. Second, limits to the size and number of chromosomes in a cell were imposed by the dimensions of the spindle and cytoplasm and the mechanics of cell division. Third, the internal architecture of the chromosome and the distribution of satellite DNA, heterochromatin and ribosomal DNA, may impose restrictions on chromosome form. Fourth, regularities in the architecture of the interphase nucleus may have adaptive effects and so influence the types of rearrangements which were selected against. Fifth, White (1973) also argued that the location of chiasmata was an orthoselective process, for their distribution could modify the types of rearrangements which were produced after breaks were formed. All, or each, of these selective criteria could be applied to a random array of mutations and produce a non-random array of chromosomal rearrangements.

White (1973) proposed that karyotypic orthoselection was the process responsible for the production of *symmetrical* karyotypes, in which all chromosomes were metacentric and of the same size, or *asymmetrical* karyotypes, where there may be two distinct size classes, macro and microchromosomes, as in birds and many reptiles. White also argued for the continued reappearance of particular karyotypes within lineages, not as being evidence for a 'primitive' karyotype, but for selection of an 'equilibrium' karyotype. This is a relatively bizarre possibility not generally supported by phylogenetic reconstruction.

An alternative hypothesis for the retention of particular chromosomal complements, based on the non-random nature of chromosomal evolution, was offered by King (1981, 1985, 1990). That is, that the numerous selective criteria proposed by White were unnecessary and that structural characteristics of the genome restricted the position and number of breaks that could occur and the type of rearrangements which could form. For example, King argued that once metacentricity had been attained in a lineage of species evolving by chromosome fusion, chromosomal evolution may have reached a dead end in terms of gross structural changes. That is, unless that metacentric karyotype had the internal structure (molecular organization) which permitted its reorganization by inversion, hetero-chromatin addition, transformation, genome amplification or polyploidization, it could remain metacentric in perpetuity. Constraints controlling the continuity of a metacentric karyotype were:

1 The propensity of the karyotype to initiate particular types of chromo-somal rearrangements.
2 The structural organization of the karyotype which permited the fixation of these changes.

The formation of acrocentric or telocentric elements by the fissioning of a meta-centric karyotype could not occur if that karyotype lacked the structure to establish breaks in the paracentromeric area. Similarly, if chromosome break-points do not occur in precise regions of a metacentric element, chromosomal evolution by pericentric inversion resulting in the formation of acrocentric, telocentric or sub-metacentric elements could not occur. Thus, the numerically most common forms of gross structural rearrangements are no longer available for the action of selec-tion. These forms are entombed by their metacentricity, unless they had the chromosome structure and appropriate mutations which ensured evolution by other means.

In some organisms with a metacentric karyomorph, the chance formation of telocentric elements by a structural rearrangement, such as an inversion, could lead to a change in chromosome number and a re-establishment of a metacentric complement, since the newly created telocentric elements may display a propensity to fuse together. This is most apparent in the inversion–fusion cycle in many groups of amphibians (King, 1990). Concomitantly, in terms of high levels of chromosomal repatterning associated with speciation, it appears that acrocentric-ity/telocentricty is a positive advantage. In the frogs of the genus *Eleutherodactylus*, once the constraints of the metacentric karyotype had been broken, and multiple acrocentric/telocentric elements had evolved, a Pandora's box of possible future karyotypic reorganization was released. Most *Eleutherodactylus* species are distin-guished from each other by fixed chromosomal differences involving an array of different types of structural rearrangement (King, 1990). This provides an inter-esting contrast to many other anuran species, or higher taxa, which retained metacentricty.

Distinguishing between the multitude of selective regimes and non-randomness as a mechanism is fraught with difficulty, for selection can apply at any level. It is most difficult to make the distinction between the absence of a particular change and selection against that change. With these qualifications in mind, Peters (1982) has provided significant support for the differential origin hypothesis. In his analysis of the recurrence of chromosome fusion in the interpopulation hybrids of *Atractomorpha similis*, he found a heritable tendency for particular chromosomal fusions between a number of different linkage groups. The specific changes induced by heritable factors could go in one direction and not in the opposite direction. Peters found that the multiple fusions were induced within, or adjacent to, paracentromeric heterochromatin, in full-sib embryos, in particular crosses. Consequently, these structural mutations were non-random.

Additional forms of evidence can be brought to bear against the formation of chromosomal mutants being a purely random process. The first, comes from the well-documented recurrence of particular chromosomal rearrangements within species. Most apparent among these are the numerous mutations which plague the human genome, producing the well-known syndromes such as Down's with trisomy 21. The occurrence of this mutation is non-random and favours a particular translocation.

Second, grossly elevated chromosomal mutation rates together with a non-random pattern of restructuring, appear to occur in certain hybrid zones, or in laboratory hybrids. Thus, in *Caledia captiva* the high incidence of novel chromosomal rearrangements in backcrosses between two cytotypes were examined in 53 surviving embryos (Shaw *et al.*, 1983). Chromosome pairs 5, 10 and the X showed a much higher mutation rate with over 20 forms of pair 10 being identified. Three important phenomena were observed: first, the same fusion (5, 10, X) occurred repeatedly among backcrosses. Second, the same fusion was present among several progeny of a single female. Third, different rearrangements were produced within the same germ line. A single individual was capable of generating an array of the same, or different, rearrangements among its descendants. Similar observations have been made in hybrids between *Chironomus thummi thummi* and *Chironomus thummi piger* by Hägele (1984). Chromosomal breaks occurred preferentially at specific polytene bands. Hägele pointed to the fact that a 120-base-pair repetitive DNA element had been located in a number of specific bands, suggesting that this mutator activity could be associated with transposable elements. The specificity of break-points forms the third area of evidence against karyotypic orthoselection being a random process. In *Drosophila buzzatii/serido* hybrids, the elevated chromosomal mutation rate which Naveira and Fontdevila (1985) detected, involved inversions, translocations and duplications. The number of translocations was significantly lower than the other rearrangements. The distribution of aberrations was evenly dispersed throughout the complement, although the total number of break-points was significantly greater in chromosome 4 than the other members of the complement.

The third form of evidence comes from the study of transposable elements. Two examples are provided here, but a more detailed explanation is found in

Section 10.2. Collins and Rubin (1984) analysed the unusual foldback transposable elements which induced chromosomal rearrangements in *Drosophila*. The ten rearrangements that they examined, indicated that sequences located between two closely linked elements were often deleted. Rearrangements and recombination events occurred within, or adjacent to, these foldback elements. The type and frequency of rearrangements depended on the element structure and DNA organization. This suggests that if the insertion sites of mobile elements are not random, then the site of chromosomal rearrangements and their frequency is also non-random.

Similarly, McClintock (1956, 1978, 1984) demonstrated that mobile elements which were present in the *Zea mays* genome in an inactivated state, could be activated by genomic stress. By using a chromosome breakage–fusion–bridge cycle to produce ruptured chromosome ends on chromosome pair 9, McClintock was able to activate the transposable elements. This resulted in a high level of chromosomal repatterning on pair 9. However, a series of additional and non-random rearrangements, involving other members of the complement, were also detected. These encompassed readily observed reciprocal translocations, inversions and duplications to short DNA segments containing transposable elements. Much of this non-random restructuring involved centromere/centromere, centromere/knob and knob/knob attachments, and the attachment of segments to new breaks, or to other chromosomes (Fig. 6.1).

The three forms of evidence outlined above place the concept of karyotypic orthoselection as a selective processes operating on random mutation in jeopardy. Evidence from the continual reoccurrence of specific mutations within genomes, particular mutants produced in hybridization studies, and the action of transposable elements, indicates that chromosomal mutations are not only non-random, but are constrained by the chromosome structure to the type of change which can be produced.

A clue to the mechanisms responsible for the non-randomness of chromosomal rearrangements may come from the cytogenetic analysis of *Mus*, in particular, the degree and direction of chromosomal evolution in *Mus domesticus* compared to *M. musculus*. The karyotype of *M. musculus* consist of 40 acrocentric chromosomes and this format occurs throughout the species range. In contrast, in *M. domesticus* populations differ by numerous fixed chromosome fusions resulting in chromosome numbers of from 2n = 22 to 2n = 40 in a huge variety of combinations. Indeed, over 60 chromosome races are now known (see Section 7.3.1), with over 122 translocations of independent origin in wild populations and laboratory strains (Mahadevaiah *et al.*, 1990). Clearly, the question as to why most of the rearrangements established in *M. domesticus* were fusions, and why these should often be multiple events within the *Mus* genome, is of some significance. Why these

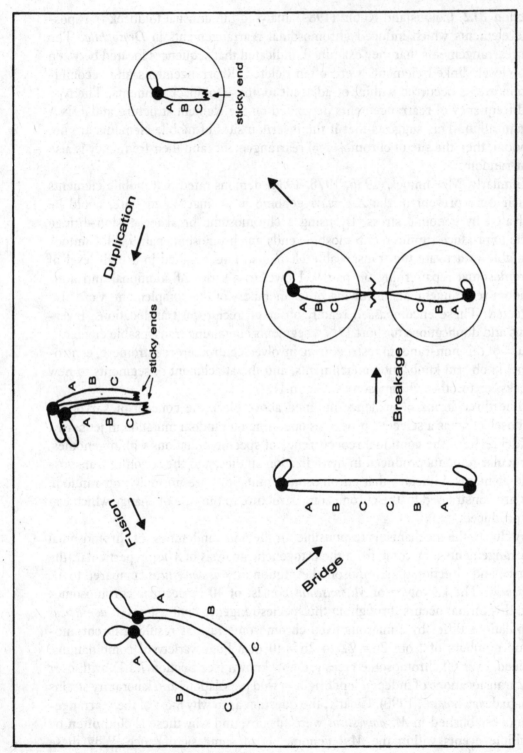

Fig. 6.1. The breakage–fusion–bridge cycle. A dicentric chromosome can be formed by chance. At anaphase the two centromeres of that chromosome may move to opposite poles producing a chromosome bridge between the separating nuclei. This will break, producing chromosomes with broken and 'sticky' ends. After chromosomal replication, the sticky ends fuse again, reforming a dicentric and restarting the cycle.

changes have not occurred in *M. musculus*, a species closely related to *M. dom-esticus*, has received a considerable level of investigation.

The constitution of satellite DNA in the two *Mus* species was analysed by Redi *et al.* (1990a) using restriction enzymes. They found that approximately 60% of *M. domesticus* satellite DNA was retained in the *M. musculus* genome, although *M. musculus* satellite DNA lacks most of the GAATTC restriction sites. The AT-rich sequences have a more clustered array in *M. domesticus* satellite DNA and occur at the paracentromeric region. The quantity of C-positive heterochromatin is larger in *M. domesticus* than *M. musculus*. Redi *et al.* (1990a) suggested that the increase in the amount of satellite DNA, with the clustering of AT-rich sequences, could have rendered *M. domesticus* chromosomes prone to whole-arm translocation.

In an earlier study, cytophotometric analysis had been used by Redi *et al.* (1986) to determine whether paracentromeric heterochromatin had been lost in the numerous fusions found in *M. domesticus*. There was no evidence that such a loss had occurred. Redi *et al.* (1990b) proposed the following model (Fig. 6.2).

1 Mouse centromeres cluster during pachytene, due to recognition of similar sequence (Rattner and Lin, 1985).

2 If chromosomal breakage occurred in adjacent centromeres, the two strands of DNA could rejoin, because of their gross sequence similarity and close proximity.

3 During discontinuous DNA synthesis, unpaired single strand regions located at the telemetric ends of two different chromosomes could occasionally base-pair. That is, a 5′ parental strand could base-pair with an homologous sequence on a 3′ parental strand of another chromosome, to give a DNA double helix with the canōpic structure reconstituted.

4 By rotating the chromosome, the proper angular position for joining the two chromosome backbones by forming a four-stranded structure could be achieved.

5 Subsequently, one of the functional kinetochores can be deactivated. Evidence for deactivation has been established by Haaf *et al.* (1989) by using immunocytogenetic attachment of fluorochromes to synaptonemal complexes and kinetochores.

Clearly, this model remains untested although it appears to agree with our understanding of the structure of *Mus* chromosomes.

6.3 Simultaneous multiple chromosome rearrangements

The possibility that the genome could simultaneously produce numerous chromosomal rearrangements of the same type, in the one individual, was a product of McClintock's (1956, 1984) investigations in *Zea mays*. King (1982) reviewed the evidence for group change, and the incidence of rare multiple mutants in comparative population studies. In earlier publications Capanna *et al.* (1977) and White (1978a) had argued against this possibility, favouring the fixation of sequential rearrangements established in numerous individuals or populations.

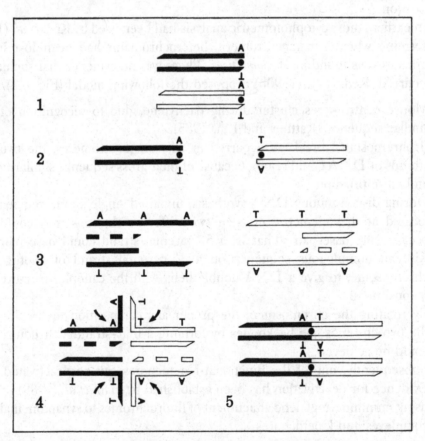

Fig. 6.2. A model for the mechanism of Robertsonian fusion proposed by Redi *et al.*, (1990b). If two telocentric chromosomes align in an anti-trans-position (1 and 2), during DNA synthesis (3), the chromosomes can fuse by the base-pairing of homologous DNA sequences (4). The DNA heteroduplex is cut and rejoined by a nick-closing enzyme (5). The paracentromeric heterochromatin maintains the fused chromosomes' polarity. (Redrawn from Redi *et al.*, 1990b.)

The primary theoretical arguments used by White against the simultaneous proposals were:

1 Such an hypothesis was not plausible on molecular grounds since no mechanism is known, or easily conceivable, for producing a large number of breaks at corresponding positions in different chromosomes of the same nucleus.
2 A simultaneous multiple mutation hypothesis is not plausible from the standpoint of population genetics, since individuals simultaneously heterozygous for many rearrangements will have their fecundity very drastically reduced and would consequently be subject to severe selection.

However, the landmark studies of McClintock (see 1978, 1984 reviews), on the activation of transposable elements by genomic stress, have demonstrated that simultaneous multiple rearrangements are an evolutionary reality. The triggering of these transposable elements resulted in varying degrees of reorganization, ranging from complex chromosomal translocations to small segments of DNA. A similar level of response to genomic stress has been discussed in regard to elevated mutation rates in inter- and intraspecific hybrids. Studies by Peters (1982) on *Atractomorpha similis*, Shaw *et al.* (1983) on *Caledia captiva*, Hägele (1984) on *Chironomus thummi* and Naveira and Fontdevila (1985) on *Drosophila buzzatii* × *serido* hybrids, all detected multiple simultaneous rearrangements in the mutant animals they produced. In two of these cases (Hägele, 1984, and Naveira and Fontdevila, 1985), transposable elements have been implicated in the high incidence of specific chromosome breaks at multiple sites. The effects of transposable elements are described in more detail in Section 10.2.

There are a number of reports in the literature of rare mutant plants and animals which carry multiple simultaneous rearrangements (King, 1982). These include translocations in the grasshoppers *Moraba scurra* (White, 1963), *Tolgadia infirma* (John and Freeman, 1976), *Atractomorpha bedeli* (Sannomiya, 1978) and *Stauroderus scalaris* (John and Hewitt, 1968). Multiple mutations (fissions) have also been encountered in rare mutant *Sceloporus grammicus* lizards (Porter and Sites, 1987). The incidence of multiple simultaneous mutants in natural populations is not high, although there is little doubt that the process has occurred in all cases. The juxtaposition between the low incidence of naturally occurring rare mutants with simultaneous mutations and the extraordinarily high incidence in stressed genomes (*sensu* McClintock, 1984), suggests that bursts of reorganization are an uncommon event and that such events are confined to particular situations that the organism encounters. It is of some interest that Lewis (1962) suggested that genomic stress could be induced by inbreeding in a founding population, whereas McClintock (1978, 1984) suggested that a variety of environmental components ranging from the accidental hybridization of two species, to particular mutagens, could activate transposable elements and provide a burst of mutator activity.

The high incidence of specific site breakage induced by transposable elements, together with the concerted nature of evolutionary changes observed in the genomes of many vertebrate and invertebrate species, implicates a direct molecular basis for the simultaneous changes. While the significance of concerted evolutionary processes and molecular drive will be discussed in detail in Section 10.1, their potential for change is apparent here.

6.4 Meiotic drive

One of the problems associated with chromosomal speciation theory is in establishing the means by which moderately or severely negatively heterotic rearrangements, which are a direct disadvantage to the heterozygous carrier, can reach fixation in the homozygous state. Of the possible mechanisms which White (1968, 1978a) suggested would be sufficient to explain this phenomenon, meiotic drive has been one of the most controversial and least explained.

It is interesting to examine the views of some of the critics of chromosomal speciation in this regard. For while Key (1968, 1979, 1981) and Lande (1979) suggested that meiotic drive was an unnecessary complication for which there was no real need and less than significant evidence, others simply ignore meiotic drive as a possible mechanism (Chesser and Baker, 1986; Sites and Moritz, 1987; Sites *et al.*, 1988a), or regard it as an improbable *deus ex machina* (Futuyma and Mayer, 1980), without really considering the available evidence. Templeton (1981) went to the other extreme; he supported the possibility of meiotic drive facilitating the fixation of negatively heterotic rearrangements, and argued that this would reduce the chance of speciation, since the very isolating barrier that is supposed to prime the speciation process (negative heterosis) is overcome. In so doing, Templeton has assumed that segregation distortion is not a transient phenomenon (see below).

Meiotic drive simply refers to a distortion in the transmission rate of a particular chromosomal rearrangement at meiosis, i.e. when segregants show a rate of transmission greater than 0.5 in one direction. Even a very small amount of meiotic drive can lead to an increased probability for the fixation of negatively heterotic chromosomal rearrangements in populations with the correct structure (Walsh, 1982). The mechanism responsible for meiotic drive usually occurs in the female meiotic cycle. Here, the derived rearrangement preferentially passes into the egg nucleus, rather than into the three polar body nuclei which are eliminated, thus elevating the transmission rate and enhancing the spread of the rearrangement throughout the derived population. However, meiotic drive can also occur in males.

Lyttle (1989) pointed to two additional forms of meiotic drive. The first of these was gamete competition. Here the chromosome which was driven, caused some form of lesion in its homologue during meiosis, resulting in dysfunction of gametes carrying that chromosome. Gamete competition would lead to effective meiotic drive for a chromosome or allele, only if it resulted in an increase in the absolute number of progeny carrying the element. This form of drive predominates in males where gamete wastage occurs. In most cases, gamete competition is due to a specific gene locus. Consequently, it cannot physically promote the fixation of structural chromosomal rearrangements, even though one of the breakpoints may be near the locus in question. Lyttle (1989) argued that gamete competition could possibly influence the fixation of rearrangements involving sex chromosomes, B chromosomes, or certain interacting inversions.

Asymmetrical gene conversion was also regarded by Lyttle (1989) as an example of meiotic drive. The concept of gene conversion supposes that the DNA sequence of one member of a pair of heterozygous alleles is altered during meiosis to produce a copy of the other. It is unlikely that this form of molecular drive could affect anything but a specific sequence set or locus. Structural rearrangements would not be assisted and neither would other associated genes. For this reason, asymmetrical gene conversion is discussed in Section 10.1.1.2.

6.4.1 Sex chromosome drive

The concept of meiotic drive was introduced by Sandler and Novitski (1957), who found that male-specific meiotic drive had been induced by certain translocations in *Drosophila*. Subsequently, Novitski (1967) found that the occurrence of non-random disjunction in *Drosophila* females, caused by preferential segregation in the egg nucleus, drove certain rearrangements towards fixation. There are also numerous reports of X-linked meiotic drive in *Drosophila* species including members from the *D. obscura* group, *D. simulans* and *D. melanogaster* (see Curtsinger, 1984). Mechanisms ranging from heterochromatic deficiencies of the X chromosome in *Drosophila melanogaster* (McKee, 1984), through transposable elements (Engels and Preston, 1981) to the impairment of spermiogenesis (Matthews, 1981) are all implicated in producing this effect.

In Section 5.1.1.3 we briefly considered the relationship between X-autosome translocations and male sterility in *Drosophila melanogaster*. McKee (1991) extended this relationship to include meiotic drive, because of the striking parallels between the genotypes responsible for meiotic drive and male sterility. Thus, X-autosome translocations caused either sterility, or meiotic drive, depending on the autosome involved. X-heterochromatic deficiencies caused meiotic drive in a normal genetic background, but caused sterility in the presence of at least two types of rearranged Y chromosomes. These parallels, in addition to the reduced fertility associated with meiotic drive genotypes, suggested a common mechanism to McKee. As it turned out, there is a close functional relationship between the X heterochromatic loci responsible for X–Y pairing and those responsible for meiotic drive, and also between segregation distortion in the X-4 translocation in males and X–Y pairing. That is, meiotic drive is a consequence of the separation of the X-euchromatin from the pairing site.

X-heterochromatin also plays a significant role in chromosomal sterility and, like meiotic drive, normal spermiogenesis depends on the continuity of the X-heterochromatin with the euchromatin. McKee (1991) argued that both cytogenetic and molecular evidence pointed to the X pairing site as also being the locus responsible for male sterility, since all X-heterochromatin deficiencies that caused non-disjunction and segregation distortion with the normal Y, also caused sterility with Y mal +. Molecular confirmation of the X pairing site and fertility locus was obtained by using transposable P-element transformation of cloned rRNA genes to rescue fertility. The NOR on the X was the locus responsible for X–Y pairing and normal sperm development. Failure of XY pairing was directly responsible for non-disjunction, meiotic drive and at least one of the forms of chromosomal sterility.

The significance of these findings with regard to pairing failure, X-autosome association and meiotic drive of structural rearrangements is paramount (see Chapters 5 and 7).

Lyttle (1981, 1982) examined the consequences of sex chromosome drive in *Drosophila melanogaster*. Since the drive for a particular sex chromosome would lead to the fixation of one sex and the extinction of that form, mechanisms which overcame this would have a significant selective advantage. Lyttle found that laboratory populations of *D. melanogaster* with Y-drive overcame this by accumulating additional aneuploid X or Y chromosomes to form XXY, XYY genotypes (see also Lyttle, 1989). A similar system of meiotic drive was detected in the sex chromosome system of the varying lemming, *Dicostonyx torquatus*, by Gileva (1987). Here, fertile XY females occurred as a result of an X-linked mutation X*. The frequency of these females in wild and laboratory populations was higher than expected due to a preferential segregation of the Y-chromosome in males. Meiotic drive was essential for the spread and maintenance of the X* mutation.

In the mosquitoes *Aedes aegypti* and *Culex quinquefasciatus*, a Y-linked gene was detected which caused a change in sex ratio favouring males. Although both

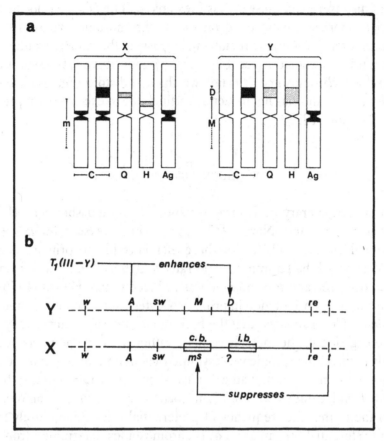

Fig. 6.3. a. The location of *m*, *M* and *D* loci within the sex chromosomes of the mosquito *Aedes aegypti* in relation to C-banding (C), Hoechst 33258 fluorescence banding (H), quinacrine banding (Q) and silver staining (Ag). b. The major genetic influences on meiotic drive at the *D* locus. The following markers are shown. *w* (white eye), *A* (factor *A*), *sw* (short wing), *M* (male-determining gene), *D* (meiotic drive factor), *re* (red eye), *t* (tolerance of meiotic drive); *m*ˢ (female determining factor, sensitive to *MD*), *c.b.* (C-band), *i.b.* (intercalary C-band). (Both a and b are from Wood and Newton, 1982, 1991, and are redrawn.)

systems were relatively similar, most of the work has been done on *A. aegypti*, and populations possessing this gene had been detected in Africa, America, Australia and Sri Lanka. Hickey and Craig (1966) found that the distortion of sex ratio was controlled by the distorter gene (*D*) found near the sex-determining locus (*M*) on the Y chromosome. Meiotic drive only occurred when *D* was coupled with *M* on the Y, since it was male specific. Nevertheless, sensitivity to the *MD*-haplotype was controlled by the female gene *m* on the X chromosome, which is the responder. The *m*-haplotype is polymorphic for sensitivity to *MD*. Two other genes are present: *t* (tolerance to distorter which influences sensitivity), and *A* (enhancer of distorter), and these are also sex linked. The *D* gene is located within an interstitial heterochromatic band on the Y chromosome, whereas the *m/M* complex occupies a considerable region encompassing the paracentromeric heterochromatin and euchromatin (see Fig. 6.3). Meiotic drive was associated with breakage in the X chromosome. It is noteworthy that distorter males have massive sperm depletion, but this has no impact on fertility due to excess sperm production (Wood and Newton, 1991).

6.4.2 Supernumerary chromosome drive

Studies on supernumerary or B chromosomes, in the grasshoppers *Melanoplus femur-rubrum* (Lucov and Nur, 1973) and *Myrmeleotettix maculatus* (Hewitt, 1973), detected an accumulation of these elements in the primary oocytes of female meiosis, which had a transmission rate as high as 0.9. However, in males, meiotic drag occurred and transmission was reduced to 0.3. Hewitt (1976) examined the chromosomal orientation in primary oocytes using bright field microscopy in *M. maculatus*. He established that the B chromosome was preferentially distributed on the egg side of the metaphase plate rather than on the polar body side. Interestingly enough, the spindle was asymmetrical in this species and the nucleoplasm surrounding the spindle had a conical shape. The Bs tended to be found in the pointed end, rather than the blunt polar body end of the spindle. Hewitt (1973) also found that the frequency of preferential orientation closely matched the level of preferential transmission of B chromosomes determined from breeding experiments. Kayano (1957) detected a similar anomaly in *Lilium callosum*, where at meiotic metaphase I the B chromosomes lie outside of the metaphase plate on the micropylar side of the spindle in the egg mother cell. They orientate in this position because of the unusual cell shape, and pass to the micropylar pole at anaphase.

In addition to this true meiotic drive, two other forms of B-chromosome drive were recognized by Jones (1991), involving nuclear divisions before and after meiosis. In *pre-meiotic drive*, mitotic non-disjunction occurred in the spermatagonial mitoses in the testes of some grasshoppers. *Post-meiotic drive* occurred in flowering plants; it was also based on mitosis and took place almost immediately after meiosis, during the development of the male gametophyte. For example, in grasses, the direct non-disjunction of the Bs into the generative nucleus occurred during the first pollen-grain mitosis.

In *Zea mays*, post-meiotic drive occurred in males and involved non-disjunction at the second pollen mitosis, coupled with preferential fertilization of the B-containing sperm. Dissection of the maize B by translocation indicated that it was

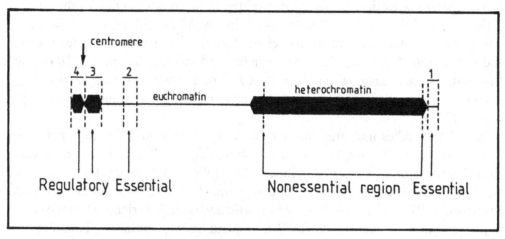

Fig. 6.4. The organization of the maize B chromosome in terms of the control of nondis-junction. (Redrawn from Jones, 1991.)

subdivided into heterochromatic and euchromatic regions (see Fig. 6.4) and has both essential control and regulatory regions. The regulatory regions were found within the paracentromeric heterochromatin (Jones, 1991).

6.4.3 Meiotic drive for autosomal rearrangements: examples from animals and plants

Meiotic drive systems are now known from a variety of plant and animal species including vertebrates and invertebrates. In this section, examples will be discussed in some detail, illustrating both the type of mechanism involved and the level at which research has proceeded in that group.

6.4.3.1 Segregation distortion in Drosophila melanogaster

The segregation-distortion (SD) system found in *Drosophila melanogaster* is one of the most intensively studied meiotic drive mechanisms. It is a male drive system and is controlled by three major loci on chromosome two (see Fig. 6.5). The segregation-distorter gene (*Sd*) was found at the base of the left chromosome arm of pair two in a euchromatic area. An enhancer gene *E(SD)* and responder gene *Rsp*, were found within heterochromatin on that chromosome. *E(SD)* occurred in the centre of the left arm of pair two, whereas *Rsp* occurred in the centre of the right arm. *Rsp* occurred in the allelic forms sensitive and non-sensitive. Males heterozygous for an *SD*-bearing chromosome and a sensitive SD^+-bearing homologue could transmit *SD* to 95% of their progeny (Temin *et al.*, 1991).

Segregation distortion in males involved the induced dysfunction of the sperm which received the SD^+ homologue. This involved both the failure of condensation of chromatin in spermatid nuclei and failure of spermatid individualization and maturation. The most recent theory advanced was that Sd and $E(SD)$ on the SD chromosome, somehow acted jointly at Rsp^s on the sensitive SD^+ homologue, to cause dysfunction of the sperm that received the SD^+ chromosome (Temin et al., 1991).

Molecular studies indicated that the Sd locus mapped to 37D2 on a polytene map. This is a tandem duplication of a 5 kb region of DNA and was present on all SD chromosomes examined. Thus, if the duplication itself was the mutant Sd locus, it was likely to have had a single origin. The responder locus found on chromosome 2R, occurred in heterochromatin and is an AT-rich 120 bp repeated sequence and its sensitivity appeared to depend on copy number (Temin et al., 1991).

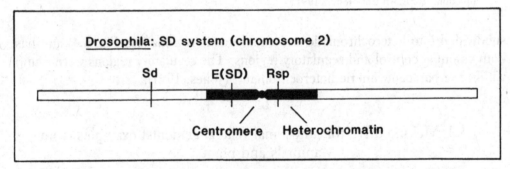

Fig. 6.5. The organization of the segregation-distorter (SD) complex in *Drosophila melanogaster*. The three major loci are Sd (segregation distorter), $E(SD)$ (enhancer of segregation distorter) and Rsp (the responder). Note that both $E(SD)$ and Rsp occur within the centromeric heterochromatin of chromosome 2. (Redrawn from Temin et al., 1991.)

6.4.3.2 Grasshoppers

In terms of fixed chromosomal differences which distinguish cytotypes, Mrongovius (1979) made an analysis of segregation distortion in laboratory-reared and natural hybrids which occurred between chromosome races of the *Vandiemenella* complex of eumastacid grasshoppers. She found that the structurally different X chromosomes of female hybrids of P24 (XY) and *Viatica* 17 exhibited a distorted ratio in embryos. The distortion was slight and favoured the parental rather than the derived chromosome. This assumed that the phylogenetic direction previously determined was correct. White (1978a) regarded this as a major finding since this was a relatively old speciation event and to find any type of segregation distortion, regardless of direction, was of value after such a long time.

An unusual heterochromatic deletion polymorphism in the grasshopper *Chor-*

thippus vagans was described by Cabrero and Camacho (1987). Embryos had twice the frequency of the M4 deletion than that found in adults sampled from wild populations. Cabrero and Camacho suggested that meiotic drive operated in heterozygous females, so that ova only yielded M4 deletion chromosomes, with the M4-B segment being lost in the polar bodies. Statistical estimates supported this hypothesis.

6.4.3.3 The sheep blowfly (Lucilia cuprina)

This organism has been extensively studied because of its great impact on the Australian sheep industry. A byproduct of the genetical research was the isolation of numerous inversions, some of which displayed non-disjunction in females. Foster and Whitten (1991) analysed a series of pericentric inversions for meiotic drive. Three of these were expected to produce drive because of their particular structure. They possessed grossly unequal terminal segments and had large differences between markers within the inversion. After crossing-over, the inversion would be the larger member of one unequal dyad, and the standard chromosome would be the shorter member of the other dyad. As expected, a lower recovery of the inverted chromosome occurred in females, whereas normal inheritance occurred in males.

In each case of distorted inheritance, the chromosome recovered in highest frequency was the one with the more central centromere. This was explained by Novitski's (1967) drag hypothesis, where differential mechanical drag operated on the unequal-sized dyads during anaphase 1, and orientated the spindle so that at anaphase 2 the shorter member had a higher probability of inclusion in the egg nucleus. In *Lucilia*, crossing-over within the inverted region between homologous heterozygotes for a pericentric inversion, increased the recovery of the chromosome with the more centrally located centromere. This suggested to Foster and Whitten (1991) a mechanism whereby metacentric chromosomes could become fixed in populations, at the expense of more acrocentric homologues, simply because they enjoyed a transmission advantage at meiosis.

6.4.3.4 The common shrew (Sorex araneus)

The chromosomally polytypic common shrew (*Sorex araneus*) is widely distributed throughout the British Isles, continental Europe and Scandinavia. A population at Oxford is polymorphic for chromosome fusions. Searle (1986) made an analysis of the complements of both foetuses from wild animals and young from laboratory crosses using wild parents, and was able to demonstrate a preferential transmission of two of the metacentric elements *Kq* and *no*, but not a third combination *pr*, in preference to acrocentric elements. In both cases, transmission was at about twice the rate of acrocentrics. However, the sample sizes were not large and Searle could not provide a precise mechanism for the nature of the drive system.

6.4.3.5 *Blue foxes* (Alopex lagopus)

A series of studies have been made on farm-bred blue foxes used for fur production (see Mäkinen and Lohi, 1987). This is a particularly interesting situation since there is great potential for mating experimentation and sample sizes are large. Blue foxes are polymorphic for a Robertsonian translocation between acrocentric pairs 23 and 24. Chromosomal complements have either $2n = 48$ or 50 or the heterozygous $2n = 49$. Mäkinen and Lohi (1987) have now confirmed previous studies which showed a relatively minor (2.4%) reduction in litter size attributed to the translocation. However, these authors also confirmed the studies of Switonski (1980) and Christensen and Pedersen (1982) which showed a positive effect on fertility, associated with the $2n = 48$ karyotype, i.e. the homomorphic fusion. That is, the number of pups in $2n = 48$ chromosome groups were larger than in $2n = 50$ groups. Moreover, in parental $2n = 49$ matings the production of $2n = 48$ offspring was favoured in a 16:18:8 ratio rather than 1:2:1. This finding agreed with the results of Christensen and Pedersen (1982) and suggested a distinct case of meiotic drive. It also suggested that the derived homomorphic rearrangement was at an adaptive advantage.

6.4.3.6 Mus domesticus

A number of studies on *Mus domesticus* specimens which were heterozygous for Robertsonian fusions, have demonstrated segregation distortion in favour of the derived form. Harris *et al.* (1986) investigated the transmission of a spontaneous Rb (9.12) fusion found in a laboratory colony of genetically wild Peruvian mice. Meiotic analysis revealed a non-disjunction rate of 10% which corresponded to the proportion of trisomic and monosomic embryos produced by heterozygous females (12–16%). There was a significant excess of normal embryos heterozygous for the chromosome fusion (2:1 ratio), which suggested that there had been a preferential retention of fused chromosomes in the egg nucleus. Harris *et al.* argued that the preferential retention of a Robertsonian chromosome in the egg nucleus could be countering the deleterious effects of non-disjunction produced by the fusion heterozygote.

The study by Harris *et al.* (1986) was most significant in three respects. First, it unambiguously demonstrated a significant level of meiotic drive in a spontaneous chromosome fusion. Second, the animals used in this study were derived from wild populations. Third, segregation distortion was in the direction of the derived form.

Gropp and Winking (1981) found segregation distortion using wild mouse translocations introduced into laboratory mice. Five out of the ten heterozygotes that they examined (Rb (1.3) 1 Bnr/+, Rb (16.17) 7 Bnr/+, Rb (10.11) 8 Bnr/+, Rb (6.13) 3 Rma/+, and Rb (4.15) 4 Rma/+, showed a pronounced deficiency in the number of fusions transmitted. That is, rather than a 1:1 ratio of heterozygous to homozygous progeny, the ratio was in favour of homozygous acrocentrics by as much as 3:1. There was meiotic drive for the standard acrocentric chromosome and not for the fusion. This profound distortion in segregation was generally only observed in females and not in males. However, in the Rb (16.17) 7 Bnr fusion, segregation distortion occurred in both sexes. Gropp and Winking (1981) concluded that the loss of the metacentric chromosome in the oocytes was so great that it could be attributed to a preferential segregation of this element to the first polar body. They also suggested that distortion in the male could have been due to a mild *t*-mutation effect (see below).

The segregation pattern in the same chromosome fusion Rb (16.17), was examined by Britton-Davidian (1990) using both wild mice and their laboratory-bred progeny, which had been obtained from the Lombardy hybrid zone between the 2n = 38 and 40 chromosome races. The breeding data showed that there was no distortion of the transmission ratio in either sex. Clearly, these data suggested that artificially produced structural heterozygotes using wild chromosomes introduced into laboratory strains, could differ from the situation pertaining in wild populations, in so far as minute chromosomal differences may exist between the two genomes. In this regard mtDNA has been analysed in several strains of laboratory mice (Yonekawa *et al.*, 1982). The cleavage patterns of most strains (50 of 55) were identical to the European *Mus domesticus*. However, *M. brevirostris* and *M. molossinus* mtDNA was also identified in some strains, whereas others were of hybrid origin (*M. domesticus* × *M. molossinus*). No *M. musculus* characters were detected. Thus, the introduction of chromosomes from wild populations of *M. poschiavinus*, or *M. brevirostris*, into any of the above strains which have genomic differences (see Gropp and Winking, 1981), could produce results which are suspect in their generality. Nevertheless, these studies unambiguously demonstrated that the process of meiotic drive was occurring in *Mus*. The work of Harris *et al.* (1986) indicated that similar mechanisms are present in the wild.

6.4.3.7 t-*haplotype in* Mus

Meiotic drive has been extensively studied in *Mus musculus* using a segregation distortion gene complex found in chromosome 17 and which is known as the *t*-haplotype. This is because the proximal third of chromosome 17 exists in two forms in wild populations: wild type and *t*-haplotype. In the latter, the proximal region occupying about 20 centimorgans of the chromosome, is readily distinguished by its G-band pattern (Lyon, 1991) and has a series of four inversions which suppress recombination in that area (Hammer, 1991).

Males that are heterozygous for a complete *t*-haplotype complex transmit it to over 90% of their offspring. However, when homozygous, *t* complexes are lethal. On the other hand, segregation is normal in females. Breeding studies have shown that the abnormal transmission ratios observed in heterozygous male mice were due to three or more segregation-distorter genes *Tcd-1*, *Tcd-2* and *Tcd-3*, acting on a single responder gene *Tcr* (Lyon, 1991) (see Fig. 6.6). Although the segregation distorters were transmitted by heterozygotes, homozygosity for any distorter in conjunction with heterozygosity for one other results in male sterility. The distorters also appear to differ in their strength. The presence, absence or homozygosity of a responder, does not lead to sterility, but can protect against sterility (Lyon, 1991).

Lyon suggested that the high transmission ratio may account for the persistence of *t* complexes in wild populations. Since the four genetic factors are distributed over a large area of chromosome 17, it is important that recombination suppression due to the four inversions has locked the region up. Hammer (1991) argued that a stepwise evolution of the *t*-haplotypes had occurred, wherein the proximal and distal inversions were established at different times in different lineages and the event leading to the spread of the *t*-haplotype was the linking of the centromeric and middle inversions. Thus, rather than a simple linear progression having occurred, the proximal inversion originated in the lineage leading to *M. domesticus* and *M. abbotti*, whereas centromeric, middle and distal inversions were established in the lineage leading to the *t*-haplotype. Hammer suggested that an interspecific hybridization event was necessary to complete the inversion set.

6.4.3.8 *Flowering plants*

Studies have been made by Parker *et al.* (1988) and Wilby and Parker (1988) on a number of plant species, but particularly on *Rumex acetosa*, a dioeceous flowering plant. Here, screening of natural populations of *R. acetosa* revealed the presence of seven centric shifts and three reciprocal translocations. Inheritance of the rearrangements was examined in the individual in which the change was identified, in backcrosses and in certain cases in crosses involving the offspring. Since the transmission data were from rearrangements tested in the generation in which they arose, or within a very few generations of their origin, genetic modification of meiotic behaviour was unlikely. The inheritance of only four rearrangements corresponded to Mendelian expectations, while others exhibited either meiotic drive (4), or drag (1).

Meiotic drive occurred through both egg and pollen, suggesting that accumulation plays a most significant role in fixing deleterious chromosomal variants. In

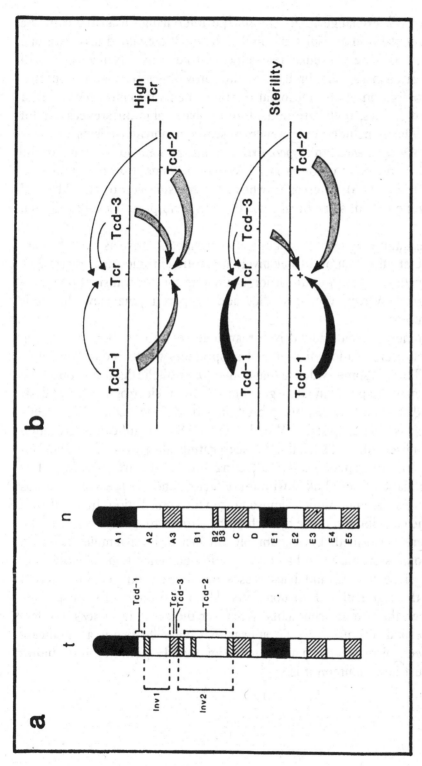

Fig. 6.6. a. G-banded chromosome 17 showing the wild type (*n*) and *t*-haplotype (*t*) configuration. The two major inversions are indicated as are the three segregation-distorter genes (*Tcd*-1 to 3) and the responder (*Tcr*). b. An operational model for male sterility and segregation distortion in the *t*-haplotype system. When the distorters are heterozygous (top), their effects on the wild type allele of the responder are strong, whereas they are weak on the *t* form of the responder (*Tcr*). When homozygous (lower) the distorters have a deleterious effect on *Tcr* and *t*, which leads to sterility. (Figures are redrawn from Lyon, 1986, and Lyon, 1991.)

the case of the X/2 exchange in *R. acetosa*, the interchange was found in the progeny of two females in an isolated population which contained less than one hundred individuals. The particular rearrangement has drive through eggs and pollen and has arisen in a small inbred deme, thus providing a fascinating opportunity to study the fixation of chromosomal rearrangements in population isolates.

In other plant species, both rare mutant and polymorphic differences exhibit meiotic drive. These include supernumerary segment polymorphisms in *Scilla autumnalis* and *Rumex acetosa*. In *Hypochoeris radicata* spontaneous mutants in the form of the centric fission of chromosome 1 were preferentially inherited through pollen (Parker *et al.*, 1988), whereas in wheat, the transmission of an additional chromosome was ensured, since breaks occurred in meiospores lacking it (Finch *et al.*, 1984).

All of these examples point to the significance of meiotic drive as a mechanism which can facilitate the fixation of negatively heterotic chromosomal rearrangements. Whether this segregation distortion is a continuous or transient phenomenon is unclear. However, recent studies add support to the view that it is sometimes transient.

From the moment a meiotically driven rearrangement arises there is a strong selection for the accumulation of unlinked suppressors which reduce the drive (Lyttle, 1989). There is some evidence that non-random allelic distortion observed in some plants may disappear in latter generations. In an electrophoretic analysis of segregation distortion between two inbred lines of *Zea mays* some 1900 seedlings of an F2 cross were analysed by Wendel *et al.* (1987). Significant segregation distortion was observed in 11 of the 17 segregating allozymes. The aberrant ratios arose from five genetic factors affecting pre-zygotic transmission. The electrophoretic markers were linked with these factors and the meiotic drive was a product of multilocus and multichromosome genetic competition. In an attempt to determine the contribution made by each of the parents to gamete competition, backcrosses were performed. Unaccountably, all loci in both families failed to show segregation distortion, i.e. it had been lost in one generation. Wendel *et al.* suggested that since the F2 and backcrosses had been produced in different environments, the segregation distortion detected was a product of selection differences between the two environments. Whatever the cause of its loss, the fact of its loss is beyond dispute. It would appear that meiotic drive is a significant evolutionary mechanism which can no longer be ignored. Hopefully, future studies may show us just how common it is.

6.5 Theoretical perspectives on fixing deleterious structural rearrangements in founding populations

Empirically, it would seem unlikely that a chromosomal variant which was negatively heterotic would have much chance of fixation in a population because of the reduced fertility it carries. Nevertheless, such rearrangements do reach fixation and it is simply a matter of expanding our concept of population genetics to account for these phenomena. Since negatively heterotic rearrangements are the basis for models of chromosomal speciation, understanding the means by which they reach fixation is crucial to our view of that process.

The spread and fixation of negatively heterotic chromosomal rearrangements can be achieved in two ways. First, particular mechanisms, such as meiotic drive, position effects, or linkage with selectively advantageous gene combinations, may overcome the deleterious fertility effects of the chromosomal rearrangements. Second, stochastic processes, such as random genetic drift, accompanied by migration and the colonization of new territory, may ensure the spread and fixation of such rearrangements.

To account for the fixation of negatively heterotic rearrangements in his model of stasipatric speciation, White (1968, 1978a) borrowed from both approaches. He suggested that the rearrangement must have selective advantage in the homozygous condition, and also argued that the species should have low vagility, with the capacity to form population isolates. Genetic drift, meiotic drive and inbreeding were factors agreed as important in assisting the fixation of structural changes.

White's inclusion of both stochastic and mechanical processes was at odds with a large body of the theoretical modellers, who argued that purely stochastic processes were sufficient to account for the fixation of negatively heterotic rearrangements, thus following the views of Wright (1941) on genetic drift. Lande (1979) adopted this purely stochastic stance. However, he qualified it to some extent by insisting that a chromosomal rearrangement with a substantial heterozygote disadvantage could reach fixation by genetic drift, only if the population was subdivided into small, nearly isolated demes. Lande (1979) argued that since the effects of drift alone could explain the fixation of structural rearrangements in small isolated populations and because there was no evidence for rearrangements showing meiotic drive (but see Section 6.4.3), then the significance of that process was negligible. Lande (1984) emphasized that strongly subdivided populations with effective deme sizes in the order of 100 or less, and high rates of local extinction and colonization, were characteristics of many present-day species.

The population structure most favourable to the fixation of negatively heterotic rearrangements was contrasted to those criteria required for the fixation of rearrangements with a heterozygous and homozygous selective advantage by

Lande (1984). He concluded that large stable populations with high interdeme migration and low rates of local extinction and colonization, so that N_e (the effective population size) was very large, were the preferred population characteristics for positively heterotic, or adaptive, homozygous changes.

Evidence has also been provided by Hedrick and Levin (1984) that kin-founding may also be a potent factor in the establishment of chromosomal variants in populations. In particular, when the individuals were very closely related, such as selfed progeny, or a full-sib group, the probability of fixation of a chromosomal variant was much higher than in a group of unrelated individuals. Indeed, this is exactly the situation one might expect if individuals with a chromosomal variant are present in an isolated deme, or are chance survivors in a population isolate. Hedrick and Levin emphasized that these effects should be most important in highly structured populations, such as plants and animals with low dispersal, or in social animals.

Wright's (1931, 1932, 1940, 1977, 1982a) 'shifting balance theory' described the stochastic processes involved in fixing deleterious rearrangements and how such novel adaptations could arise and spread in geographically subdivided populations. Chief among these was the process of 'selective diffusion'. Here, the number of emigrants and colonizers which dispersed from a deme, increased with the average fitness of individuals in that deme. Since genetic variability between geographically separated demes was maintained by migration and mutation, fitness in a particular deme could be increased if individuals in it moved from an old adaptive peak to a new and more adaptive peak by mutation and random genetic drift. Once such an improved adaptation had been achieved, more migrants and colonizers would be produced by this than neighbouring less adaptive demes, by the process of 'interdemic selection'.

In a major theoretical study in which all of Wright's parameters, including selective diffusion, were incorporated for the first time, Lande (1985) modelled Wright's shifting balance theory when applied to the fixation of major negatively heterotic chromosomal rearrangements. Lande (1985) made an analogy between the shifting balance theory and the fixation of major chromosomal rearrangements in a strongly subdivided population with local extinction and colonization. His points of comparison were that every spontaneous rearrangement with a heterozygote disadvantage, creates a new adaptive peak for the population which can be attained by random genetic drift across an adaptive valley, selection up the new peak and spread by colonization and migration.

In essence Lande's studies indicated that:

1 Interdeme selection during the spread of a mutation depended more on the capacity of the mutant to invade and become established in other

demes than on selective diffusion, unless there was rapid local extinction and colonization.

2 The intensity of interdeme selection was reduced by random local extinction and colonization. When these processes were rapid (with no selective diffusion), the expected fixation rate of spontaneous mutations with a heterozygous disadvantage approached that in a single isolated deme.

3 Local extinction and colonization, and selection on the homozygotes, accelerated the spread of chromosomal mutations which were destined to be fixed.

Thus, the act of colonization appeared to be of major significance in the spread of deleterious rearrangements in a derived population.

A computer simulation prepared by Chesser and Baker (1986), attempted to determine what stochastic conditions were necessary for the fixation of chromosomal mutations in small isolated demes. Parameters such as litter size, age-dependent mortality, overlapping generations and varying sex ratios, which had not been previously considered, were included in these studies. Chesser and Baker did not include inbreeding between close relatives and varied the degree of fertility effect of the heterozygous rearrangement from 0 to 0.5.

These authors found that random processes alone could explain the frequency of fixation of chromosomal rearrangements when the number of initial founders was small (five to ten individuals), when meiotic problems had a small impact on fertility and when the number of offspring per mating was high. However, they found that the fixation of a mutation was unlikely when the number of karyotypic rearrangements was low and when survivorship was high. With population bottlenecks involving more than 20 individuals, few offspring and substantial fertility effects, random processes could not explain the fixation of such changes. Meiotic problems which may reduce fecundity were relatively unimportant to either the fixation or extinction rates when the number of founders was small, but these became increasingly important with higher numbers of founders. Mutations did not reach fixation when fecundity reductions of 0.25 or greater were encountered, or when founding population sizes were 20 to 50 for either model. Chesser and Baker (1986) concluded that, in cases where species karyotypes have been massively repatterned, factors other than stochastic processes and population bottlenecks were probably important to the fixation of chromosomal mutations.

Chesser and Baker (1986) argued that larger demes have the greatest probability of survival and that high rates of extinction within small demes made it unlikely that chromosomal mutations in small demes would reach fixation. However, they also found that meiotic problems which may reduce fecundity are relatively unimportant to the fixation or extinction rates when the number of founders was small, but became increasingly important with larger numbers of founders.

A slightly different approach to the question of the fixation of chromosomal rearrangements was taken by Sites *et al.* (1988a). The basic programme of Chesser and Baker (1986) was modified to include several population genetic correlates (F-statistics), based on electrophoretic parameters which Sites and Moritz (1987) considered significant, and these were applied to the analysis of a single iguanid lizard species, *Sceloporus grammicus*. Sites *et al.* (1988a) then tested their findings against Hall's (1983) cascading model of chromosomal speciation. This model is very similar to White's (1978b) chain processes model which is dealt with in Section 9.1.3.

When 'bottlenecked' population sizes had been reduced to five individuals, extinction rates were high, but from 1 to 2.5% of populations fixed new chromosomal rearrangements. As population size increased, the probability of fixing novel rearrangements decreased. Sites *et al.* (1988a) argued that the cascading model of chromosomal speciation was unlikely in *Sceloporus grammicus*, since the empirical and simulated electrophoretic profiles were far too different. Another interpretation can be made. That is that the electrophoretic parameters and genetic correlates proposed by Sites and Moritz (1987) as being critical determinants to chromosomal speciation are, in reality, quite irrelevant to that process, and that Sites *et al.*'s computer simulation study supports that conclusion. The assumptions and conclusions advocated by Sites and Moritz (1987) are dealt with in more detail in Sections 8.2.8 and 8.4.1.

All of the studies described above ignored the possibility of a role for mechanistic processes in the fixation of deleterious rearrangements. Lyttle (1989) offered the following three arguments to rebuff this purely stochastic approach. First, meiotic drive reduced the frequency interval over which drift must act to ensure that a rearrangement was established in the face of selection. It also increased the minimum population size which was needed to generate drift. Meiotic drive increased the probability that a rearrangement would reach fixation and relaxed constraints on population size which Lande (1979) considers necessary without meiotic drive. Second, in regard to how common drive is, Lyttle pointed to the likelihood for the selection for the suppression of meiotic drive as soon as it arose. Thus, rearrangements established initially with the assistance of drive, may no longer exhibit this mechanism. Third, it was unreasonable to assume that populations would remain small, or that interdemic migration rates would stay lower than extinction rates, as Lande's drift model does. Only a small drive advantage was necessary to ensure the fixation of a rearrangement, and this was even less if drift and other stochastic conditions were operating.

Mathematical modelling was used by Hedrick (1981) to test the effects of meiotic drive, the selective advantage of the homokaryotype, inbreeding and genetic drift, both in isolation and in combination. He concluded that meiotic drive, when associated with a heterokaryotypic disadvantage, could lead to a substantial increase in the probability of fixation. Most significantly, Hedrick found that inbreeding in concert with a selective advantage of the new homokaryotype, or with genetic drift, was important in the fixation of a rearrangement in a population isolate. Walsh (1982) took a slightly different approach, analysing the rate at which reproductive isolation occurred. He considered negatively herotic rearrangements by themselves, those with a homozygous advantage and those with meiotic drive. The results are of interest.

1 In populations with small to moderate effective population size, unassisted rearrangements could only produce reproductive isolation if they were numerous and only weakly deleterious.
2 With moderate population size, strong homozygote advantage or drive can fix a few strongly deleterious rearrangements, resulting in reproductive isolation.
3 With a large population size, only meiotically driven rearrangements can become fixed.

Walsh pointed out that chromosomal speciation can only result from mildly to strongly negatively herotic rearrangements inducing hybrid sterility, if the populations are very small (less than 50), with a high chance of extinction and inbreeding depression. These rearrangements could only reach fixation in moderately large populations with a high mutation rate and with driven rearrangements.

In summary, most of the population modelling data support the concept that negatively herotic chromosomal rearrangements can be established by purely stochastic processes in small, isolated, inbred demes of low-vagility species. The significance of population size, or the particulars of a variety of other variables, are all open to debate and suggestion. After all, the data obtained from computer modelling studies are only as good as the assumptions that are made and the weighting given to any variable. Conclusions can at best only be regarded as tentative. Genetic drift, selection and the capacity of an organism to colonize are all key features. Equally, the capacity of the population to retain its isolation and resultant genetic integrity through inbreeding and kin-founding, increase the likelihood for the fixation of negatively herotic rearrangements.

Unless these stochastic processes were available, negatively herotic rearrangements which occurred as spontaneous mutants within species distributions would be swamped and presumably eliminated. Whatever their precise fate, it is unlikely that they could reach fixation. Nevertheless, there is little doubt that an undue emphasis has been placed on purely stochastic mechanisms, when it is now apparent that deterministic mechanisms such as meiotic drive have a major and significant role in the fixation of chromosomal rearrangements.

6.6 Concluding remarks

The role of chromosome change in speciation has been challenged in two distinct areas. The first of these supposes that negatively heterotic chromosomal rearrangements cannot be established in derived populations because of their heterozygous disadvantage, and thus, by inference, the chromosomal differences that we observe between species are not negatively heterotic and not involved in speciation. Second, that the fertility effects of structural rearrangements are not sufficient to prevent gene flow between parental and daughter populations and that, consequently, these rearrangements cannot be effective post-mating isolating mechanisms.

In this chapter we have dealt with the first of these areas of contention, whereas the second, defining the fertility effects of structural rearrangements, will be detailed in the following Chapter 7. For the moment it should be stated that **there is no longer any room for debate as to whether profoundly negatively heterotic chromosomal rearrangements can reach fixation in derived populations, the fact of their fixation is undeniable. It is up to the opponents of chromosomal speciation to explain how these rearrangements have reached fixation. That is, our concepts of population genetics must be modified to account for the existing phenomena.**

The two broad areas considered in the present chapter impinge directly on the capacity of plant and animal populations to fix negatively heterotic chromosomal rearrangements. First, the innate mechanisms, such as chromosomal mutation rate, the type of structural rearrangement which can occur, whether changes are fixed simultaneously or sequentially, or whether these can be driven to fixation, all have a direct impact on the nature and capacity for chromosomal evolution.

Second, the inherent difficulty of explaining how negatively heterotic rearrangements can ever reach fixation in plant and animal populations requires an array of stochastic and mechanistic assumptions. These parameters have been considered by theoretical geneticists using computer modelling, although the assumptions these authors have added have further complicated a most complex situation. Nevertheless, their conclusions have guided our thinking as to the population structure necessary for the fixation of such deleterious changes.

The conclusions that one might reach from an examination of more-recent research findings, indicate that the basic innate mechanisms may have a far greater impact on the fixation of negatively heterotic rearrangements than was previously considered likely.

1 Estimations of the chromosomal mutation rates for plants and animals made by White (1978a) at 1 in 500 would appear to be a substantial underestimation and oversimplification. Considerable variation in the mutation rate is known between species and this might extend from as high in 1 in 6 to lower than 1 in 3000. Moreover, there appears to be substantial variation in the mutation rate between chromosomes within complements. When chromosome breakage induced by transposable elements is considered, the mutation rate may increase enormously involving up to 70% of cells within an individual. Nevertheless, at this stage this cannot be regarded as a general phenomenon. The apparent chromosomal uniformity displayed by entire families of plants and animals is difficult to reconcile with the mutability we are now confronted with. The conclusion seems unavoidable that particular genomes have a high potential for change, whereas others remain frozen in time. Nevertheless, the possibility of substantial chromosomal evolution induced by a high mutation rate is far greater than would have been dreamed of a decade ago.

2 There is now little doubt that chromosomal evolution in terms of the site of chromosome breakage and the type of chromosome rearrangement established is a non-random process. Characteristics of the structure of the genome and the potential for transposable elements to cause consistent breaks at particular sites, can lead to the formation of certain types of rearrangements, involving specific chromosomes and in some cases the whole genome. The explanation of karyotypic orthoselective processes as being the result of selection on a random array of rearrangements, no longer appears tenable. It is now possible to explain the uniform pattern of chromosomal reorganization found in particular species as being based on the molecular structure of that species genome.

3 Equally, there appears to be evidence to support the view that some of the more substantial incidences of genome reorganization attributed to structural rearrangements may have been due to a series of simultaneous changes rather than a cascade of sequential rearrangements. This provides for the possibility of rapid chromosomal evolution and the formation of decisive reproductive isolation, rather than the gradual accumulation of individual differences.

4 There is now no doubt that meiotic drive is a significant evolutionary

mechanism. Not only have numerous instances of drive been documented over the last few years, but these have been established in a wide range of organisms and have implicated different types of structural rearrangements. While it is clear that not every chromosomal rearrangement is being driven to fixation, the possibility that many of the negatively heterotic changes have been established with the assistance of drive can no longer be discounted.

The four areas considered above have a direct impact on the nature of chromosomal evolution and the possibility of fixing powerfully negatively heterotic rearrangements. *In toto*, they tell us that such rearrangements can be rapidly formed and established in population isolates, at a rate which is greater than previously thought possible.

Despite these findings, there is an apparent reluctance of theoretical geneticists to incorporate such innate mechanisms in their estimations. There is no doubt that computer simulations examining the fixation of chromosomal rearrangements in isolated populations are only as good as the assumptions introduced. There is equally no doubt that many of the simulations bear little relationship to reality because of these assumptions. Let us take an extreme case. Meiotic drive has been ignored as a process by many modellers despite the evidence for its existence in some organisms (see Lande, 1979, 1985; Chesser and Baker, 1986; Sites *et al.*, 1988a). When this mechanism is included in computer simulations, the probability that negatively heterotic rearrangements will reach fixation increases substantially (Hedrick, 1981; Walsh, 1982). Unnecessarily constrained models which argue against the likelihood of deleterious rearrangements being established, appear to reflect a personal position rather than an approximation of reality. Surely the aim of modellers is to explain how powerfully deleterious mutants have been established, not why they cannot be established.

The following population characteristics have been assessed as being important to the fixation of negatively heterotic structural rearrangements:

1 Purely stochastic processes, such as random genetic drift, migration and the colonization of new territory.
2 Genetic drift alone if the rearrangements are established in a deme size of from 10 to 100 individuals, in the case of vertebrates.
3 Inbreeding, steep selective gradients, population bottlenecks, selection for the homozygous rearrangement and the kin-founding in populations of low vagility, or social animals.
4 Meiotic drive in animals with a moderate population size, which can fix several strongly deleterious rearrangements resulting in reproductive isolation.

It would appear that negatively heterotic rearrangements capable of forming post-mating isolating mechanisms can be established in an array of situations, provided that isolation, inbreeding, steep selective gradients, small population size and colonization are included. Fixation would appear to be automatic if the change is driven.

7

The impact of structural hybridity on fertility and viability

Muntjac deer and equids:

> What neither of these two mammalian cases resolve is the actual cause of meiotic blockage which leads to hybrid sterility. The probability is that it does not depend on the chromosome differences, but, *rather, results from faulty genic interaction, though this has not as yet been established.*
>
> (John and Miklos, 1988, p. 260)

and

Mice:

> Thus, even when numerous structural differences exist between the two chromosome sets these differences do not produce complete reproductive isolation, *and the extent to which the genotype is involved in the fertility produced has yet to be clarified.*
>
> (John and Miklos, 1988, p. 267)

and

> Indeed, there are still too many biologists who, in the name of hypothesis, fail to appreciate the difference between evidence and assumption, or else who are only too ready to ignore the one in favour of the other.
>
> (John, 1981, p. 48)

7.1 Fertility effects induced by chromosome changes

Two approaches have been used to determine the impact of structural chromosome hybridity on organismal viability and fertility. The first of these is the purely mechanistic analysis of meiosis in hybrids. Chromosomal behaviour, including pairing and segregation patterns, may be used to estimate the effects of chromosomal rearrangements on gametogenesis and thus hybrid fertility. The second way of estimating the fertility effects of structural rearrangements is to analyse the number, condition and type of hybrid progeny produced. Indeed, in some cases the total inviability of zygotes, or reduction in litter size, gives a direct assessment of the incidence of aneuploidy and production of unbalanced gametes. The two sets of data often match remarkably well (see below).

The two basic sources of chromosomal hybrids are laboratory crosses and hybrid zones. The first option is to produce laboratory hybrids between animals from different wild populations. By doing this the direct effect of chromosomal changes can be estimated by meiotic investigation and then cross-referenced to the fertility of hybrids. Alternatively, structural rearrangements found in wild populations can be introduced into laboratory-reared strains of known genotype and the fertility effects of these changes may then be estimated. Naturally occurring rare mutants found in laboratory populations can also be investigated in terms of their fertility effects and in some cases segregation patterns can be scored in the animal in which they occurred (Evans et al., 1967). Second, naturally occurring hybrids between species or cytotypes can be removed from the hybrid zones and investigated in the laboratory. The difficulty of establishing the status of hybrid zone hybrids (i.e. whether they are F1, F2 or backcross), is a substantial drawback for this form of analysis and can only be overcome by using chromosomal banding markers, or genotypic markers. The additional drawback of lack of knowledge of the precise age of individuals, or the number of eggs which they may or may not have laid, creates difficulties in estimating the effects of chromosomal hybridity on morphology and fecundity.

Two major constraints are associated with hybrid studies and the validity of conclusions which can be drawn from them. The first of these, although equally relevant to several issues raised above, is the genetic integrity of the forms hybridized. While the effects of chromosomal hybridity can be determined in organisms which have genotypic differences established between cytotypes, it is not possible to make a distinction between the fertility effects caused by the genetic and chromosomal differences. Unless the genotypic component of both parental genomes is known, it can be claimed that the fertility effects are due to genotypic differences.

The second constraint which must be imposed on hybridization/structural hybrid analyses involves the nature of the chromosomal differences investigated. The critical distinction between investigating heterozygotes for balanced chromosomal polymorphisms and structural hybrids between cytotypes with fixed chromosomal differences has often been ignored. There is little value in analysing the meiotic segregation patterns of heterozygotes for a balanced polymorphism, and then arguing that there are no meiotic ill-effects of the particular type of structural rearrangement involved, and then concluding that rearrangements of this type cannot be involved in chromosomal speciation. Equally, it is not reasonable to extend observations on polymorphisms to fixed differences distinguishing cytotypes, even when both occur within the same species and involve similar types of rearrangements and similar chromosomes. There is no good reason why certain fixed differences might not be profoundly negatively heterotic as structural heterozygotes, whereas similar rearrangements present as chromosomal polymorphisms might be neutral. Surprisingly, numerous analyses of segregation patterns in balanced polymorphisms have been used as a basis for argument against chromosomal speciation.

Porter and Sites (1985, 1987) and Davis *et al.* (1986) analysed a number of naturally occurring pericentric inversion or fission polymorphisms and scored the populations for the frequency of chromosomal phenotypes. Not surprisingly, they found that the frequency of chromosomal heterozygotes did not depart significantly from Hardy–Weinberg equilibria. Clearly, one would expect to find this situation in a balanced polymorphism where the heterozygote was not selected against, i.e. it was neutral. Davis *et al.* (1986) argued that the chromosomal polymorphisms they examined were not negatively heterotic: 'These data are interpreted to indicate that in some cases, structural rearrangements of chromosomes do not result in the generally assumed reduction of heterozygote fitness' (p. 645). Further on they concluded that 'Meiotic selection against chromosomal heterozygotes may be generally less restrictive than has been previously assigned in the context of vertebrate evolution' (p. 648). Much of Davis *et al.*'s (1986) discussion revolved around the role of negatively heterotic rearrangements in speciation and the implications of their findings to this process. Similarly, Porter and Sites (1985), when considering fission polymorphisms, emphasized in a much more direct manner that 'Both the frequency of the polymorphisms in this sample and the meiotic analysis should be considered in the light of the chromosomal speciation hypotheses in general . . .' (p. 253). Sites and Moritz (1987) argued 'yet these are the same types of rearrangements suggested by Hall (1983) to have initiated speciation in this complex' (p. 165). To restate the point unambiguously, **the only means of determining whether a putatively negatively heterotic rearrangement which distinguishes one cytotype from another is deleterious to fertility, is to examine meiosis in hybrids between the cytotypes or species, not by examining balanced polymorphisms for similar rearrangements.**

7.2 Spontaneous mutations

A great many reports are present in the literature describing the fertility effects of spontaneous chromosomal changes which have arisen in domesticated animals. However, it should be noted that many of these rearrangements are effectively balanced polymorphisms, because of the unusual mechanisms which have been developed which overcome the otherwise substantial fertility effects associated with structural heterozygosity. Nevertheless, some of these are worth examining from the perspective of repetitive rearrangements, fidelity to the type of change, and the avoidance mechanisms which have been established which mitigate against the more deleterious effects of structural heterozygosity.

7.2.1 Pigs *(Sus scrofa)*

Chromosomal differences detected between many of the extant pig species involve chromosome fusions. Species as morphologically diverse as the giant forest hog (*Hylochoerus meinertzhageni*), the warthog (*Phacochoerus aethiopicus*) and bush pig (*Potamochoerus porcus*) (all of which are African species), the European wild boar (*Sus scrofa*) and the domestic pig (*Sus scrofa*), are chromosomally similar and distinguished by Robertsonian fusions (Bosma, 1978; Melander and Hansen-Melander, 1980). Whereas wild boars are characterized by chromosomal polymorphisms for two independent fusions producing chromosome numbers of 2n = 36, 37 and 38 (Tikhonov and Troshina, 1975), domestic pigs are chromosomally monomorphic. Exceptions to this statement include numerous spontaneous reciprocal translocations which have been detected in domestic pig populations. Interestingly enough, chromosomal fusions which distinguish pig species have not been encountered among these spontaneous mutants.

A total of 22 reciprocal translocations (Mäkinen and Remes, 1986; Popescu and Boscher, 1986) and one X-autosome translocation are known in the domestic pig (Gustavsson *et al.*, 1989). The great majority of these have been detected through the analysis of lineages exhibiting a decrease in litter size. Such fertility effects vary from a 25% reduction to absolute sterility.

The general effect on litter size suggested that chromosomally unbalanced zygotes were eliminated before implantation in the uterus. However, some cases are known where stillborn and malformed piglets have been produced (Gustavsson *et al.*, 1988). Offspring which appeared as normal pigs were comprised of chromosomally balanced individuals with both normal and translocation chromosomes present. Most boars which were encountered had a normal semen profile.

In the case of the X-autosome translocation, male meiosis was arrested and no stages later than pachytene were encountered. That is, total sterility resulted from the translocation. Nevertheless, no impact on fertility was encountered in females carrying the translocation (Gustavsson *et al.*, 1989) (see also Section 5.1.1.3).

Gabriel-Robez *et al.* (1988) and Jaafar *et al.* (1989) analysed synaptonemal complexes in boars heterozygous for reciprocal translocations. They determined that a substantial degree of heterologous pairing occurred at pachytene. This could well avert a deficiency in pairing due to translocation heterozygosity and so increase the chances of germ cell survival. Nevertheless, heterosynaptic pairing would minimize chiasma formation and thus increase the incidence of aneuploidy due to the resultant non-disjunction. However, it would also increase the incidence of embryonic loss by producing unbalanced chromosome sets. Gustavsson et al. (1988) also detected non-homologous pairing in one of the four reciprocal translocations they analysed. Nevertheless, ring quadrivalents formed in meiosis gave 2:2 disjunction with alternate and adjacent-1 segregation, whereas chain multiples found in 70% of cells gave rise to adjacent-2 disjunction (Figs. 7.1 and 5.3). Thus, offspring consisted of chromosomally balanced individuals with both normal and translocated chromosomes. Translocations can therefore persist in pig populations, lowering both viability and the fertility and providing a substantial economic encumbrance to farmers.

7.2.2 Cattle *(Bos taurus)*

Chromosomal analyses of over 50 species of Bovoidea have been made (Wurster and Benirschke, 1968). These reveal substantial interspecific and intergeneric chromosome differences with numbers ranging from 2n = 30 to 60. The predominant mode of differentiation has been by chromosome fusion, although inversion and both reciprocal translocations and tandem fusions have been found. Changes in the quantity of paracentromeric heterochromatin have also occurred (Buckland and Evans, 1978a, b; Wallace, 1978; Iannuzzi and Di Berardino, 1985).

In domestic cattle *(Bos taurus)*, large numbers of spontaneous chromosomal mutants have been established. Unlike pigs, these are generally not reciprocal translocations, but chromosome fusions. Over 30 different chromosome fusions have been detected, although occasional pericentric inversions and translocations are known (see Berland et al., 1988). Most of the fusions occur as single spontaneous heterozygotes, although several double fusion heterozygotes have been detected (Roldan et al., 1984). A single tandem fusion was recorded by Hansen (1969) which produced a 10% reduction of fertility in bulls. In the majority of cases, the fertility effects of the chromosome fusions are minor. For example, Logue and Harvey (1978) investigated spermatogenesis and meiotic behaviour in *Bos taurus* bulls which were heterozygous for spontaneous 1/29 Robertsonian translocations. There was no evidence for any impairment to fertility in terms of mating behaviour, semen quality and spermatogenesis. The rate of non-disjunction was 6.4% in heterozygotes. Gustavsson (1969) made a major analysis of the 1/29 fusion polymorphism in Swedish red and white cattle with a sample

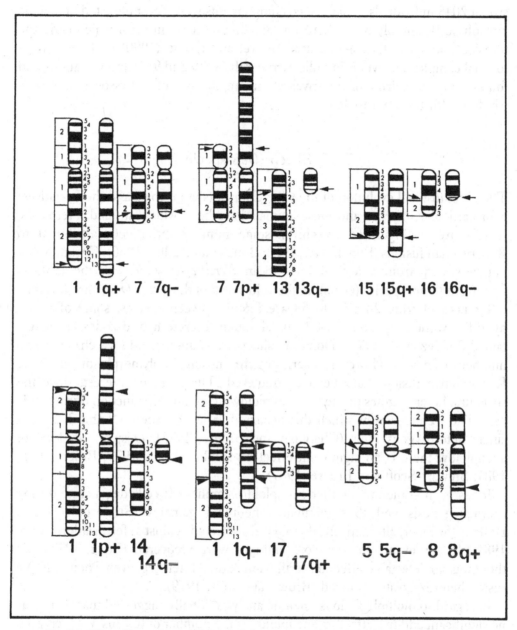

Fig. 7.1. These illustrations of G-banded chromosomes show six of the spontaneous translocations which have been detected in domestic pig populations. The arrowheads represent
the breakage and rejoining points in the normal and translocated chromosomes. (Redrawn
from Gustavsson, 1988, and Gustavsson *et al.*, 1988.)

size of 2045 individuals. Unbalanced complements were never detected in meiotic metaphase II, and although there was a subtle increase in embryonic death, the translocation was effectively neutral. Bouvet and Cribiu (1990) analysed synaptonenal complex behaviour in bulls carrying the 1/29 and 9/23 translocations and found complete pairing of the trivalent and an absence of any association of this trivalent with the sex vesicle.

7.2.3 Sheep (*Ovis* species)

The chromosomal evolution of the tribe Caprini (sheep, goats and their relatives) is of particular interest, and these animals have been widely studied. Here, too, species are distinguished by chromosome number differences attributed to Robertsonian fusions (Bunch *et al.*, 1976; Bunch and Nadler, 1980). Chromosome numbers vary from a low of 2n = 48 in *Hemitragus jemlahicus*, the tahr, to 2n = 60 in *Capra* species (goats). In the wild sheep of the genus *Ovis*, chromosomal differences of from 2n = 52 to 58 are found between species, some of which hybridize without apparent ill-effects, although discrete hybrid zones are maintained (Valdez *et al.*, 1978). Domestic sheep are characterized by a chromosome number of 2n = 54. However, a series of chromosomal polymorphisms involving Robertsonian fusions have been encountered. These are of interest, since the structural heterozygotes produced aneuploidy due to non-disjunction during meiosis. The frequencies at which this occurred varied between 6.9 and 20.2% in single heterozygous rams (Chapman and Bruere, 1975). Despite this level of aneuploidy, there is no significant effect on fertility (Long, 1977; Bruere *et al.*, 1981; Stewart-Scott and Bruere, 1987).

It has been argued that either aneuploid metaphase II cells degenerate during spermatogenesis and thus eliminate aneuploid gametes (Stewart-Scott and Bruere, 1987), or, alternatively, that they are selected against at fertilization (John, 1981). In any event, no aneuploid embryos were recovered (Long, 1977) and there was no detectable effect on fertility, including litter size, even when multiple fusion heterozygosity occurred (Bruere and Ellis, 1979).

In regard to multiple fusions, meiotic analysis initially suggested that there was no significant additive effect on aneuploidy as the number of fusions was increased (Bruere *et al.*, 1981). However, more recent studies by Stewart-Scott and Bruere (1987) found that aneuploid frequencies of 3.9, 4.3 and 5.3% were present in rams heterozygous for 1, 2 and 3 translocations respectively. This suggests an additive effect despite the low levels of aneuploidy. It appears that the chromosome fusions in sheep are essentially balanced polymorphisms with no deleterious fertility effects due to the elimination of aneuploid cells. However, the demonstrated fact of additive fertility effects, suggests that multiple fusions may provide a significant fertility barrier in those organisms which lack aneuploid elimination mechanisms.

7.2.4 Domestic fowl *(Gallus domesticus)*

Bonaminio and Fechheimer (1988) investigated the relationship between the frequency of meiotic products of spermatocytes, heterozygous for reciprocal translocations involving chromosome pair 1 and a microchromosome, in the domestic chicken. These were compared to the frequencies of chromosomally balanced and unbalanced zygotes in a sample of embryos. In the secondary spermatocytes, neither balanced nor unbalanced chromosomal types had complementary rearrangements in the expected 1:1 ratio. An excess of spermatocytes with long-arm deficiencies suggested lagging of this element at metaphase I, although 52.5% of secondary spermatocytes contained a balanced complement. In embryos, the balanced products of segregation occurred in the expected 1:1 ratio, but the complementary products did not. Some 49.6% of embryos carried a balanced complement, i.e. a higher proportion of duplications and deficiencies were present. Bonaminio and Fechheimer (1988) presented evidence which indicated that balanced and unbalanced complements were being transmitted to embryos at different frequencies than those produced in secondary spermatocytes. The authors attributed these observations to differential fertility and viability in spermatogonia with different unbalanced genomes, rather than to the different sperm being produced in unequal numbers. These results provide another example of meiotic drive and in this case for a reciprocal translocation (see Section 6.4).

7.2.5 The importance of investigation into spontaneous mutations

The examples shown here illustrate the relatively high rate of spontaneous mutation in domesticated animals. They also indicate that chromosomal rearrangements of specific types recur in particular species and that these often involve a biased representation of the chromosomes involved, i.e. they are non-random. In a number of instances simultaneous chromosome changes have occurred, and in those cases where several rearrangements are present in the one organism, their effect on fertility is additive.

Nevertheless, a large number of these chromosome changes will remain as balanced polymorphism, either because they lack any dramatic effect on viability or fertility, or because mechanisms are present which effectively disarm them.

Thus, potentially deleterious mutations are tolerated. This even extends to reciprocal translocations which persist and lower the viability and fertility for the population as a whole, but not necessarily for certain translocation carriers. In this regard, some reciprocal translocations are parasitic, whereas the effects of others are so profound that their progeny are rendered sterile, or the individuals are themselves sterile. Many of the chromosome fusions, on the other hand, had very minor fertility effects.

The degree to which the deleterious side-effects of potentially damaging structural changes are eliminated, modified or by-passed is related to the area at which the effect is expressed. Thus:

1 The co-orientation of reciprocal translocation multiples ensures that balanced individuals with both normal and translocated elements are produced.
2 Chromosomally unbalanced zygotes produced by reciprocal translocations may be eliminated before implantation (*Sus scrofa*).
3 Chromosomally unbalanced and balanced gametes may be transmitted at different frequencies due to the differential viability of spermatogonia (*Gallus domesticus*), i.e. male segregation distortion by meiotic drive.
4 Non-homologous pairing at pachytene in reciprocal translocation heterozygotes, could increase germ cell survival, but also increases the incidence of aneuploidy by changing the position of chiasmata (*Sus scrofa*).
5 Aneuploid cells produced by non-disjunction of chromosome fusion heterozygotes may degenerate during spermatogenesis (*Ovis*).
6 Aneuploid gametes may be selected against at fertilization (*Ovis*).
7 Heterozygotes for negatively heterotic changes with reduced litter size may be driven. This may take the form of selection for the derived homozygote which is associated with increased litter size and is thus adaptive (*Alopex lagopus*) (see Section 6.4.3.5).

One might validly conclude from these observations that the prediction of viability and fertility effects of spontaneous structural rearrangements is a most precarious exercise, which might be directly influenced by the reproductive physiology and cellular dynamics of the organism involved. Clearly, conclusions relating to chromosomal speciation based on chromosomal mutants in domesticated animals may be misleading, since most are effectively balanced polymorphisms.

7.3 Hybridization studies: simple and complex systems

In this section, the effects of structural rearrangements on the viability and fertility of artificial hybrids between chromosomally distinct forms will be explored. Rather than use an exclusively mechanical approach and deal with the different types of rearrangements found in different species under their broad structural categories, they will be dealt with in case by case scenarios. As it turns out, the outcome is not very different, since most chromosome changes which distinguish species, or chromosome races, are of the same type in the one complex. Thus, in mice, lemurs and rats most rearrangements encountered are Robertsonian fusions, or multiple Robertsonian fusions, with brachial homologies.

7.3.1 Fertility effects of fusion heterozygosity in *Mus domesticus*

Chromosomal evolution within the genus *Mus* is enormously complex, with species differing by numerous fixed rearrangements (Jotterand-Bellomo, 1984, 1986, 1988). Within *Mus domesticus* a total of 22 different chromosome races have been characterized in Western Europe and Africa and a total of 60 chromosome races are now known throughout the world (Winking *et al.*, 1988; Mahadevaiah *et al.*, 1990).

Chromosomal variation detected between monomorphic chromosome races of *Mus* involve Robertsonian fusions. A total of 77 different chromosome arm combinations are now known in *Mus* (Winking *et al.*, 1988) (see Fig. 7.2 and Table 7.1). These may be divided into two categories. First, independent fusions which are present as one or more entities in the genome, i.e. some races have one fusion, others ten fusions. They are independent in the sense that separate chromosome arms are involved in the fusion and in structural heterozygotes a trivalent is formed at metaphase 1 (see Fig. 5.2). The second category are chromosome fusions which share monobrachial homologies. Here, when at least two fusions occur in a genome, both fusions share common chromosome arms. The result is that chain multiples or rings are formed at metaphase 1. Large numbers of fusions sharing brachial homologies may be present, and these may involve most elements in the genome (Fig. 7.3), producing chains or super-rings.

The viability and fertility effects of these types of rearrangements have been extensively investigated using three distinct approaches. In the laboratory, crosses have been made between different cytotypes found in the wild to estimate meiotic segregation and fertility effects on F1 hybrids. Second, spontaneous chromosome fusions present in laboratory mouse strains have been examined in the animals in which they were found, or, alternatively, particular fusions which were only known

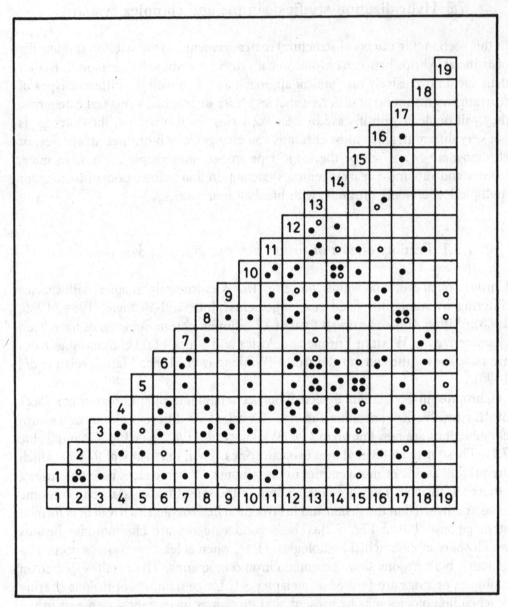

Fig. 7.2. A summary of the different Robertsonian fusions which have been detected in wild populations (●) and laboratory stocks (○) of *Mus domesticus*. (Redrawn form Redi et al., 1990c.)

Table 7.1. Composition and frequency of Robertsonian translocations in karyotypically distinct populations of Mus domesticus. Robertsonian translocations of laboratory origin are marked by an asterisk. (From Winking et al, 1988)

Composition	Frequency	Composition	Frequency	Composition	Frequency	Composition	Frequency
1.2	2	3.10	1	6.10	2	10.11	2
1.2*	1	3.12	1	6.12	2	10.12	8
1.3	3	3.13	1	6.13	2	10.14	5
1.5	1	3.14	1	6.13*	1	10.14*	1
1.6	1	3.15*	1	6.14	1	10.15	1
1.7	1	3.16	1	6.15*	1	10.17	1
1.10	1	3.X*	1	6.16	2		
1.11	3			6.19*	1	11.13	7
1.15*	1	4.6	4			11.14	1
1.18	1	4.8	1	7.8	2	11.14*	1
		4.10	2	7.13*	1	11.15	1
2.3*	1	4.11	1	7.18	4	11.16	1
2.4	2	4.12	3			11.17	1
2.5	1	4.13	1	8.9	1	11.18	1
2.6*	1	4.14	1	8.10	1		
2.8	4	4.15	1	8.12	2	12.13	1
2.14	1	4.15*	1	8.14	3	12.13*	1
2.15	1	4.17	1	8.15	1	12.14	1
2.16	1	4.18*	1	8.17	4		
2.17	1			8.17*	1	13.14	1
2.17*	1	5.12	1	8.19*	1	13.15	1
2.18	1	5.13	3			13.16	2
		5.14	3	9.11	1		
3.4	3	5.15	10	9.12	2	15.17	2
3.5*	1	5.17	1	9.13	1		
3.6	2	5.19*	1	9.14	8	16.17	7
3.8	4			9.16	3		
3.9	2	6.7	3	9.19*	1	17.18	1

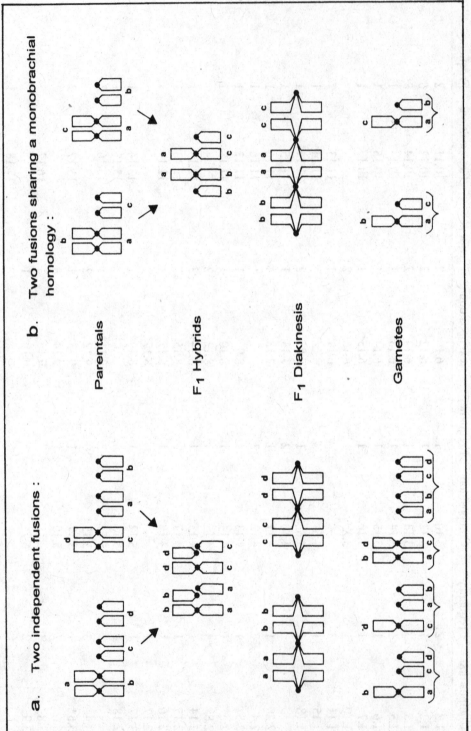

Fig. 7.3. A schematic representation of meiotic configuration and segregation pattern in structural hybrids for Robertsonian fusions which are heterozygous for two independent fusions (a) and for two fusions sharing a monobrachial homology (b).

in wild *Mus* populations were introduced into laboratory strains by selective breeding and then the effects on viability and fertility effects were studied. Third, fertility effects were investigated in naturally occurring hybrid zones.

Following an extensive series of hybridization studies in *Mus domesticus*, which were largely based on the introduction of wild Robertsonian rearrangements into laboratory mouse genomes, Gropp *et al.* (1982) were able to provide a broad perspective on the fertility effects of fusion heterozygosity. Two principal areas of effect were identified by Gropp *et al.* First, the meiotic non-disjunction of multivalents in structural heterozygotes which led to the production of unbalanced gametes and in turn produced aneuploid zygotes which were eliminated. Second, a mechanism was found in males which affected the process of gametogenesis as a whole and caused the breakdown of spermiogenesis leading to maturation arrest, or sterility. This secondary mechanism also occurred in females (Setterfield and Mittwoch, 1986). Both non-disjunction and gametogenic impairment can provide powerful post-mating isolating mechanisms; however, the situation is complex and the levels of effect are correlated with the type of Robertsonian rearrangements involved. A third mechanism has also been implicated in fertility effects. That is, pairing failure in fusion multivalents can lead to the association of unpaired autosomal segments with sex chromosomes. This form of association can have profound fertility effects. Thus, fertility effects in *Mus* can be broadly considered in terms of the type and number of rearrangements present:

7.3.1.1 Single Robertsonian fusions

In those animals which were heterozygous for a single Robertsonian fusion producing a trivalent at diakinesis, the rates of malsegregation of the trivalent could be readily calculated by scoring metaphase II cells. Gropp and Winking (1981) and Gropp *et al.* (1982) made an analysis of 16 separate fusions involving chromosomes of different size combinations (Table 7.2) in male mice. The results showed non-disjunction rates of from 2 to 28%. These are significant fertility effects for single fusions.

In seven cases, the effects on both males and females were compared for the same fusion. This was done by scoring the rates of meiotic anaphase 1 malsegregation (Table 7.3). The difference was striking, for in some females segregational errors shared a tenfold increase resulting in a 33 to 61% non-disjunction rate.

Redi and Capanna (1988) point to the fact that chromosomal fusion combinations involving the same arms, vary in their non-disjunction rates when these are derived from different areas. Thus, Rb (10.11) 8 Bnr/+ males from Val Bregaglia, northern Italy, have 5%, whereas Rb (10.11) 5 Rma/+ males from central Italy have 13.8%. They also note that malsegregation rates are much higher in older females and are also higher when wild rearrangements are intro-

Table 7.2. *Malsegregation rates in male single Robertsonian metacentric heterozygotes of* Mus domesticus. *(From Gropp and Winking, 1981)*

Designation (arm composition)	Chromosome arm counts in male meiotic metaphase II (%)					Non-disjunction rates calculated on the base of >20 × 2	No. of M II cells scored
	<19	19	20	21	>21		
I Feral origin							
Rb (1.3) 1 Bnr/+		8	85	7		14	400
Rb (4.6) 2 Bnr/+		10	79	11		22	300
Rb (5.15) 3 Bnr/+		13	73	14		28	300
Rb (11.13) 4 Bnr/+		12	74	14		28	400
Rb (8.12) 5 Bnr/+	2	3	93	2		4	300
Rb (9.14) 6 Bnr/+	1	5	89	5		10	300
Rb (16.17) 7 Bnr/+	1	2	95	2		4	300
Rb (10.11) 8 Bnr/+		4	95	1		2	600
Rb (4.12) 9 Bnr/+		5	91	4		8	400
Rb (1.10) 10 Bnr/+		2	96	2		4	400
Rb (3.8) 2 Rma/+	0.5	9	81.5	9		18	200
Rb (6.13) 3 Rma/+		4.3	92.3	3.3		6.6	300
Rb (4.15) 4 Rma/+		11	78	11		22	300
Rb (10.11) 5 Rma/+		6.6	86.2	7.2		14.4	600
Rb (2.18) 6 Rma/+		7	84.7	8.3		16.3	300
Rb (9.16) 9 Rma/+		11.7	80	8.3		16.6	300
II Laboratory origin							
Rb (9.19) 163 H/+		3	94	3		6	300
Rb (6.15) 1 Ald/+		8	89	2		4	600
Rb (8.17) 1 Lem/+		6	92.5	1.5		3	300

duced into laboratory genomes than when spontaneous rearrangements are detected within laboratory genomes. Baranov (1980) found essentially similar results, where post-implantation embryonic deaths ranged from 12 to 15% in laboratory fusions compared to 16 to 25% with wild fusions in laboratory mice. Disturbances in gametogenesis were not encountered with single Robertsonian fusion heterozygosity.

In certain mouse strains, particular heterozygotes for one or more Robertsonian translocations are sterile or infertile. Grao *et al.* (1989) analysed synaptonemal complexes in male mice heterozygous for three independent Robertsonian translocations, none of which had been implicated in sterility. They found that the non-homologous centromeric regions of the acrocentrics were unpaired in the trivalents and heterosynaptic pairing occurred at pachytene. Associations between

Table 7.3. *Comparison of rates of meiotic anaphase I malsegregation (M II evaluation) in male versus female single Robertsonian metacentric heterozygotes of* Mus domesticus. *(From Gropp* et al., *1982)*

Spermatocytes		Rb/+	Oocytes	
n	Rate of non-modal M II* (%)		Rate of non-modal M II* (%)	n
400	14	Rb (1.3) 1 Bnr	58	103
400	28	Rb (11.13) 4 Bnr	34	77
300	4	Rb (16.17) 7 Bnr	34	83
300	7	Rb (6.13) 3 Rma	34	69
300	22	Rb (4.15) 4 Rma	50	85
300	16	Rb (4.17) 13 Lub	61	66
300	4	Rb (8.17) 1 Lem	33	102

*Calculated by doubling the hyperhaploid M II counts.

the unpaired region in one of the three trivalents, and the sex vesicle were found in 30% of the nuclei. In this study the very limited association between autosomes and sex chromosomes had no impact on fertility. However, Forejt (1979) examined male meiosis in Rb 1 Lem/+ and Rb 7 Bnr/+ structural hybrids, where fertility was normal. In the 1185 diakinesis and metaphase I cells scored, multivalent/X-chromosome associations were due to chance alone. Since the degree of X-autosome association may be directly related to the level of fertility, it would be appropriate to analyse pairing behaviour in those single Robertsonian translocations which are known to be sterile. Luciani *et al.* (1984) and Rosenmann *et al.* (1985) have found cases of heterozygote sterility correlated with the association of Robertsonian trivalents, or unpaired autosomes, with the sex vesicle in man. Here it is thought that the association interferes with the X-inactivation process.

7.3.1.2. Multiple independent Robertsonian fusions

Gropp and Winking (1981) found that mice which were heterozygous for several or many independent Robertsonian fusions (i.e. where metacentrics other than homologues did not share homologies and thus only formed trivalents at meiosis), non-disjunction rates rose to very high levels. In such cases, females still had higher non-disjunction rates than males (52 and 51% compared to 68 and 77% in tobacco mouse F1 Rb (1.7) Bnr/+ and CD-F1 Rb (1.9) Rma/+). These profound effects were not purely arithmetically additive for the single values of each fusion. However, a contrasting result was found by Baranov (1980) who found low incidences of trisomy and unaffected fertility in laboratory mice with two or three independent fusions. Redi and Capanna (1988) have encountered gametogenic disturbances in mice where multiple trivalents are present, although this was not a severe primary impairment of gametogenesis.

7.3.1.3 Multiple Robertsonian fusions with monobrachial homologies

In those animals where double Robertsonian heterozygotes share a chromosome arm (a monobrachial homology), quadrivalent chains, or rings, are formed at meiosis. Where larger numbers of homologies and fusions are present, chains or super-rings of up to 18 chromosomes can be formed (see Figs 7.3 and 7.4).

Baranov (1980) found that in double heterozygotes with a monobrachial homology, such as Rb (8.17) 1 Lem and Rb (16.17) 7 Bnr, non-disjunction effects are massive and males are completely sterile, whereas females had post-implantation loss of 30 and 25% of early resorbed embryos. That is, the situation had been reversed when compared to single or multiple independent fusions where the female had enhanced sterility. Here, males were sterile and females could be 'fully' fertile (Redi and Capanna, 1988). Although Gropp et al. (1982) were less dogmatic in this respect, they noted that in 42 cases of double heterozygotes with monobrachial homologies, female fertility was only reduced by 25 to 50%. In contrast, in 15 of the 42 cases, male heterozygotes showed complete sterility with spermatogenic/spermiogenic breakdown, and in 12 cases subfertility. These authors argued that the lower levels of sterility in females than males were due to the fact that in females only a single mechanism was responsible for fertility impairment, i.e. malsegregation resulting in unbalanced gametes and pre- and post-zygotic loss. In males, malsegregation along with a male-specific depression of gametogenesis also occurred. However, in none of these cases was any information provided on the possibility of X–autosome association, which could explain what could be interpreted as a Haldane effect.

Forejt (1979) found that sterile male mice which were doubly heterozygous for the two Robertsonian translocations Rb (16.17) 7 Bnr and Rb (8.17) 1 Lem (i.e. they shared a monobracial homology), formed non-random associations with the X chromosome. In addition, the XY chromosomes showed impaired condensation when associated. The data strongly support the hypothesis that interference with X-chromosome inactivation is a possible cause of spermatogenic breakdown.

This hypothesis also received support from Mahadevaiah et al. (1990) who examined pachytene pairing and sperm counts in mice with both single Robertsonian translocations and those which had compound chromosomes with monobrachial homologies. In the compound heterozygote Rb (6.15) 1 Ald/Rb (4.6) 2 Bnr, which is male sterile, sperm were sparse, whereas in Rb (6.15) 1 Ald/ Rb (4.15) 4 Rma, which is male fertile, the sperm count was 50% of the parental

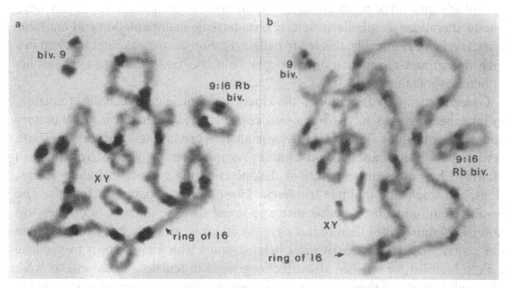

Fig. 7.4. a and b. C-banded diakinesis cells taken from the testis of *Mus domesticus* hybrids between Cittaduccale and Camp Basso populations in central Italy. Both cells show a ring involving 16 Robertsonian translocations. (These photographs were generously donated by Dr H. Winking.)

value. The heterozygotes for single fusions formed trivalents, of which 20–40% had unsynapsed ends. The quadrivalents had 85% unsynapsed ends in the male-sterile heterozygote and 75% in the male-fertile heterozygote. In both cases, quadrivalents associated with the XY in about 70% of cells examined. X–autosome association also occurred in 25% of trivalents.

Gropp *et al.* (1982) found that complete sterility regularly occurred when chains with pentavalent, or longer, composition were formed in meiosis. Gropp *et al.* (1982) were able to construct animals with either rings, or complex chain multivalents, in their breeding trials. They found that ring heterozygotes did not show the spermatogenic failure encountered in chain heterozygotes (Fig. 7.4). Sperm were present in abundance; nevertheless, pre- and post-implantation losses were very high when crosses were analysed. This was due to non-disjunction and unbalanced foetal complements, the number of which increased proportionally with the greater ring size, with up to 100% non-disjunction in super-rings. However, this is segregationally induced sterility rather than gametogenic sterility.

Eichenlaub-Ritter and Winking (1990) used immunofluorescence to examine non-disjunction, spindle structure and chromosome alignment, in maturing oocytes in CD/cremona mice hybrids which were heterozygous for multiple and different fusions, resulting in super-rings of 18 at meiosis. They found that pairing and chiasma formation occurred in the heterozygotes and that high levels of aneuploidy and variation in chromosome number in metaphase II oocytes were due to alterations in spindle structure, chromosome malorientation and chromosome lagging. Presumably, this led to malsegregation at anaphase and resulted in a high rate of non-disjunction and aneuploidy in oocytes, leading to total sterility in heterozygous females.

Garagna *et al.* (1990) examined the kinetics of female germ cell differentiation in mice heterozygous for Robertsonian rearrangements derived from laboratory and wild populations. This most extensive analysis constructed heterozygotes with seven separate chromosomal constitutions ranging from $2n = 23$ to $2n = 38$, each of which had different numbers of diakinesis configurations including chains, rings, and single and multiple trivalents. These were compared to controls. The results are of undoubted significance and reveal that the number of oocytes was lower in heterozygotes than in homozygous controls, and this was irrespective of the genetic background. Structural heterozygotes which are known to produce male subfertility, or sterility, also affect oogenesis in females. These effects were often substantial and there was some evidence that the impact on meiosis included the earliest stages, with up to an 87% reduction in heterozygotes at pachytene/ diplotene.

These authors determined that the effects on oocyte numbers were greater with chain configurations (chains of five, or nine, or seven independent trivalents), than with ring configurations, thus confirming data already available in males. The data show that chromosomal heterozygotes producing chain multiples have dual-level effects in females, just as they do in males. The impact on gametogenesis shortens the reproductive life span of the heterozygous female by reducing the number of oocytes available for reproduction. The second-level effects of aneuploidy and chromosomal imbalance lead to zygotic loss. Although it is now clear that gametogenic impairment is present in females as well as males, it is also clear that the degree of gametogenic impairment differs between the two sexes, with male sterility effects being far more profound.

It appears that the massive effects on fertility and viability found in heterozygotes for as few as two chromosome fusions sharing monobrachial homologies, implicates three mechanisms as the causative agents. These are: gametogenic impairment, aneuploidy and malsegregation, and the association of sex chromosomes with the multivalent.

7.3.1.4 *Reciprocal translocation in* Mus

Reciprocal translocations are well known in laboratory strains of *Mus domesticus*, but are virtually unknown in the wild. Autosomal translocations which cause male sterility were first described by Forejt (1974) and by Forejt and Gregorová (1977). They are generally characterized by a break-point in a proximal position on one chromosome and in a telomeric position on the other. Forejt found that male-sterile translocations had a non-random pachytene association with the X chromosome which was thought to be responsible for male sterility (Forejt et al., 1981). Subsequently, de Boer et al. (1986) established that in all male-sterile chromosome mutants, pachytene association of the X and unpaired autosomal segments occurs.

Two possible explanations have been proposed for this sterility. First, Miklos (1974) argued that the breakdown in gametogenesis was caused by the failure of saturation of pairing sites and thus the presence of unpaired regions in pachytene. Second, Lifschytz and Lindsley (1972) and Forejt and Gregorová (1977) proposed that the association of the autosomal material with the XY bivalent interfered with the X-chromosome inactivation cycle which was necessary for normal spermatogenesis. That is, it reactivated a normally silenced X chromosome. A corollary to this is that oogenesis would be unaffected in females, if autosomes were associated with the XX, since both X chromosomes remain active.

de Boer et al. (1986) analysed four reciprocal translocation heterozygotes and two double heterozygotes, each of which had a different level of sterility. By comparing synaptic behaviour in pachytene, they had hoped to test the two hypotheses for male sterility. In the case of reciprocal translocations, spermatogenic impairment was correlated with proximity of the sex chromosomes and the translocation multivalent, and to the lack of meiotic pairing in the multivalent. In the double heterozygotes, the same correlations were found.

A similar result was obtained by Mahadevaiah and Mittwoch (1986) using tertiary trisomic $Ts(5^{12})$ 31 H chromosomes derived from the reciprocal translocation $T(5.12)$ 31 H. The extra chromosome was associated with the XY bivalent in most spermatocytes, whereas in 50% of the oocytes, the 5^{12} chromosome was associated with unpaired chromosomes, with heterochromatin, or was self-synapsed. These authors concluded that the errors in pairing bore no relationship to the breakdown in spermatogenesis and impairment of oogenesis that they encountered. Perhaps the most significant finding of this study was the indication that anomalies which resulted in the breakdown of spermatogenesis also effected oogenesis. Setterfield and Mittwoch (1986) carried out oocyte counts on 3- to 5-day old tertiary trisomic $Ts(5^{12})$ 31 H mice and found a 71% reduction in the number of oocytes when compared to normal litter mates. In comparison, the sperm count of trisomic males was less than 1% of normal, i.e. a 99% reduction. Most surviving sperm were abnormal. Clearly, the effects of trisomy on male and female gametogenesis differ in degree rather than manner.

A number of studies have been directed at assessing the direct effect of X–autosome association on fertility in translocation heterozygotes. Hotta and Chandley (1982) found increased levels of X-linked enzymes in late prophase spermatocytes of X–autosome translocations, or reciprocal translocations associated with the sex vesicle. This suggested a reactivation of the X-linked genes, thus supporting Lifschytz and Lindsley's (1972) hypothesis of interference of autosomes in X inactivation.

Speed (1986) failed to detect an increase in the amount of incorporation of H^3 Uridine, i.e. there was no increase of RNA synthesis in the sex vesicles of tertiary trisomic mice in which the extra chromosome was always associated with the sex vesicle. Jaafar *et al.* (1989) examined the distribution of H^3 Uridine incorporation in primary spermatocytes of mice carrying the X-16 X–autosome translocation. This translocation produced sterility in males and semi-sterility in females. Here, too, there was no evidence for the reactivation of the X as Lifschytz and Lindsley's model would suggest. Rather, there was a spread of X inactivation over normal and translocated autosome, suggesting partial inactivation of autosome 16. This technique was less sophisticated in terms of resolution than the analysis of Hotta and Chandley (1982). Nevertheless, Jaafar *et al.* (1989) felt that it was possible that partial reactivation of the X chromosome may have been occurring as well as an inactivation of the autosome (see Fig. 7.5).

These data in combination, indicate that X–autosome associations have a powerful effect on sterility and suggest that this interferes with the X-inactivation process and may also result in autosome inactivation. Pairing failure does, however, make a contribution to these processes, in that unpaired regions of the synaptonemal complex will preferentially associate with the X.

7.3.1.5 *Genic variation in* Mus

The argument has been advanced by Sites and Moritz (1987) and John and Miklos (1988), that fertility effects found in structural hybrids of organisms such as *Mus* are not necessarily due to the chromosomal rearrangements which impair meiotic segregation, but to unspecified 'genic effects'. This notion generally remains without support in those studies where genic and chromosomal analyses have been made. There is little doubt that this is the case in *Mus*.

For example, Ferris *et al.* (1983) made a major study on the mtDNA (300 mapped restriction sites), in eight subspecies of *Mus*. This analysis included six chromosome races of *Mus domesticus* distinguished by from two to nine pairs of chromosome fusions. The chromosome races investigated shared great homogeneity in their mtDNA. Only one abnormal type of mtDNA was encountered in specimens with two pairs of fusions, the others were indistinguishable. The authors concluded that the mtDNA data indicate a very recent origin for the populations (20000 to 40000 years), and suggests that chromosomal differentiation may have been established by a few founders. However, Moriwaki *et al.* (1984) detected greater levels of divergence between chromosomal forms and

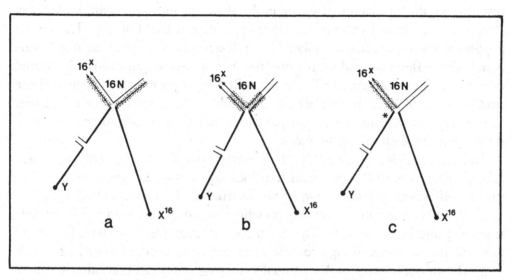

Fig. 7.5. Three hypotheses were advanced by Jaafar *et al.* (1989) to explain male sterility in *Mus domesticus* specimens possessing an X–autosome (pair 16) translocation. These were based on the distribution of H³ Uridine incorporation. In all cases the *unlabelled* segments are stippled.

a. An extension of inactivation only occurs on the translocated chromosome 16 segments.

b. An extension of inactivation to both segments of bivalent 16 which are nearest the X.

c. An extension of inactivation to the segments of bivalent 16 in contact with that part of the X which has the inactivation centre (*).

The results indicate that hypothesis b is most likely with X inactivation extending over both normal and translocated chromosomes 16 near the X break-point. It is likely that the autosomal inactivation leads to sterility. (Redrawn from Jaafar *et al.*, 1989.)

subspecies of *Mus*. This is not surprising since most forms analysed were of a greater level of systematic divergence than those analysed by Ferris *et al.* (1983).

Britton-Davidian *et al.* (1989) surveyed 28 populations of *Mus domesticus* in both western Europe and north Africa for the proteins encoded by 34 nuclear loci. This electrophoretic analysis was conducted on populations which were geographically adjacent and which included all-acrocentric populations and those characterized by from one to nine pairs of fixed fusion differences. The results were striking in their consistency. The mean values for heterozygosity at each of the polymorphic loci followed the same pattern of variability in all-acrocentric mice as it did in the Robertsonian populations ($H = 0.09$). There were no fixed electrophoretic differences attributed to any of the chromosome races. Genetic distances varied from $D = 0.01$ to $D = 0.07$ over all populations analysed. Most significantly, two of the parapatric Italian populations which failed to show genic differentiation were totally reproductively isolated by monobrachial homology resulting in the formation of chain multivalents in meiosis. Hybrids between Bergamo and Binasco chromosome races produced a chain of five. The Ovada population was reproductively isolated from all-acrocentric mice by eight independent fusions; Bergamo and Ovada differed by nine fusions and Binasco differed from Ovada by eight fusions. These three Robertsonian populations yielded the smallest levels of genic differentiation. The authors attributed this genic similarity to the very recent common ancestry, rather than to gene flow between reproductively isolated chromosome races.

Saïd *et al.* (1986) investigated *M. domesticus* populations in Tunisia, which carried nine pairs of Robertsonian translocations (2n = 22), when compared to the 2n = 40 acrocentric karyomorph. An electrophoretic analysis of both cytotypes used 36 presumptive loci. The only genetic differences observed involved allele frequency and heterozygosity. The mean Nei distance was $D = 0.037$. Results of this type do not mean that genic differences have not been established between other chromosome races of *Mus*. Equally, they do not mean that other forms of differentiation have not occurred. Capanna *et al.* (1985) analysed two sympatric chromosome races which had diverged to a considerable level in terms of their behaviour and had established pre-mating isolating mechanisms. Because of the pre-mating isolating mechanisms, these 2n = 24 and 2n = 26 chromosome races did not form hybrids. There were no fixed genetic differences between the chromosome races and they only differed by allelic frequencies in the 26 loci investigated ($D = 0.024$).

In those instances where genic divergence has been investigated in chromosome races of *Mus*, which are distinguished by profound chromosomal reproductive isolating mechanisms, substantial genic differences have not been detected with either mtDNA or electrophoretic analysis. There is no evidence that reproductive failure in these forms can be attributed to genetic change and much evidence to suggest that it is purely chromosomal in origin.

7.3.2 *Rattus sordidus* complex

Three species of Australian rodents *Rattus sordidus* (2n = 32), *R. colletti* (2n = 42) and *R. villosissimus* (2n = 50) are distinguished by multiple chromosome fusions. Baverstock *et al.* (1983b) produced a series of hybrids between these species in the laboratory using wild caught animals. F1 hybrids of *R. colletti* × *R. villosissimus* survived, but had reduced viability and fertility. Male meiosis revealed the presence of three trivalents and a chain of five. In backcrosses of F1 hybrids to the parental forms, litter size was reduced by 70%. F1 males of the *R. sordidus* × *R. colletti* cross were completely sterile and in meiotic metaphase cells four trivalents, a ring of four and chain of seven were produced. Similarly, in *R. sordidus* × *R. villosissimus* crosses, F1 hybrid males were totally infertile, as were backcrosses (Mahony, pers. comm.). These authors attributed sterility, or reduced fertility, in structural hybrids to the fusions sharing monobrachial homologies disrupting segregation. However, Mahony (pers. comm.) has indicated that X–multivalent associations are also present and are directly implicated in fertility and viability effects.

Baverstock *et al.* (1986) were able to discount the possibility that hybrid infertility was a product of genic differences between the species. In a most extensive electrophoretic analysis using 55 loci, Baverstock *et al.* found that only one fixed electrophoretic difference had been established between *R. sordidus* and *R. colletti* and *R. villosissimus*. The genetic distance between *R. colletti* and *R. villosissimus* was $D = 0.001$ and between this pair and *D. sordidus* it was $D = 0.08$. A microcomplement fixation analysis of serum albumin was used to assess the extent of structural gene divergence. None had occurred. Only very subtle pellage differences separate these forms which, until the chromosomal analyses had been made, most regarded as subspecies.

Despite this convincing data, John and Miklos (1988) concluded that 'none of these cases offer convincing evidence that the fixed fusion differences observed between related mammal species are likely to have played a causative role in the speciation process as a result of the malsegregation they produce in hybrid combinations' (p. 268).

One can only deduce from this statement that John and Miklos (1988) failed to consider the biochemical data, even though it had been alluded to in Baverstock *et al.* (1983b) and described in detail in Baverstock *et al.* (1986). It is clear that these authors misunderstood the backcross fertility data in regard to litter size, since they comment that litter size was reduced 'to about 70% that found in the parental species' (p. 268) rather than litter size being reduced by 70%, which is the case. The absolute F1 sterility of *R. sordidus* × *R. colletti* F1 hybrids also appears to have been ignored. All evidence suggests that viability, fertility and indeed speciation in the *R. sordidus* complex is based on chromosomally induced reproductive isolating mechanisms.

7.3.3 Muntjac deer

The tiny muntjac deer of the subfamily Muntiacinae are one of the most cytogenetically exciting groups. The genus *Muntiacus* comprises five species, the Indian muntjac (*M. muntjak*), Fea's muntjac (*M. feae*), the black muntjac (*M. crinifrons*), the Chinese muntjac (*M. reevesi*) and Roosevelt's muntjac (*M. rooseveltorum*). One of the subspecies, the red muntjac (*M. muntjak vaginalis*), has the lowest chromosome number known in any vertebrate 2n = 6 (♀), 7 (♂) (Wurster and Benirschke, 1970). M. *muntjak muntjak* has 2n = 8 (♀) (Wurster and Atkin, 1972), and M. *rooseveltorum* from Laos has 2n = 6 (♀) (Wurster-Hill and Seidel, 1985), whereas, the rare *M. feae* has 2n = 13 (♀) (Soma *et al.*, 1983). In contrast, M. *reevesi* has 2n = 46 chromosomes (Wurster and Benirschke, 1967), a much higher chromosome number which is characteristic of many other Cervidae (2n = 50 to 70, Mayr *et al.*, 1987).

The extraordinary difference in chromosome number between what are morphologically similar species (*M. muntjak* and *M. reevesi*), is thought to have been due to a series of tandem fusions and Robertsonian fusions of a M. *reevesi*-like complement. Indeed, Liming *et al.* (1980) were able to reconstruct a matching G-banded M. *muntjak vaginalis* complement from an F1 *M. reevesi* × *M. muntjak vaginalis* hybrid karyotype. It is noteworthy that this series of structural rearrangements was correlated with a 20% difference in DNA content between the two species (Wurster and Atkin, 1972). Schmidtke *et al.* (1981) and Johnston *et al.* (1982) determined that this difference was due to a loss of middle repeat DNA. Loh-Chung *et al.* (1986) used restriction enzymes, *in situ* hybridization and DNA sequencing analysis on both species. Their analysis of seven DNA clones isolated from highly repeated DNA of Indian muntjac cells, indicated that a satellite specific to the paracentromeric regions of that species was also present in the Chinese muntjac. Blot hybridization revealed a great homology of the satellite DNA component despite the massive karyotypic differences between the species.

Hybridization studies between *M. muntjak vaginalis* and *M. reevesi* by Liming *et al.* (1980) and Liming and Pathak (1981) determined that both male (2n = 27) and female (2n = 26) F1 hybrids could be produced. Gametogenesis in the male suggested a low level of pairing homology, as indicated by an absence of synapsis at pachytene, despite the broader molecular similarity. Moreover, the hybrid male was sterile, since spermatogenesis had been arrested at early prophase and spermatocytes were degenerative.

It would be most interesting to examine the chromosomes of *M. crinifrons* and gametogenesis in *M. reevesi* and the structurally intermediate *M. feae* and its hybrids. At the moment the data suggest that the profound chromosomal differences between the morphologically most similar Indian and Chinese muntjacs are providing an absolute reproductive barrier.

7.3.4 Horses (*Equus* species)

The horses of the family Equidae provide an interesting example of speciation involving substantial chromosome change followed by a high level of morphological and presumably some genic divergence. The seven surviving species are unmistakably different in morphology: compare the ass (*E. asinus*) to Hartmann's mountain zebra (*E. zebra hartmannae*). There seems to be little doubt that considerable morphological divergence has occurred subsequent to their speciation. It may be reasonable to conclude that most members of the genus *Equus* are survivors of a less recent cladogenic process.

George and Ryder (1986) made a molecular analysis of the mtDNA of seven extant *Equus* species producing restriction-endonuclease cleavage maps. Sequence divergence between *E. przewalskii* and *E. caballus* ranged between 0.27 and 0.41%. There was greater divergence within *E. caballus* lines (0.55%). The former is in the same order of magnitude as between human racial groups (0.36%).

The degree of divergence varied between *E. hemionus onager* and *E. h. kulan* (0%) and *E. przewalskii* and *E. hemionus* at 7.8%. George and Ryder (1986) estimated that the lineage to extant horses was present 3.9 MYBP and the most recent divergence was between the two zebras *E. grevyi* and *E. burchelli* 1.6 MYBP. These findings suggest an older radiation when compared to *Mus*.

With this in mind, the very extensive chromosome differences found between the species suggest that structural rearrangements have played a major role as a post-mating isolating mechanism. Chromosome numbers include *E. przewalskii* (2n = 66, FN = 92), *E. caballus* (2n = 64, FN = 92), *E. assinus* (2n = 62, FN = 102), *E. hemionus onager* (2n = 56, FN = 102), *E. grevyi* (2n = 46, FN = 80), *E. burchelli* (2n = 44, FN = 80), *E. zebra hartmannae* (2n = 32, FN = 62). While undoubted Robertsonian rearrangements have had a significant impact on the evolution of this complex, the G and C-banding investigation of Ryder *et al.* (1978) indicated that numerous pericentric inversions and changes in the amount of heterochromatin have also been established.

F1 hybrids have been produced between all species. All but two of these hybrids have been totally sterile, showing a breakdown of gametogenesis and general failure of meiosis at pachytene. The first of the exceptions is the hybrid between the horse *E. caballus* and *E. przewalskii*. Here, a single chromosome fusion distinguished the species (Short *et al.*, 1974). During the course of meiosis the heterozygous chromosomes form trivalents and segregate regularly. F1 hybrids and backcrosses to *E. prezewalskii* are totally fertile. If there are any genic differences which distinguish these species they have no impact on fertility. It is noteworthy that *E. przewalskii* is thought to be the direct ancestor of *E. caballus*.

Interspecific hybrids between *E. asinus* and *E. caballus* produce male and female mules; male and female hinny are produced by the reciprocal cross. Chandley *et al.* (1974) examined meiosis in male mules and hinny and found that spermatogenesis breaks down at pachytene due to a substantial disruption of pairing, although on some rare occasions spermatozoa were encountered in the hinny. Chandley *et al.* attributed infertility to a block in meiosis at the primary spermatocyte stage, due to the dissimilarity between maternal and paternal chromosome sets preventing synapsis. This same form of blockage, due to karyotypic incompatibility, was present in other *Equus* hybrids. However, the fact that occasional sperm could be produced by the hybrids, indicates a close homology between horse and donkey despite the significant chromosomal reorganization. Chandley *et al.* suggested that quasi-haploid sperm with entire horse complements may have been formed by chance and that these could account for the viable sperm. They also encountered occasional complete and unexplained pairing in male hinny pachytene cells.

In a most remarkable study, Zong and Fan (1989) convincingly demonstrated that occasional mule and hinny F1 hybrids were in fact fertile and that these could be backcrossed to either of the parental forms. Zong and Fan analysed the karyotypes of eight of these rare backcross progeny and found that no two individuals were chromosomally the same. Chromosome numbers included 2n = 60, 62, 63 and 64, and very different proportions of acrocentric and metacentric chromosomes were present in the complements in each case. Only one karyotype was horse-like. This finding indicates that Chandley *et al.*'s (1974) quasi-haploid sperm hypothesis, which was used to account for viable sperm in hybrids, is untenable. It is probable that those hybrid backcrosses which survived had complements balanced enough to provide the genic constitution necessary for viability. The authors argued that there are varying degrees of fertility and sterility in the F1 and backcross hybrids.

The chromosomes of the horse and the donkey are distinguished by multiple pericentric inversions and other complex rearrangements, despite the fact that their chromosome numbers are so similar (2n = 62 and 64) (Ryder *et al.*, 1978). The evidence of variable infertility in F1 hybrids between these species indicates that sterility cannot be solely attributed to faulty genic interactions between the species, as John and Miklos (1988) proposed. It seems most likely that reproductive incompatability is the product of erratic meiotic pairing failure, caused by the gross chromosomal differences established between the species. In such a situation, most pachytene cells would remain unpaired, occasional cells might show partial synapsis, whereas very rare cells may show complete synapsis. Chandley *et al.* (1974) observed this rare pachytene pairing in male hinny meiosis and found that it was correlated with the production of occasional spermatozoa. One might not expect to see cases of partial or complete pairing in meiosis if fertility differences were due to genic barriers, or genic interactions. Rather one might expect a genic blockade of particular parts of the meiotic cycle.

7.3.5 Interspecific hybridization in the Bovidae

7.3.5.1 Dik-diks

Among the smallest members of the Bovidae, the five species of African dik-diks are characterized by both isolated and overlapping distributions. The two species from the subgenus *Rhynchotragus* have a distinctly elongated proboscis. These are *Madoqua kirki* and *M. guentheri*; the former occurs in eastern and western geographic isolates. Both *M. kirki* and *M. guentheri* have been subdivided into a series of subspecies, but these are of uncertain status.

The failure of breeding trials in North American zoos has bought to light a complex system of chromosomal variation. In the case of *M. guentheri*, two karyotypes, $2n = 48$ and $2n = 50$, were identified in the three females examined. These are distinguished by a (5:17) chromosomal fusion/fission difference. The $2n = 49$ chromosome hybrids between the $2n = 48$ and $2n = 50$ forms reproduce in captivity (Ryder *et al.*, 1989).

A contrasting situation exists with *M. kirki*. Here two distinct cytotypes were found and these were distinguished by multiple rearrangements including an X–autosome translocation, inversions and tandem fusions. Both cytotypes have $2n = 46$ in the female, but the $2n = 46b$ cytotype has $2n = 47$ in the male due to an XX/XXY sex chromosome system (Ryder *et al.*, 1989). Hybridization studies revealed that only female offspring of matings between $2n = 46a$ and $2n = 46b$ cytotypes could produce progeny. The mature hybrid male analysed had extensive meiotic activity, but spermatogenesis had been arrested and the animal failed to produce spermatozoa.

F1 hybrids between female *M. kirki* ($2n = 46b$) and *M. guentheri* ($2n = 50$) males produced both male and female progeny. Complete male sterility was obtained. Once again meiosis was completely arrested, but in these hybrid males, testis size had been reduced and meiotic disturbance was far greater than in the hybrids between the cytotypes of *M. kirki*. Female F1 hybrids failed to produce progeny.

The preliminary studies by Ryder *et al.* (1989) suggest that the multiple rearrangements distinguishing cytotypes of *M. kirki* are responsible for the chromosomally induced reproductive isolation evident between cytotypes and species. Indeed, in the examples provided, the more profound the chromosomal differences were between species, or cytotypes, the greater was the degree of hybrid inviability, or sterility.

7.3.5.2 Bos *hybrids*

A number of studies have been made on hybrids between chromosomally indistinguishable species of *Bos*, i.e. indistinguishable in terms of their G-banded complements (Pathak and Kieffer, 1979). Results were unusual in that genetically closely related species such as F1 and backcross progeny of *Bos indicus* × *B. taurus* which have inconsequential fertility impairment, showed disturbance of their pachytene pairing. Dollin *et al.* (1989) found a 9.0% level of pachytene pairing abnormalities in full blood parental strains of *B. indicus* and *B. taurus* using synaptonemal complex spreads examined by electron microscopy. Dollin *et al.* (1991) found significantly higher levels in the F1s and backcrosses (11 to 30%), which were largely due to partial pairing failure, although other abnormal configurations were encountered. A diminution of these effects was observed in the backcross generation. It is noteworthy that XY–autosome associations were encountered with unpaired regions in from 2 to 11% of the cells, although individual bulls went as high as 18%. An individual bull was also observed to have 53% asynapsis (see Fig. 7.6)

Hybrids between the more distant *Bos taurus* and *B. javanicus*, which were also chromosomally indistinguishable in terms of G-banding (Pathak and Kieffer, 1979), produced highly fertile F1 cows with a level of 90%, whereas F1 bulls proved to be sterile with sperm development ceasing at the secondary spermatocyte stage (according to Kirby, 1979). Pathak and Kieffer (1979) found that F1 hybrids exhibited degenerating pachytene and there were no later stages present. Liming and Pathak (1981) reasoned that a meiotic blockade formed at pachytene was probably a physiological disturbance rather than mechanical error and that such mechanisms were common in the Bovidae and Cervidae. Dollin and Murray (pers. comm.) used synaptonemal complex analysis to investigate pairing in an F1 sterile hybrid bull. They found extensive pairing failure, thus suggesting a mechanical disturbance.

Our understanding of the evolution of chromosome form and fertility effects in the genus *Bos* is inhibited by the absence of information on the degree and timing of genetic divergence in the species we are interested in. However, comparisons which have been made between *Bison bison*, *Bos grunniensis* and *Bos taurus* using mtDNA sequencing, indicate that the degree of divergence varied from 2.6 to 9.5% in the species compared. Since it is estimated that 2% of nucleotide substitutions occur per million years (George and Ryder, 1986), the divergence of the above species ranged from 1.3 to 4.7 MYBP. This is a similar degree of divergence to that found in *Equus* (Miyamoto *et al.*, 1989). These results are comparable to an analysis of allozyme divergence in 27 species of the Bovidae, using 40 enzyme systems made by Georgiadis *et al.* (1990), which suggested a third wave of speciation events from 5 to 2.5 MYBP.

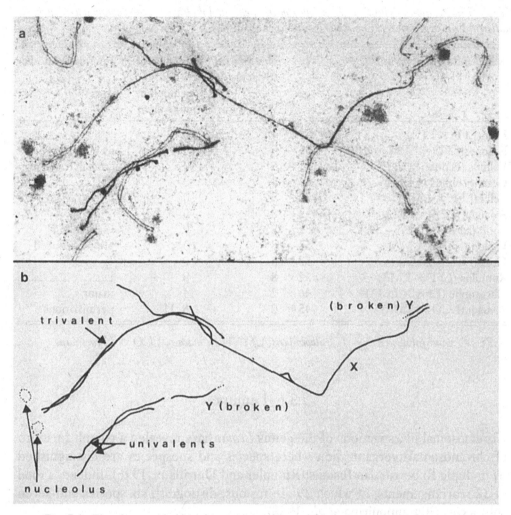

Fig. 7.6. The photograph (a) and interpretive diagram (b) show associations between the X and Y chromosomes, two univalents and a partially paired Robertsonian trivalent in a synaptonemal complex pachytene spread from an Africander bull. The Robertsonian translocation is between the smallest NOR-bearing pair 28 and a larger pair. In this cell the Y chromosome has been broken at a fragile site distal to the X–Y pairing region. (This plate was generously donated by Dr A. E. Dollin.)

Brown *et al.* (1989) detected intraspecific mtDNA variation of 0.01 to 0.5% of sequences in *Bos taurus*. When combined, this information suggests that *Bos* species represent older speciation events and indicate the likelihood of internal chromosomal divergence since speciation.

Table 7.4. *Relationship between the number of trivalents and multivalents and spermatogenesis in different* Lemur *hybrids. (From Ratomponirina* et al., *1988)*

Name and kind of hybrid[a]	2n	Number of trivalents	Number of chromosomes included in multivalent	Spermatogenesis
Loft (LFF × LFA)	54	4	0	
Isidore ([LFF × LFA] × LFF)	54	6	0	as in pure
Volmus (Isidore × LFF)	55	5	0	animals
Cesar (Volmus × LFF)	47	3	0	————
Felix (LFF × LMA)	52	8	0	
Aviateur (LFF × LMA)	52	8	0	
Potin (LFF × LMA)	52	8	0	variable but
Espion (LFF × LMA)	52	8	0	always reduced
Gros (LFF × LMA)	52	8	0	
Romulus (LFF × LMA)	52	8	0	————
Christophe (LFA × LMA)	46	2	14	major
Bassam (LCO × LMA)	45	0	6, 11	perturbations

[a]LFF = *Lemur fulvus*, LFA = *L.f. albocollaris*, LMA = *L. macaco*, LCO = *L. coronatus*.

7.3.6 Lemurs

Chromosomal investigations of the genus *Lemur* have revealed a complex pattern of chromosomal reorganization where species and subspecies are distinguished by multiple Robertsonian fusions (Rumpler and Dutrillaux, 1976). Indeed, a total of 32 rearrangements, of which 29 are fusions, distinguish six species and seven subspecies (Ratomponirina *et al.*, 1988).

In an analysis of hybrid fertility and spermatogenesis of *Lemur*, Ratomponirina *et al.* (1988) crossed *L. fulvus* (2n = 60), *L. f. collaris* (2n = 52), *L. f. albocollaris* (2n = 48), *L. macaco* (2n = 44) and *L. coronatus* (2n = 46). Three separate levels of fertility effect were observed in the fertile hybrids (Table 7.4), and these were directly correlated with the complexity and number of trivalents, or multivalents, formed at meiosis. Thus, hybrids between *L. fulvus* and *L. f. albocollaris* (2n = 54) and between individual specimens with chromosomal variation 2n = 54, 55 and 47, all reproduce in captivity. Spermatogenesis was normal despite the presence of three to six trivalents. Synaptomenal complex analysis in hybrid meiosis revealed irregular and unsynchronized pairing.

Sterile hybrids were produced between *L. macaco* and *L. f. albocollaris* (2n = 46) and *L. macaco* and *L. f. coronatus* (2n = 45). In these, gametogenesis had been

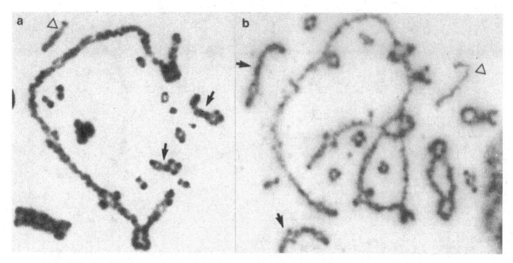

Fig. 7.7. a and b. Diakinesis cells from a hybrid between *Lemur fulvus albocollaris* and *L. macaco*. Note the chain multivalent and two trivalents (arrowed) and XY bivalent (hollow triangle) in each cell. (These photographs were generously donated by Professor Y. Rumpler.)

profoundly disturbed, and when the early stages of meiosis were present most cells degenerated after pachytene (see also *Mus* and *Bos*). Some spermatozoa were produced, but very few. The 2n = 46 hybrid produced bivalents, two trivalents and one multivalent involving 14 chromosomes, whereas the 2n = 45 hybrid had two multivalents (a chain of 11 and a ring of 6) (Fig. 7.7). Synapsis at pachytene was never observed to be complete.

The third class of hybrid displayed irregular fertility (*L. f. fulvus* and *L. macaco*, 2n = 52). One out of six animals was capable of reproduction. Eight trivalents were present and these had delayed pairing. Heterosynapsis was often observed and in occasional cells unpaired chromosomes associated with the sex chromosomes (Fig. 7.8).

In the hybrids which showed chain multivalent formation there was a direct association with infertility. This suggested to Dutrillaux and Rumpler (1977) and Ratomponirina *et al.* (1988) that Robertsonian rearrangements which formed multiples were a powerful reproductive barrier and associated with speciation in lemurs. Conversely, small numbers of rearrangements which resulted in trivalents at meiosis could in some situations be tolerated. Ratomponirina *et al.* (1988) argued that the complexity in multiple formation was accentuated by pairing failure, with the more complex chain multiples and multivalents showing more pronounced failure. These authors suggested that a delay in synapsis may have led to germ cell degeneration.

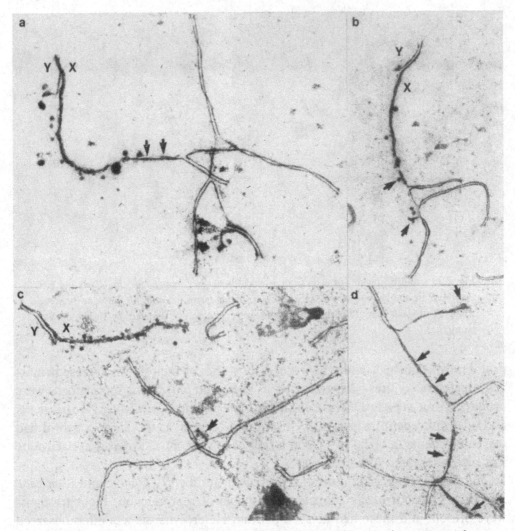

Fig. 7.8. a. Multivalent synaptonemal complex of a chain involving six chromosomes of a hybrid between *Lemur fulvus albocollaris* and *L. f. collaris* (this hybrid presents two chains; one involving six chromosomes, the other four). One of the asynapsed extremities of the chain is associated with the XY bivalent, the other one shows a dense appearance.

b. Multivalent synaptonemal complex of a chain involving four chromosomes of a complex hybrid 'lionel'. This hybrid resulted from successive crosses between *L. f. fulvus*, *L. m. macaco* and *L. albocollaris*. (This complex hybrid presents only one chain involving four chromosomes.) In this spermatocyte the XY bivalent is also associated to the quadrivalent.

c. Multivalent synaptonemal complex of a quadrivalent from the same *Lemur* hybrid as in b. Note that the asynapsed segments are more dense (arrow), but that there is no association with the XY bivalent, although they are in close proximity.

d. Multivalent synaptonemal complex of a chain of four from the same *Lemur* hybrid. Note that the asynapsed segments are more dense (arrows).

(These photographs were generously donated by Professor Y. Rumpler.)

7.3.7 Rock wallabies

The marsupials remain as one of the most intensively studied animal taxa with 178 species having been karyotyped (Hayman and Martin, 1974; Rofe and Hayman, 1985; Hayman, 1990). The techniques which have been applied include standard chromosomal analysis through to the most sophisticated DNA sequencing (Peacock et al., 1981).

The chromosomes of Australian and South American marsupials, which share the putative ancestral $2n = 14$ karyomorph common to most marsupial higher taxa, were G-banded by Rofe and Hayman (1985). The complements were virtually identical, thus supporting the hypothesis that the $2n = 14$ karyomorph was ancestral and that the great many species which had higher chromosome numbers, such as the Macropodidae, Potoroidae, Tarsipedidae, Petauridae, Didelphidae, Notoryctidae and Burramyidae, had all attained their higher chromosome numbers ($2n = 16$ to 24) by fission. In many cases, additional forms of rearrangement had been superimposed on this format.

Numerous interspecific hybrids between species had been either deliberately, or fortuitously, produced by zoos (Smith et al., 1979; Close and Lowry, 1990). Fertility effects ranged from total sterility to partial fertility in both sexes, although in a sample of 33 hybridizations examined by Close and Lowry (1990), fertility was more commonly disturbed in males than in females (Haldane's law?).

The remarkable levels of hybrid infertility in macropod matings suggested that many of the species have accumulated sufficient fixed genic differences between them to become reproductively isolated. Indeed, in certain crosses the parental species were chromosomally indistinguishable and still produced F1 male sterility. Most interspecific macropod hybrids were totally infertile (Smith et al., 1979; Close and Lowry, 1990).

One of the most interesting marsupial groups, which show particularly high levels of chromosomal evolution associated with speciation and chromosome race formation, are the rock wallabies of the genus *Petrogale*. Sharman et al. (1990) analysed 21 taxa of *Petrogale*, 11 of which are recognized as species, and found that only two of these showed the same karyotype. Chromosomal fusion and fission differences predominated and, in some cases where species were parapatrically distributed, hybrid zones occurred between them. The species/race distributions are highly complex involving relic populations on particular mountain ranges (Fig. 7.9) and what appear to be actively speciating complexes. In certain situations where species have parapatric distributions, they may be distinguished from each other by particular fusions. However, the two forms may be distinguished from their direct ancestor by a shared rearrangement, suggesting a sequential fixation of rearrangements.

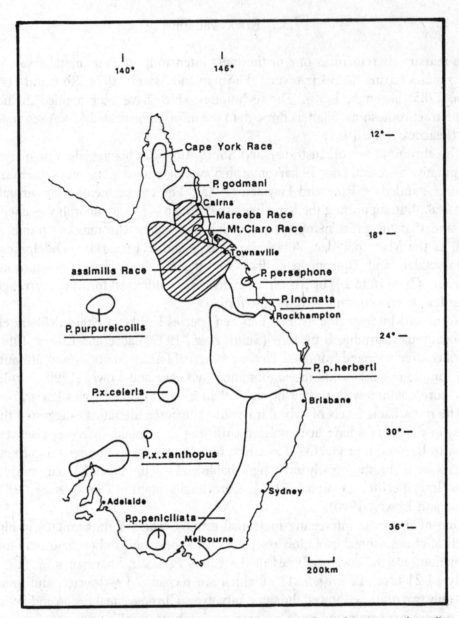

Fig. 7.9. The distribution of the rock wallabies of the genus *Petrogale* in eastern Australia. The shaded areas denote the distribution of members from the *Petrogale assimilis* complex. (Redrawn from Eldridge *et al.*, 1988.)

Sharman *et al.* (1990) suggested that the rock wallabies had all of the lifestyle requirements necessary for stasipatric speciation. They are found as localized colonies of up to several hundred animals, although individual rock outcrops may be occupied by as few as ten. A small rock wallaby colony may have a single dominant male which fathers most of the offspring. Sharman *et al.* argued that such a male, heterozygous for a single rearrangement, may have access to several females over a long period, including daughters and granddaughters, and recruitment of competing males may be infrequent.

Three of the chromosome races of the *Petrogale assimilis* species complex (*assimilis*, Mareeba and Mt Claro) are morphologically similar, but can be distinguished by unshared electrophoretic polymorphisms (Briscoe *et al.*, 1982). mtDNA analysis also indicates that they are most similar (Cathcart, 1986). The *assimilis* race (2n = 20) differs from the ancestral format by a 6:10 fusion, the Mt Claro race (2n = 20) by a 5:10 fusion and the Mareeba race (2n = 18) by 5:10 and 6:9 fusions.

Interracial hybrids between these forms produce totally infertile males in the *assimilis* × Mt Claro cross, where a quadrivalent is formed in meiosis, and *assimalis* × Mareeba cross, where a pentavalent is formed in meiosis. Females were subfertile. In the Mareeba × Mt Claro cross, a simple trivalent is formed in meiosis, but males carrying this are sterile with numerous abnormal spermatids (Eldridge *et al.*, 1990; Sharman *et al.*, 1990).

A synaptonemal complex analysis of these hybrids made by Eldridge *et al.* (1990) indicated that both non-homologous pairing and association of unsynapsed autosomes with the sex chromosomes occurred at pachytene. A high incidence of association occurred with the trivalent and X in the Mareeba × Mt Claro cross and the quadrivalent and X in the *assimilis* × Mt Claro cross (Close, pers. comm). This association may account for the reproductive failure of male F1 hybrids. Once again, females had impaired fertility, but were not sterile like the males.

7.3.8 *Caledia captiva*

In the majority of interspecific and interracial hybrids discussed thus far, inviability and infertility appear to have been associated with malsegregation in heterozygotes, gametogenic breakdown, pairing failure and the association of multivalents with sex chromosomes, due to mispairing, or heterochromatic association. The Australian grasshopper, *Caledia captiva*, provides a meaningful contrast, for the taxa are distinguished by multiple pericentric inversions involving 8 of the 12 pairs in the complement. Shaw *et al.* (1986) elevated two of the cytotypes to subspecific level; these were previously referred to as the Moreton and Torresian chromosome races. These two taxa meet in southeastern Queensland where they form a narrow hybrid zone. In addition to the eight fixed inversions, all chromosomes have C-band differences and the two subspecies are electrophoretically distinct ($D = 0.18$ using 26 loci).

The situation is complicated by the fact that the distinctive complement of the Moreton 'subspecies' is in fact a chromosomal cline, which at the southern end of its distribution, at Lakes Entrance, is chromosomally indistinguishable from the Torresian karyotype in terms of the centromere positions, but not C-banding pattern (see Fig. 7.10). The Lakes Entrance cytotype is electrophoretically, and also in terms of its highly repeated DNA profile, very similar to the Moreton taxon ($D = 0.06$).

Crosses between the Moreton and Torresian subspecies have the greatest chromosomal and high genic differences ($D = 0.18$), the F2 generation is inviable and the backcross is 53 to 64% inviable. Shaw et al. (1986) have attributed this inviability to the redistribution of chiasmata in F1 hybrids leading to recombination of the genome not normally exposed to the effect, thus breaking up significant co-adapted gene complexes.

By experimental manipulation Shaw et al. have been able to separate the effect of genic and chromosomal components on F2 inviability (see Fig. 7.10). Thus:

1 In the Lakes Entrance × Moreton cross where chromosomal differences were greatest and genic differences minimized, F2 inviability was improved by 58%.
2 In the Lakes Entrance × Torresian cross where the chromosomal differences are removed, but genic differences remain high ($D = 0.25$), F2 viability increases by 46%.
3 The portion of inviability attributed to the genic differences in the three F2 generations can be calculated in terms of the Moreton and Torresian subspecies. This is 54%. Therefore the chromosomal component must be 42%.
4 By removing the chromosomal component in the backcross generation, i.e. by using a Torresian × Lakes Entrance F1 as a parent to Torresian × Moreton, inviability improves by 46 to 65%.

These results are compelling. They provide a clear demonstration that the multiple pericentric inversions which distinguish the Moreton, Lakes Entrance and Torresian cytotypes play a major role in hybrid inviability, irrespective of the genic differentiation which is present. Indeed, the possibility that the remaining hybrid breakdown could be attributed to 'genic' differences by Coates and Shaw (1984) was criticized by Marchant et al. (1988). These authors found that genic correlates of observable allozyme differences which they encountered between Torresian and Moreton taxa, did not contribute to the lack of hybrid fitness.

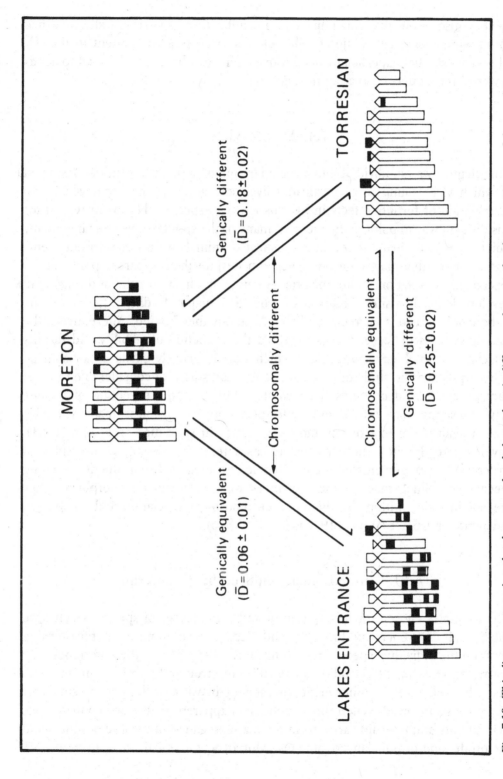

Fig. 7.10. This diagram summarizes the chromosomal and electrophoretic differences between the chromosome races of the grasshopper *Caledia captiva*. (Redrawn from Shaw *et al.*, 1986.)

They also show that the impact of recombination on structurally modified chromosomes can have major fertility effects in crosses subsequent to the F1. This suggests that superficially benign inversions can have the potential to act as powerful reproductive isolating mechanisms.

7.4 An overview

This chapter was subdivided into a consideration of spontaneous mutations, most of which were found in domesticated livestock, and was then followed by an examination of fertility effects in hybrids between species. The main reason for this division was to point to the fact that many of the spontaneous rearrangements which have been encountered in domesticated animals were not rearrangements that could readily play a role in speciation despite their apparent potential for reproductive isolation. The meiotic systems of sheep, cattle and pigs had developed mechanisms which overcame the major fertility effects of the Robertsonian fusions, or reciprocal translocations they carried. Alternatively, the genetic systems of these species tolerated the reduction of viability and fertility associated with the rearrangement which now effectively exists as a balanced polymorphism. The study of these innate mechanisms is of considerable value in our appreciation of chromosomal evolution but, with the exception of meiotic drive, these may have little relevance to speciation. Nevertheless, many who have argued against the role of chromosome change in speciation (see John, 1981), have incorrectly cited these particular cases (e.g. the sheep), as examples of structural rearrangements not being capable of providing a reproductive isolating mechanism. Since most of these changes were balanced polymorphisms they are not included within the theory of chromosomal speciation and are but an independent artifact (White, 1978a; Mayr, 1982b).

7.4.1 Chromosomally induced hybrid infertility

The examples chosen for this section describe complexes of species which have substantial differences in the type and degree of chromosomal repatterning between the individual species in each lineage. The impact of these chromosomal rearrangements on the viability and fertility of structural hybrids can, in some cases, be isolated from other genic interactions because of the genetic similarity of the forms involved. When this is done, it is apparent that major contributions to inviability and infertility arise from the malsegregation of meiotic products and the production of unbalanced gametes. Moreover, the impairment, or failure, of pairing in multivalents, combined with the association of unpaired segments with sex chromosomes, provides a powerful sterility effect in males. The nature of chromosomally derived hybrid infertility and hybrid inviability is discussed under these broad areas.

In those instances where species were distinguished by one or two Robertsonian translocations (dik-dik, *Lemur, Petrogale, Equus, Mus*), most fertility effects attributed to malsegregation were minor and fertile F1 hybrids could be formed. However, in those single Robertsonian heterozygotes where the unpaired segments of the synaptonemal complex had a high level of association with the sex vesicle, significant levels of infertility, or in some instances sterility, obtained (man). When closely related forms within the same species complex had multiple rearrangements of the same type or of different types, fertility effects were enhanced and sometimes absolute. These effects included F1 sterility, or F1 viability and F2 sterility. That is, there was a clear relationship between an increased level of infertility associated with an increased degree of chromosomal differentiation in related forms.

In those examples (Muntjac, *Caledia, Rattus sordidus*), which were characterized by multiple changes alone (this includes both multiple independent translocations, multiple pericentric inversions and multiple translocations sharing monobrachial homologies), the same pattern followed, with high levels of inviable F1 or F2 progeny and profound reductions in fertility in F1, F2, or both, or only F2.

The areas of effect targeted in spermatogenesis or oogenesis were consistent across groups sharing the same types of structural rearrangements. However, a number of unusual features were involved. For example, in *Mus*:

1 Heterozygotes for single fusions had a 2 to 28% impairment to fertility in males, but this level was amplified to 33 to 61% in terms of non-disjunction in females with the same rearrangements.

2 When additional independent fusions were present this sexual dichotomy continued, with effects being greater in females than males. With additional fusions, fertility effects increased greatly, but they were not simply additive.

3 When multiple fusions with monobrachial homologies were present, males were either inviable, or completely sterile, if chain multiples were formed, but females were not. Some females were fully fertile. However, significant effects on oocyte production were detected.

4 Where unpaired elements in reciprocal translocations, or trisomics, associated with sex chromosomes, male sterility and impaired female fertility occurred.

A second level of sterility effect was also observed in both males and females of *Mus domesticus*. This affected gametogenesis and resulted in a breakdown of spermatogenesis before pachytene in males, whereas females showed a reduction in oocyte number. The degree of effect was also more profound in males than females. There was no evidence for a genic effect since the chromosome races in question were often indistinguishable both electrophoretically and in terms of mtDNA. The degree of gametogenic effect on males was not observed when single fusions were present, was occasionally observed when there were several independent fusions present, and was profound when multiple fusions (as few as two) shared monobrachial homologies.

A most similar set of stages were observed in Lemurs and *Petrogale*, where one or two fusions were coincident with irregular or unsynchronized pairing. Structural hybrids with multiple independent fusions had reduced fertility, with delayed and heterosynaptic pairing, whereas those hybrids with chain multivalents were sterile and pairing failed. Profound pairing failure was also observed in muntjac hybrids with numerous tandem fusion differences producing sterile F1 males. Dik-diks, which had multiple rearrangements, exhibited a similar breakdown in spermatogenesis in males, whereas this was not the case in those forms which were hybridized and which had a few chromosomal differences. The same pattern was observed in horses, where in those species with a few fusions there was no impact on fertility in hybrids, in terms of malsegregation, or pairing, whereas pairing failure was characteristic of all species distinguished by multiple rearrangements. It appears that a second-level effect on pairing was present in all species hybrids which were characterized by multiple and complex structural arrangements. In most cases, no studies on oogenesis had been made. This second-level effect was independent of malsegregation effects.

It is of some note that similar pairing breakdown occurred in hybrids between chromosomally indistinguishable species of *Bos*. The degree of failure increased with the distance between parental species. A number of authors have attributed this second-level pairing breakdown to genic differences between the parental species. While it is not known whether this is the case in *Bos*, dik-diks, muntjac or *Lemur*, in *Mus*, pairing breakdown occurred where no discernable genic differences were detected between the hybridizing parental forms.

Pairing breakdown in chromosomally identical (in terms of G-banding) but evolutionarily divergent species such as *Bos*, may be due to some genic perturbation of meiosis, but there is no evidence for this. Possible reasons for this anomaly are either a change to the structure of pairing sites at the molecular level in the parental species, or a change in the timing of cell cycles in the parental strains causing asynchrony and thus delayed, or failed, synapsis. The latter may well be a genetic perturbation, but this has not been demonstrated.

Multiple structural rearrangements might have a similar impact on pairing behaviour in structural hybrids. Evidence in support of this comes from two areas. First, in those members of species complexes which have only minor chromosomal differences established between them, second-level pairing impairment is not necessarily present in meiosis of the structural hybrids. The severity of pairing impairment in hybrids becomes progressively more apparent when additional chromosomal differences are inserted into the genome of the parental species (lemurs, muntjac deer, horses, mice, dik-diks). This also argues for a chromosomal cause to the disruption of pairing rather than a genetic cause.

Second, in the many F1 horse hybrids, males which are sterile and in which profound pairing breakdown is correlated with multiple rearrangements, rare pachytene cells do pair and produce spermatozoa. If a genetic mechanism was responsible for pairing failure, one would not expect occasional pairing success within animals and the production of totally viable and fertile hybrids. Presumably, the viable backcross progeny resulting from successful matings would be the product of rare sperm with a chance balanced complement.

In the great majority of cases where pairing breakdown occurred and was attributed to structural rearrangements, unsynapsed chromosomes often associated with the sex vesicle at pachytene in male meiosis. When high levels of this form of association occurred, either subfertility or sterility resulted. The partial failure of pairing was observed in rearrangements as diverse as Robertsonian fusions and complex reciprocal translocations, and in most of these cases a degree of sex chromosome association was encountered.

The impact of sex chromosome association on the fertility of structural rearrangements has been largely ignored by those considering such changes as post-mating isolating mechanisms. Yet this form of association could explain a number of current imponderables including the Haldane effect observed in so many *Drosophila* species hybrids (Lindsley and Tokuyasu, 1980). In particular, the great variation of fertility between different single Robertsonian fusions could be explained by complete synapsis in those trivalents which do not malsegregate and have no effect on fertility, and incomplete synapsis and sex chromosome association in those trivalents which malsegregate, are infertile or have some degree of infertility. Indeed, this has been observed in man where sterility is the rule in individuals heterozygous for Robertsonian trivalents which associate with the sex vesicle at pachytene. Here, XY–autosome associations can reach as high at 96% and result in complete infertility (Gabriel-Robez *et al.*, 1986; Johannisson *et al.*, 1987). In mice, Forejt (1979, 1982) found male sterility associated with double heterozygotes for Robertsonian translocations which shared a monobrachial homology. Here, unpaired, or incompletely paired, regions of the translocated chromosomes formed non-homologous association with the X chromosome. Similarly, in the *Petrogale assimilis* complex a single fusion was enough to provide sterility in hybrids which were genetically most similar, but in which the trivalent was associated with the X at pachytene.

In those species of *Bos* which failed to show any chromosome differences, but which had pairing impairment or failure in hybrids, X–autosome associations were observed at pachytene and the incidence was as high as 18% (Dollin *et al.*, 1991). However, it was difficult to dissociate the impact of pairing failure from sex chromosome association on the fertility of these heterozygotes.

It would therefore appear likely that four mechanisms are responsible for inviability or fertility impairment in structural heterozygotes involving translocations. First, the malsegregation of heterozygotes and the formation of aneuploid and unbalanced gametes. Second, synaptic impairment due to complex structural reorganization. Third, the association of unpaired segments, or heterochromatic segments, with sex chromosomes. Fourth, a series of minor effects which occur before pachytene and provide a direct inhibition to gametogenesis in structural hybrids. Each of these may act in concert or independently to provide profound inviability or fertility barriers in organisms distinguished by gross structural rearrangements.

7.4.2 Genic effects on fertility: more apparent than real?

The examples shown above illustrate two levels of interaction between chromosome changes and fertility effects. The first of these is that the addition of deleterious rearrangements to members of a species complex enhance the level of infertility found in species hybrids (compare single and multiple rearrangements in horses, *Mus*, lemurs and dik-diks). Second, when sex-limited sterility is present and it is male limited, there is a reasonable probability that this is a product of X–autosome association (compare *Mus* and man). It is very difficult to attribute these fertility effects to anything but chromosomal mechanisms. Yet the argument has been consistently advanced that fertility effects observed in structural hybrids are caused by genic differences rather than chromosomal mechanisms.

The elimination of the genic component in organisms such as *Rattus* or *Mus* (since the chromosome races are genically indistinguishable in terms of electrophoretic, mtDNA or immunogenetic differences), leaves nothing but a chromosomal mechanism that can be associated with induced sterility. Equally, in man there is no evidence to lead us to suspect that individuals which happen to be heterozygous for Robertsonian rearrangements and which are sterile because of X–autosome associations, induced as a byproduct of these rearrangements, also have some genic differences which cause sterility. Such a suggestion would not be taken seriously.

From another perspective, there may well be genic differences which distinguish the various species of horses, cattle or lemurs. But there is no evidence to indicate that genic differences between these species are responsible for the fertility effects encountered. Indeed, if we examine the literature, there are many examples of old, well-differentiated, allopatric, chromosomally uniform species, which have numerous genic differences established between them and their relatives, yet with which they have no difficulty in hybridizing and producing fertile progeny (see Section 3.4.5). One might then ask why it is necessary to suppose that genically indistinguishable species, or cytotypes, should have hypothetical genic effects which are responsible for sterility in structural hybrids? The answer seems to be that there is no reason to suspect that such a genotypic interaction is present and no evidence for its existence in those chromosomally divergent examples considered.

By the same token, several recent studies have suggested that the observed fertility effects of substantial structural differences in hybrids of *Mus* should be discounted because wild *Mus domesticus* chromosomes were introduced into laboratory strains of *M. domesticus*, and that these were genetically different. This is despite the fact that genetic analysis using restriction patterns of mtDNA indicates that most *Mus* laboratory strains are identical to *M. domesticus* (Yonekawa *et al.*, 1982). Nevertheless, the argument was used that genic differences may have been responsible for the chromosomally induced fertility effects despite the absence of any evidence which might suggest that this was so. This view was supported by the fact that the degree of effect on fertility differed between rearrangements which arose as spontaneous mutations in *M. domesticus* laboratory strains and those which evolved in the wild. The possibility that this was due to the fact that some fusions, even involving the same chromosomes, may have different fertility effects in different areas (Redi and Capanna, 1988), or that it may have been due to the impact or pairing failure or X–autosome association, were not considered. More surprising is the fact that the data on fertility effects of animals extracted from wild populations, and hybrid zones between genetically indistinguishable chromosome races which showed massive fertility impairment, were also ignored.

The quotations which introduced this chapter highlight a basic philosophical problem in our understanding of the role played by chromosomal rearrangements as post-mating isolating mechanisms, i.e. the introduction of notional evidence which is given the same weight as hard data. Genetic interaction has been used as a red herring, drawn across the trail to deter us from reaching the conclusion that gross chromosomal rearrangements may affect fertility in hybrids between genically indistinguishable populations. What is alarming is that this ill-considered view is given credibility.

There is little doubt that the effects of structural heterozygosity on the meiotic systems of hybrids has in the past been substantially underestimated. Most emphasis has correctly been placed on malsegregation and the production of unbalanced gametes. Nevertheless, the involvement of the X-inactivation cycle in structural heterozygotes and its association with the disruption of pairing, together with second-level effects induced by recombination of inversion products, are profound fertility barriers generally not given adequate consideration in the context of reproductive isolation. There appears to be ample evidence to suggest that chromosomally initiated reproductive isolation can occur in those forms distinguished by negatively heterotic translocations of all types and multiple inversions. These rearrangements can provide a basis for chromosomal speciation.

8
Genic change and chromosomal speciation

> We cannot measure the genetic divergence that has occurred between species (even sibling species) and conclude that this amount of genetic change has occurred during speciation – most of it, in the majority of cases, will have arisen after speciation has been completed.
>
> (White, 1978a, p. 41)

8.1 Chromosomal and genetic differentiation: the relationship

The key issue involved in establishing a relationship between chromosome change and speciation is 'to be able to distinguish between those changes which preceded the origin of the species in question, those which might have arisen subsequent to the speciation event and those which not only accompanied speciation, but were responsible for it' (John and Miklos, 1988, p. 258).

Thus, the first of the two requirements necessary for demonstrating a relationship between chromosome change and speciation is to show that the derived chromosomal rearrangements were profound enough to establish reproductive isolation between daughter and parental forms. By doing this, such chromosome changes cannot be seen as neutral events accompanying or following speciation, but are shown to be deterministic in its outcome. In Chapter 7 it was unequivocally demonstrated that many types of chromosomal differences between otherwise undifferentiated species could provide a profound and sometimes totally impermeable barrier to gene flow, thus acting as effective post-mating isolating mechanisms. Second, it is critical to determine whether the fixation of chromosomal rearrangements occurred before, or after, any other form of genetic differentiation was established. If it can be consistently demonstrated that chromosome races which are distinguished by fixed chromosome differences are established before genic or morphological changes have reached fixation, then it follows that chromosome change is the primary and causative factor in speciation and other changes are secondary.

Two assumptions are critical to an assessment of genetic diversity between lineages of chromosomal differentiated and related species:

1 In a linear array of ancestor/descendant related species, genetic distances between the forms reflect the order of their origin and their degree of relatedness.
2 Differences in the genetic distances between separate linear arrays of ancestor/descendant related species, reflect the antiquity of each radiation. That is, some radiations are older than others.

The underlying premise to both of these assumptions is that genic diversity between species accumulates progressively with the passage of time. This premise has been consistently supported in regard to the degree of genic differentiation at electrophoretic loci (see Nevo, 1983; Thorpe, 1983; see Chapter 3 for details). Moreover, comparative electrophoretic and microcomplement fixation studies on rodents and dasyurid marsupials, made by Baverstock and Adams (1987), revealed a strong correlation between albumin immunological distances and genetic distances estimated from a suite of other structural genes. A similar correlation between Nei distance and albumin immunological distance was reported in a range of vertebrates (frogs, salamanders, lizards and mammals) by Wyles and Gorman (1980). These data support the assumptions made above and would justify the following classification of the age of speciating lineages (see Fig. 8.1):

Class 1: A relic colonizing radiation in which all species in a sequence A–D have numerous fixed genetic differences between them, although the older A species would have greater genic distances than the D species (the most recently evolved).
Class 2: A linear array of forms which have distinct species at the A end, with numerous fixed genic differences, and either species, or subspecies, at the D end with very few genic differences.
Class 3: A linear array of forms ranging from species to chromosome races, with some genic differences at the A end and no genic differences between C and D at the D end.
Class 4: A linear array of very recently differentiated chromosome races with no significant genic differences between any of the races.

In the following section, a series of examples of chromosomal and genic divergence in vertebrates that are categorized as class 3 or 4 lineages are described. In each instance, a major chromosomal radiation has occurred in which a series of chromosome races has been established, each of which is distinguished from

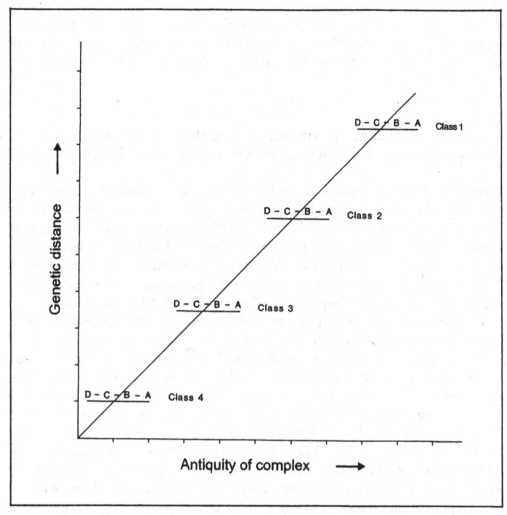

Fig. 8.1. This diagram illustrates the assumption that increased genetic divergence accumulates linearly with time. Any lineage of chromosomally speciating, or speciated forms (A–B–C–D), can be broadly classified (class 1 to 4) as to the antiquity of its formation by comparing morphological and genetic divergence. Thus, class 1 complexes will have greater genetic and morphological differences between species within the lineage, than will the most recently evolving class 4 lineages.

its ancestor by one or more negatively heterotic, or potentially negatively heterotic changes. Levels of genic diversity are seen to vary substantially between species within these colonizing radiations and also between radiations. This form of classification provides a predictive situation, for, if chromosome races are the most recently formed entities in a lineage and are genetically indistinguishable, the chromosomal changes which distinguish the races can only have been established as the first stage of differentiation. If these genetically undifferentiated chromosome races are reproductively isolated, chromosomal speciation producing sibling species has occurred. It is noteworthy that one might not expect to find class 3 or 4 levels of genic diversity in lineages which have evolved by other than chromosomal means.

8.2 Genic changes in speciating complexes distinguished by fusions, fissions and rearrangements sharing brachial homologies

When discussing case studies of speciating complexes in Chapter 7, in terms of the role of chromosomal rearrangements as post-mating isolating mechanisms, a number of examples were cited which were also appropriate to consider in this chapter. That is, both genic and molecular analyses had been made on these speciating complexes. Rather than simply reiterate the differences established, it is more important to put these complexes into the context of the classification used in this chapter and integrate the salient findings of these analyses.

The horses and cattle of the genera *Equus* and *Bos* provide excellent examples of class 2 species (see Chapter 7), for they have certain recently differentiated species and an array of well-established divergent congeners in each lineage. These are older colonizing radiations in which the species have many fixed differences and significant levels of reproductive isolation established between them.

The second group of species complexes, the *Mus domesticus*, *Rattus sordidus* and *Petrogale assimilis*, are all class 4 radiations. There is very little data available for the *Petrogale assimilis* complex other than a mtDNA analysis which shows great uniformity between the species (Cathcart, 1986), and an electrophoretic analysis which shows that certain races are distinguished by unshared electrophoretic polymorphisms (Briscoe *et al.*, 1982). On the other hand, both the *M. domesticus* and *R. sordidus* complexes have been intensively studied and provide the strongest evidence for the primacy of chromosomal speciation. The *R. sordidus* complex has chromosomally distinct species, which were elevated to species level because of the profound chromosomal differences between them, since they only have very subtle pelage differences which otherwise distinguish them. The chromosomes of these forms share complex monobrachial homology, they are genetically indistinguishable in terms of their electrophoretic phenotype (one fixed difference between *R. sordidus* and both *R. colletti* and *R. villosissimus*), and albumin immunogenetic distances are indistinguishable (see Section 7.3.2). The species are isolated from each other by the most powerful chromosomal post-mating isolating mechanisms resulting in absolute sterility, or profound fertility loss, in hybrids (Baverstock *et al.*, 1983b, 1986).

Equally, parapatric Italian chromosome races of *Mus domesticus* surveyed by Britton-Davidian *et al.* (1989) (see Section 7.3.1.5) were electrophoretically indistinguishable from each other, yet were totally reproductively isolated by the monobrachial homology of the chromosome differences which they possessed. This indicates both the primacy of chromosome change, in that the chromosome races had been established before other forms of differentiation, and the impermeable barrier to gene flow, which this form of rearrangement produces between chromosome races.

8.2.1 Class 3. The *Rhogeesa tumida–parvula* complex

The South American vespertilionid bats of the *Rhogeesa tumida–parvula* complex are now known to have seven distinct cytotypes, $2n = 30$, $2n = 32N$, $2n = 32B$, $2n = 34$, $2n = 42$, $2n = 44$ and $2n = 52$ (Baker *et al.*, 1985). A G and C-banding and electrophoretic analysis of the $2n = 30$, 32, 34 and 44 forms, identified 13 different fusions distinguishing the cytotypes. The allopatric $2n = 32B$ sample from Belize and $2n = 32N$ sample from Nicaragua, although sharing the same external karyomorph, differed by eight chromosome fusions which were revealed by G-banding (Baker *et al.*, 1985).

Each of the cytotypes differed by chromosome fusions which shared brachial homologies. Thus, although hybrids were not produced between the forms, it was possible for the authors to calculate the probable meiotic configurations which would occur and thus estimate the impact of these chromosome rearrangements on fertility. The $2n = 44$, $2n = 34$, $2n = 32B$ and $2n = 32N$ cytotypes appear to be reproductively isolated by the chromosomal differences alone.

The electrophoretic analysis of 21 isozymes revealed that fixed allelic differences had been established between all cytotypes examined. Indeed, members of the $2n = 32B$, $2n = 34$, $2n = 42$ clade, which appeared to be composed of the most recently differentiated and most closely related cytotypes, were distinguished by one or two separate fixed electrophoretic differences, and the $2n = 32B$ and 42 cytotypes had a genetic identity of $I = 0.32$. Genetic identities in the complex ranged from $I = 0.32$ to 0.81. It is noteworthy that the cytotypes were also morphologically indistinct.

These data suggest that the bats of the *Rhogeesa tumida–parvula* complex had differentiated chromosomally and subsequently established fixed electrophoretic differences, without establishing morphological distinctiveness. Baker *et al.* (1985) argued that these cytotypes, distinguished by multiple chromosome fusions, are distinct biological species.

8.2.2 Class 3. The *Proechimys guairae* species group

A linear pattern of chromosomal speciation in the Venezuelan spiny rats of the *Proechimys guairae* complex was described by Reig *et al.* (1980). These forms are very difficult to differentiate morphologically and are nominally regarded by Reig *et al.* (1980) as being two species and four subspecies. These tropical, forest-dwelling rodents have undergone extensive chromosomal repatterning with little accompanying genic divergence. Six distinct chromosomal forms are present, all of which correspond to the systematic groups: these are 2n = 42, 2n = 44, 2n = 46, 2n = 48, 2n = 50 and 2n = 62. The complements are distinguished by single or multiple Robertsonian rearrangements and also pericentric inversions. Reig *et al.* (1980) suggested that the chromosome changes appeared to be acting as a post-mating isolating mechanism. Each chromosome race was allopatrically, or para-patrically, distributed and there was no indication of introgression of one form into the next. The chromosome races are linearly arrayed and appear to be distributed in their sequential order of formation. This suggested that a series of major colonizing radiations had occurred, encompassing the lowland forests which surround the major mountain axis, linearly bisecting Venezuela.

An electrophoretic analysis of specimens used in the chromosomal study was made by Benado *et al.* (1979) and based on 22 loci. They found that increasing genetic divergence was correlated with greater karyotypic divergence. Genetic distances between 2n = 46 populations at La Trilla and San Esteban were $D = 0.014$, whereas those between higher chromosome number forms gradually increased: 2n = 48 ($D = 0.023$), 2n = 50 ($D = 0.036$), 2n = 62 ($D = 0.029$) and 2n = 62 *P. urichi* ($D = 0.134$). These authors pointed to the fact that such low levels of genetic divergence are analogous to other rodent species complexes which have established extensive chromosomal differences associated with reproductive isolation. Benado *et al.* argued that the chromosomal radiations were established before genetic differentiation had occurred.

8.2.3 The genus *Rattus*

The Australian representatives of the genus *Rattus* provide a magnificent evolutionary contrast. On the one hand are a complex of allopatric species (*R. lutreolus*, *R. tunneyi* and *R. fuscipes*), which are chromosomally most similar and have established numerous morphological and electrophoretic differences between them, and are undoubtedly good class 2 species (see Baverstock *et al.*, 1983b, 1986, and Section 3.4.5 for a detailed account). However, these produce totally viable hybrid progeny when artificially hybridized (the chromosomal differences having no impact on fertility or viability), and are thus good species which are not reproductively isolated. On the other hand, the allopatric complex of species *R. sordidus*, *R. colletti* and *R. villosissimus* have massive chromosomal differences between them, and are reproductively isolated and genically indistinguishable. These contrasting examples illustrate both the primacy of chromosome change in speciation and the efficiency of multiple rearrangements sharing monobrachial homologies as a barrier to gene flow.

8.2.3.1 Class 3. The Rattus rattus complex

The widely distributed black rat, *Rattus rattus*, has been one of most extensively analysed of any vertebrate species. Numerous studies by Yosida and his colleagues have defined the distribution of the chromosome races of this species, and this information is best summarized by Yosida (1980a). Other investigations include restriction endonuclease digests of the mtDNA of this and other *Rattus* species (Hayashi *et al.*, 1979; Brown and Simpson, 1981). The amino acid composition of the transferrin molecule was determined by Moriwaki *et al.* (1971), and most recently Baverstock *et al.* (1983c) compared the amount of genetic differentiation of the five chromosomal forms of *Rattus rattus*, to other *Rattus* species, with both an electrophoretic analysis on 45 loci and an albumin immunological analysis using the microcomplement fixation technique.

By using C and G-banding techniques and standard chromosome preparations Yosida (1980a) was able to characterize the following chromosome races of *Rattus rattus*:

1 $2n = 42$ SEA: This race occurs throughout South East Asia, and all chromosomes in it have distinctive paracentromeric heterochromatic blocks.

2 $2n = 42$ Jap: This Japanese race is the same as the $2n = 42$ SEA race in gross morphology, but differs in that seven of the 12 pairs of telocentric chromosomes lack the paracentromeric heterochromatic blocks.

3 $2n = 40$: This race is found in the highlands of Sri Lanka and differs from the $2n = 42$ SEA in that pairs 11 and 12 have been fused, thus reducing the chromosome number to 40.

4 $2n = 38$: This oceanian race is found from India west to the Middle East, and appears to have been introduced to Europe, Africa, North and South America, and Australia by man's shipping activities. The karyotype of this race has the pair 11/12 fusion of the $2n = 40$ form as well as a pair 4/7 fusion.

5 $2n = 42$ Mau: Specimens from this race are only found on the island of Mauritius. They possess the pair 11/12 and 4/7 fusion metacentrics of the $2n = 38$ race and have also established chromosome fissions involving metacentric pairs 14 and 18 of that race (Yosida *et al.*, 1979).

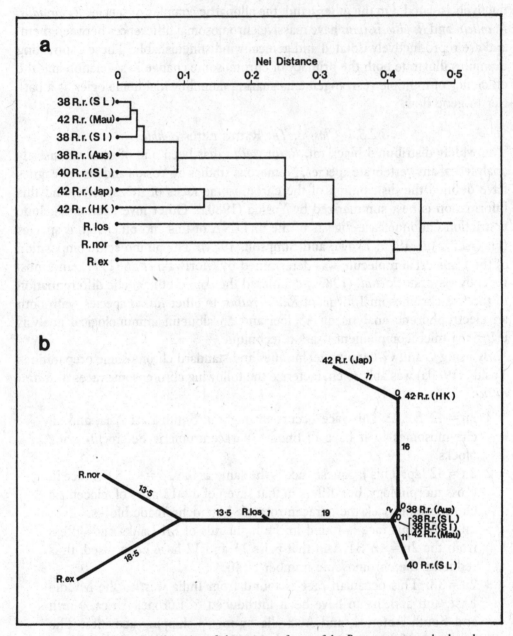

Fig. 8.2. a. A phenetic clustering of chromosome forms of the *Rattus rattus* complex based on Nei distance. The outgroup species and *R. rattus* chromosome races are: *R. exulans* (R. ex), *R. norvegicus* (R. nor), *R. losea* (R. los), and *R. rattus* Hong Kong (HK), Japan (Jap), Sri Lanka (SL), Australia (Aus), Mauritius (Mau), Seychelles Islands (SI).
b. A Wagner tree based on fixed allelic differences (%) showing the relationships within the *R. rattus* complex. (Redrawn from Baverstock *et al.*, 1983c.)

Hybridization studies made in the laboratory indicate that F1 hybrids between the 2n = 38 and 2n = 42 forms are sterile, whereas hybrids between 2n = 38 and 2n = 40 forms are fertile, but with reduced litter size. Hybrids between the 2n = 38 and 2n = 42 Mau forms have a higher level of fertility than those with the 2n = 40 animals (Yosida, 1980b).

On the basis of these data, Yosida (1980a) proposed that the karyotypic forms of *R. rattus* were related in the lineage 2n = 42 SEA, 2n = 40, 2n = 38, 2n = 42 Mau. Yosida believes that the colonization of Mauritius was relatively recent (within the last 400 years), and subsequent to the colonization of the island by the 2n = 38 chromosome race, two pairs of fissions were established in this founder population resulting in the 2n = 42 Mau form.

The electrophoretic and microcomplement fixation data of Baverstock *et al.* (1983b) are unambiguous in their findings and they fully support the model advocated by Yosida for the evolution of this complex. Unfortunately, the very recent origin (400 years BP), for the 2n = 42 Mau form, has not been confirmed by these experiments, although its origin is undoubtedly very recent (see Fig. 8.2).

The electrophoretic analysis indicated that the three 2n = 38 populations (D = 0.022 to 0.033), clustered together and are most closely related to the 2n = 42 Mau chromosome race from which they differed by one fixed difference (D = 0.020 to 0.055). Both of these chromosome races are more closely related to each other than each is to the 2n = 40 race (D = 0.072 to 0.101). The 2n = 40 form shares no alleles with the 2n = 38 race at two loci. Both of the 2n = 42 races (Jap and SEA) have five fixed differences between them (D = 0.127), and form a group which is distinct from the other chromosome races from which it differs by nine fixed differences (D = 0.25). Results from the albumin microcomplement fixation analysis support these findings. They indicate that the 2n = 42 Mau race was relatively recently and rapidly established, and that a minimal amount of genetic divergence has occurred since that time.

8.2.4 Class 3. The *Gerbillus pyramidum* complex

Extensive chromosomal variation in the great sand gerbil, *Gerbillus pyramidum*, was described by Wahrman and Gourevitz (1973). They found that the Sinai peninsula and Negev were inhabited by a southern chromosome race (2n = 63– 66), whereas a northern race with 2n = 50–52 occurs south of Tel Aviv–Yafo (Fig. 8.3). These chromosome races differ by seven fixed chromosome fusions. Wahrman and Gourevitz reasoned that the more metacentric northern race evolved from the southern race, spreading northward in a colonizing radiation of the coastal sand dunes. The two races are separated by a 150 × 150 km hybrid zone, where numerous karyotypic forms are encountered. A series of African chromosome races (2n = 38–40) were found and these are believed to be an independent offshoot from the ancestral acrocentric stock.

Nevo (1982) made an electrophoretic analysis of populations from the northern and southern chromosome races of *G. pyramidum*, and compared these to three populations from the chromosomally monomorphic (2n = 40, all metacentric) *G. allenbyi*, a close relative to *G. pyramidum*. Nevo analysed allozymic variation in 27 presumptive loci and found extremely little genetic diversity in terms of polymorphism, heterozygosity and genetic distance, within and between species.

The genetic distances suggested that there was almost no difference between the chromosome races of *G. pyramidum* (D = 0.009), whereas the genetic distance between *G. allenbyi* and *G. pyramidum* was relatively low (D = 0.11) and was similar to the difference between populations of *G. allenbyi*. Nevo argued that chromosomal mutations could have been established relatively rapidly (several thousand generations), and that these have provided the initial basis for speciation. This differentiation is associated with very little genic change. Reproductive isolation is still imperfect between the northern and southern chromosome races, but is complete between *G. pyramidum* and *G. allenbyi*.

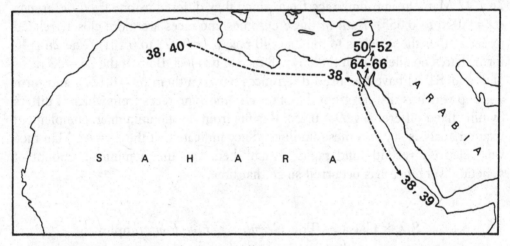

Fig. 8.3. The distribution of chromosome races in the *Gerbillus pyramidum* complex. (Redrawn from Wahrman and Gourevitz, 1973.)

8.2.5 Class 4. The *Spalax ehrenbergi* complex

The fossorial mole rats of the *Spalax ehrenbergi* complex represent one of the most extensively studied vertebrate taxa, at least in terms of their chromosomal evolution. In Israel, the four chromosome races 2n = 52, 54, 58 and 60 are morphologically indistinguishable and geographically arrayed in an ancestor/descendant alignment (Wahrman *et al.*, 1969a, b; Nevo, 1985a) (Fig. 8.4). The $2n = 60$ race also extends into western Egypt and North Africa (Lay and Nadler, 1972).

It was argued by Nevo (1985a) that this complex represented a clear-cut case of peripatric speciation, a means of colonizing chromosomal speciation (*sensu* Mayr, 1982b). The chromosomal species display progressive stages of differentiation, which have occurred subsequent to their colonizing radiation which started some 250 000 years BP. This involved radiations into the upper Galilee Mountains ($2n = 52$); the Golan heights ($2n = 54$); the lower Galilee Mountains and coastal plains ($2n = 58$); and the mountains of Samaria, Judea, northern Negev, the southern Jordan Valley and the coastal plain ($2n = 60$). All of these regions are climatically distinct and it is probable that there have been several sequential radiations of these mole rats southward ($2n = 52, 58, 60$) and eastward ($2n = 52, 54$). The radiations into increasingly arid environments were presumably accompanied by climatically co-adapted genomic adaptations at all levels (Nevo, 1985a).

Heth and Nevo (1981) and Nevo (1985a) found that post-mating isolating mechanisms appear to have preceded the development of pre-mating isolating mechanisms. There is an absence of pre-mating isolating mechanisms in the younger divergence ($2n = 60$–58), whereas post-mating isolating mechanisms exist between these chromosome races (see below). In contrast, the older divergences ($2n = 52, 54, 58$) have well-established pre- and post-mating isolating mechanisms and these involve olfaction (Nevo *et al.*, 1976), vocalization (Heth and Nevo, 1981) and aggressive interactions (Nevo *et al.*, 1975), leading to positive assortative mating of the chromosome races.

Hybrid populations in three of the hybrid zones were analysed in some detail by Nevo and Bar-El (1976), who found that they vary significantly in fitness, depending on which races are involved. Litter analysis suggested that the hybrids are partially fertile, but inferior in fitness. The hybrid zones decrease in width from north to south and Nevo and Bar-El (1976) found that the larger the chromosome differences which occurred between the karyotypes, the narrower was the hybrid zone. That is, reproductive isolation increases progressively from the southern to northern hybrid zones, thus suggesting that there is an active process of speciation occurring between the chromosome races involving lower hybrid fitness (Wahrman *et al.*, 1985) and hybrid breakdown. Nevo (1985a) argued that the chromosome changes established between the forms of *Spalax ehrenbergi* were a primary post-mating isolating mechanism which initiated speciation. That is, 'in *S. ehrenbergi* chromosome differentiation appears to be causal, not incidental, to the speciation process' (Nevo, 1985a, p. 87).

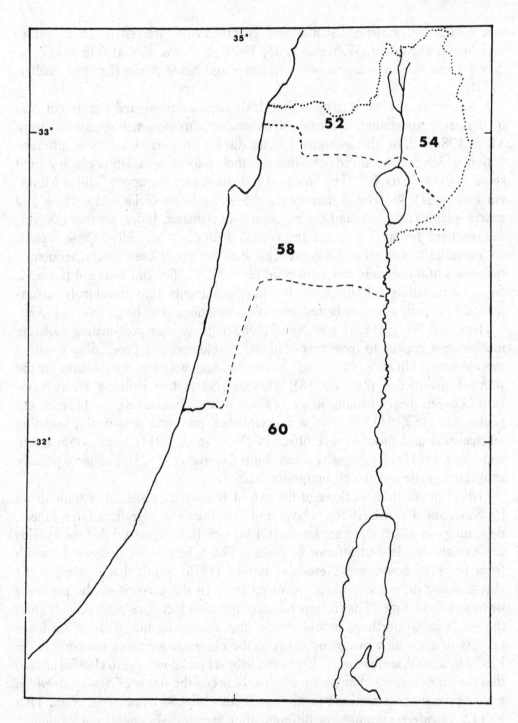

Fig. 8.4. The distribution of chromosome races in the *Spalax ehrenbergi* complex. (Redrawn from Nevo and Bar-El, 1976.)

Nevo and Shaw (1972) and Nevo and Cleve (1978) made electrophoretic analyses on some 882 specimens from the four chromosome forms. A total of 25 loci were examined in these studies and the results are most informative. Genetic uniformity is found at 24 of the 25 loci in 21 populations examined. Genetic identity between the four karyotypes is most similar, ranging from $I = 0.931$ to 0.988, and genetic distances range from $D = 0.002$ to 0.07. Heterozygosity levels and proportion of polymorphic loci are: $2n = 52$ ($H = 0.035$, $P = 0.12$), $2n = 54$ ($H = 0.016$, $P = 0.24$), $2n = 58$ ($H = 0.037$, $P = 0.16$), $2n = 60$ ($H = 0.069$, $P = 0.28$) (Nevo and Cleve, 1978). Clearly, genic changes associated with the secondary divergence of the chromosome races are minimal, despite the establishment of numerous ethological differences.

A further analysis using DNA–DNA hybridization to measure the average genomic divergence among the four chromosome race of *Spalax*, was made by Catzeflis *et al.* (1989). This produced a corroborative phylogeny to all other techniques applied and suggested that the most recently derived chromosome races ($2n = 58$ and 60), differ by a 0.2 to 0.6% base-pair mismatch. This is a particularly small degree of difference and suggested an upper Pleistocene divergence between 0.18 and 0.25 MYBP. The $2n = 54$ divergence occurred from 2.0 to 2.35 MYBP. These estimates of time of divergence are a little greater than those obtained from electrophoretic analyses. Nevertheless, the primacy of chromosomal differentiation in this complex is apparent.

8.2.6 Class 3. The *Acomys cahirinus* complex

The third major analysis involving Israeli rodents was made by Nevo (1985b) on the genus *Acomys*. The species analysed were *Acomys russatus* ($2n = 66$) and two chromosomal forms of *A. cahirinus* ($2n = 36$ and $2n = 38$). Their cytology was originally investigated by Wahrman and Goiten (1972). *A. russatus* is distributed throughout the south and west of Israel and lives in extreme desert habitats.

The golden spiny mouse, *A. cahirinus*, is widely distributed in Israel and Sinai and is restricted in its habitat preference to rocky areas. The Israeli populations have $2n = 38$ chromosomes, whereas those from the Sinai have $2n = 36$. The populations are parapatrically distributed, and a 16×15 km hybrid zone which contained $2n = 37$ animals was found. The chromosome races of *A. cahirinus* are morphologically indistinguishable and have not attained complete reproductive isolation (Nevo, 1985b). Each of these chromosome races is distributed sympatrically with *A. russatus* over at least a part of its distribution, although it should be noted that the *A. cahirinus* forms are nocturnal, whereas *A. russatus* is diurnal (Fig. 8.5).

Nevo (1985b) analysed 220 specimens from ten populations taken from throughout the distribution of all three species/chromosome races, scoring them electrophoretically for 35 loci. Genetic distances between the 2n = 36 and 2n = 38 chromosome races of *A. cahirinus* were very low (*D* = 0.023) and were not dissimilar to differences between populations (*D* = 0.012). Indeed, genetic distances within *A. russatus* were also very low (*D* = 0.006), however, this 2n = 66 species was very different from both chromosome forms of *A. cahirinus* (*D* = 0.300). These

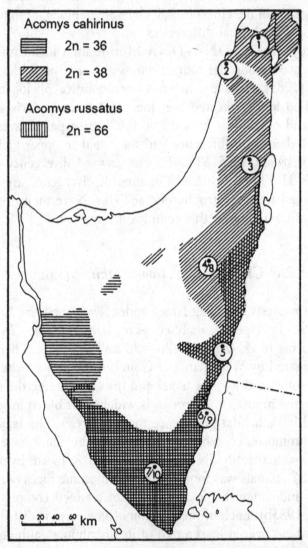

Fig. 8.5. The distribution of chromosome races in the *Acomys cahirinus* complex. (Redrawn from Nevo, 1985b.)

distances suggested an evolutionary divergence time of 1.5 million years between *A. russatus* and *A. cahirinus* and of only 115 000 years between the *A. cahirinus* chromosome races (Nevo, 1985b). Nevo argued that this was in accordance with the view that *A. cahirinus* colonized mesic Israel in the upper Pleistocene, and during this recent colonization chromosomal forms were established which were distinguished by the fixation of a single centric fission.

When analysing levels of genic diversity, Nevo (1985b) found higher levels of H and P for *A. cahirinus* ($P = 0.09$ to 0.27, $H = 0.014$ to 0.061), than for *A. russatus* ($P = 0.07$ to 0.19, and $H = 0.005$ to 0.009), although the number of populations sampled was very different (seven and three respectively).

In summary, the evidence suggests that two speciation events had occurred, the divergence of *A. cahirinus* from *A. russatus* and the very recent divergence of the chromosomal forms of *A. cahirinus*. Chromosomal differences had been established between the 2n = 38 and 2n = 36 races, before the fixation of any significant electrophoretic or morphological characteristics.

8.2.7 Class 3. The *Gehyra variegata–punctata* species complex

King (1979, 1984) described a series of chromosome races in the gekkos of the *Gehyra variegata–punctata* species complex and proposed a phylogeny for their evolution. Subsequently, Moritz (1986) reanalysed this complex using G-banding and electrophoresis and found essentially the same pattern of differentiation and produced a most similar phylogeny. One lineage in particular, occurs as a geographically arrayed sequence of ancestor/descendant related species and chromosome races 2n = 44 *(G. pilbara)*, 2n = 42a *(G. minuta)*, 2n = 40a *(G. variegata)*, 2n = 38a and 2n = 38b *G. variegata* (Fig. 8.6). These chromosome races differ from each other by unequal chromosome fusions. Each of the chromosome races is allopatrically distributed although the distributions are broken up into a series of isolates. The 'older' forms with higher chromosome numbers are morphologically defined species, whereas the most recently derived chromosome race (2n = 38b) cannot be readily distinguished from its direct ancestor (2n = 40a) by morphology alone (King, 1984). Moritz (1984) made an electrophoretic analysis on specimens from four of these chromosome races based on the variation at 36 presumptive genetic loci. As one might expect, the older chromosome races had established greater genic distances between them (2n = 44 to 42a, $D = 0.148$, 2n = 42a to 40a, $D = 0.141$) than had the most recently derived forms. Indeed, the majority of populations of the 2n = 38b race were electrophoretically indistinguishable from the 2n = 40a race ($D = 0.023$ to 0.025). Nevertheless, certain isolated populations of the 2n = 38b race had established fixed genic differences which distinguish them from other populations of that race. Members of the $2n = 38a$ chromosome race, which had presumably been established for a longer time than the $2n = 38b$ race (distance between $2n = 38$ races $D = 0.090$ to 0.102), had also established several fixed differences, distinguishing them from the $2n = 40a$ race ($D = 0.072$). It appears that a sequential series of colonizing radiations had occurred. Each chromosome race had colonized a discrete geographic territory before genic or morphological differences had reached fixation. The geographic distribution of the most recently derived chromosome race ($2n = 38b$) is adjacent to that of its direct ancestor ($2n = 40a$) in the Macdonnell ranges of the Northern

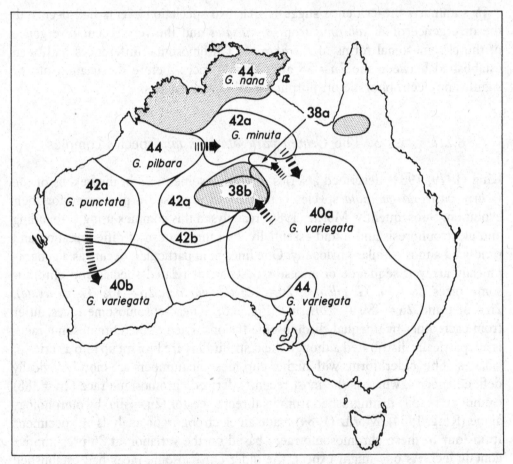

Fig. 8.6. The generalized distributions of the chromosomal forms of the *Gehyra variegata–punctata* complex. *G. purpurascens*, which is sympatric with many central Australian forms, is not included. The stippled areas refer to the three isolated populations of the relic species *G. nana*. The colonizing radiation discussed in Section 8.2.7 includes $2n = 44$ *G. pilbara*, $2n = 42a$ *G. minuta*, and the $2n = 40a$, $38a$, $38b$ forms of *G. variegata*. Note that the distributions of ancestor/descendant species are abutting (see arrows).

Territory. Since there is no evidence for the introgression of chromosomes from one race into the next, and there are no fixed genic differences between these races, it may be reasonable to conclude that the chromosome races had been established and attained their distribution as a primary function, and initiated speciation in this complex.

8.2.8 Class 3. The *Sceloporus grammicus* complex

The analysis of the *Sceloporus grammicus* complex of chromosome races has a somewhat chequered history. Hall and Selander (1973) described a geographic array of chromosome races, each of which was distinguished by a fixed chromosomal fission difference and which they regarded as the result of a chain of speciation events. The chromosome races included the ancestral $2n = 32(S)$, $2n = 34(F5)$, $2n = 34(F6)$, $2n = 36(F(5+6))$, $2n = 40(FM1)$ and $2n = 46(FM2)$. Porter and Sites (1986) described an additional form, the $2n = 38(FM3)$. The chromosome race distributions were complex, but often abutted, and narrow hybrid zones were sometimes present between them. The hybrid zones occurred between ancestor/descendant chromosome races in terms of Hall and Selander's geographic lineage of descent (see Fig. 8.7).

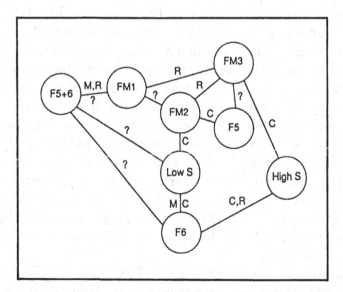

Fig. 8.7. Geographic relationships of the cytotypes of the iguanid lizards of the *Sceloporus grammicus* complex. Hybridization events between these forms were detected by chromosomal (C), mtDNA (M), or rDNA (R) analysis. (Redrawn from Sites and Davis, 1989.)

A series of papers on different aspects of the evolution of the *S. grammicus* species complex were presented by Sites and his colleagues. These used various characters to test the possibility of chromosomal speciation, and generally argued against the likelihood of this mode of speciation. In an analysis of morphological variation, Sites (1982) concluded that geographical isolation was more important than the fixation of chromosomal rearrangements in determining morphological divergence. In a comparison of allozyme variation in 19 presumptive loci in the $2n = 32$, $2n = 34$ and $2n = 36$ races, Sites and Greenbaum (1983) found little divergence between cytotypes (although they did find fixed differences occurring within populations of the $2n = 32$ and $2n = 36$ cytotypes), and concluded that 'the pattern of variation suggests that chromosomal differences between cytotypes do not reduce gene flow among cytotypes to any appreciable extent. Allopatric isolation appears to be more effective than chromosomal rearrangement in promoting or maintaining genetic divergence' (Sites and Greenbaum, 1983, p. 63–4).

Subsequently, Sites (1983) and Porter and Sites (1985, 1986) analysed the segregation pattern of chromosomal polymorphisms occurring within chromosome race distributions of *S. grammicus* and found that these were without meiotic side-effects (which is hardly surprising), and thus were not involved in chromosomal speciation. Thompson and Sites (1986) and Sites *et al.* (1988b) continued to cite these studies as evidence that the fixed differences between chromosome rearrangements have minimal meiotic side-effects (also see Sections 5.1.2.2 and 7.1 for a critique of this approach).

Electrophoretic analyses were used by Thompson and Sites (1986) and Sites *et al.* (1988b) to attempt to determine the population structure of the *S. grammicus* complex. The genetic distances between the cytotypes examined ranged from $D = 0.00$ to 0.108, and fell within the range of those for local populations. In both studies, a range of precarious assumptions were made and used to calculate what the heterozygosity and mean number of alleles per locus should be in founding populations, compared to panmictic populations. These authors concluded that *S. grammicus* did not have the population structure conducive to the fixation of negatively heterotic rearrangements. That is, it was panmictic rather than founding. They concluded that the rearrangements were neutral, or positively heterotic, and had reached fixation because they were adaptive (Thompson and Sites, 1986; Sites *et al.*, 1988b). This conclusion appears to have been reached despite the occurrence of fixed electrophoretic differences at isolated localities within chromosome races.

A re-examination of this complex using rDNA, mtDNA and the 16 most polymorphic electrophoretic loci, presented a substantially different picture (Sites and Davis, 1989). The rDNA data clearly resolve the high elevation portion of the $2n = 32(S)$ cytotype and the $2n = 34(F6)$ race. Most cytotypes have distinct markers, although considerable hybridization had occurred between the F5+6, FM3 and FM1 cytotypes. The F6 race was also distinguished by a duplicated G3 pdh-B locus and had three other fixed differences (Sites *et al.*, 1988c). In fact, this cytotype had been described as a species (*S. paluciosi*) by Lara-Gongora (1983), as had the 2n = 32 'P1' cytotype of Hall and Selander (1973), which was called *S. anahuacas*. These are both basal forms in Hall and Selander's (1973) lineage and it is not surprising that they had differentiated to this level. Patterns of mtDNA variability confirmed the integrity of several cytotypes such as the high elevation S, FM2 and FM3. The S and F6 differ by as many as 18 restriction sites.

By combining mtDNA, rDNA and electrophoretic data, Sites and Davis (1989) were able to show that the basic linear array of relationships proposed by Hall and Selander (1973) were supported. All trees placed the 2n = 32(S) cytotype basally, followed by the F6 (a single rearrangement away), and then stepwise to the F5+6 which was intermediate in terms of mtDNA and rDNA differentiation, whereas the FM1 and FM2 were most derived. The FM3 cytotype did not fit where expected, which suggested to these authors that it had a different chromosomal origin.

Hybrid zones between the chromosome races were no broader than 500 metres in all situations. Sites and Davis claimed that F1 and backcross individuals which possessed electrophoretic markers were present and insisted that significant introgression of these markers could occur. This would appear to be in contradiction to the rDNA and mtDNA data which showed the integrity of many cytotypes.

There seems to be little doubt that this recently evolved class 3 speciating complex provides a clear-cut example of the formation of chromosome races before the fixation of subsequent genic, molecular and morphometric differences. Most recent evidence supports the model of speciation presented by Hall and Selander (1973).

8.2.9 Class 3. The *Phyllodactylus marmoratus* complex

This southern Australian gekko was analysed chromosomally by King and Rofe (1976) and King and King (1977) and subdivided into four chromosome races 2n = 36ZW, 2n = 36, 2n = 34 and 2n = 32. Each of these chromosome races was allopatrically distributed and the distributions were geographically linearly arrayed in the presumed order of descent (2n = 36, 34 and 32) (see Fig. 8.8). Those with different numbers were distinguished by single chromosome fusions, whereas the 2n = 36ZW race had a characteristic sex chromosome heteromorphism. The distributions of these animals were very large and there was no evidence of chromosomal introgression of one race into the next. The distributions gave the impression of an east to west colonizing radiation, during the course of which the chromosome races had been established.

Storr (1987) conducted a morphological analysis on the Western Australian part of this complex and recognized the 2n = 32 and 34 races as the subspecies *P. marmoratus marmoratus* and the 2n = 36 race as *P. m. alexanderi*. That is, there were no adequate morphological characters to distinguish the most recently evolved 2n = 32 form from the 2n = 34 race. A preliminary electrophoretic analysis of these forms using 32 presumptive loci (S. Donnellan, South Australian Museum, pers. comm.), found genetic distances of $D = 0.05$ between the 2n = 32 and 2n = 34 chromosome races and $D = 0.08$ between the 2n = 34 and 36 races. A cryptic species distinguished by three fixed differences ($D = 0.18$) was detected in the western sector of the 2n = 36 race distribution. High levels of heterogeneity between populations suggest that a considerable degree of isolation has existed. This may have arisen since the initial radiation of the chromosome races and been a product of high aridity levels in the late Tertiary or Pleistocene disrupting a more continuous distribution. This complex provides an example of genetic differences being established within isolated populations subsequent to the primary colonization by the chromosome races.

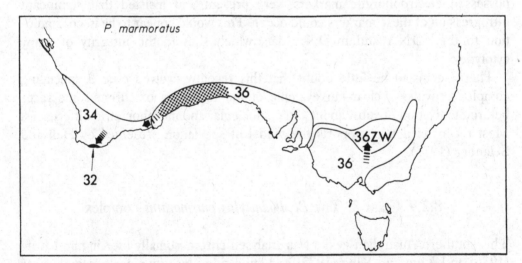

Fig. 8.8. The distribution of the chromosome races of the gekko *Phyllodactylus marmoratus* in southern Australia. The stippling refers to the electrophoretically distinct form of the 2n = 36 race.

8.3 Genic changes in speciating complexes characterized by both neutral chromosomal changes and also by negatively heterotic rearrangements

In the previous section, we examined genic differentiation in those species complexes where the forms are distinguished by known or presumptive, negatively heterotic chromosomal rearrangements. In other species complexes or genera, such as *Thomomys* and *Peromyscus*, a contrasting situation exists. Here, chromosome races or species may be distinguished by negatively heterotic rearrangements, whereas others in the same complex may have neutral or polymorphic chromosomal differences present. Further still, some chromosome races may have both fixed negatively heterotic differences and neutral monomorphisms or polymorphisms also present, one overlaying the other.

8.3.1 The genus *Thomomys*

The pocket gophers of the genus *Thomomys* are one of the demographically best known of the vertebrate groups. Their similarity of form and lifestyle provides us with a unique opportunity to examine genic changes which have occurred in speciating complexes exhibiting diverse forms of chromosomal reorganization.

The three most chromosomally variable species complexes in *Thomomys* appear to have differentiated by unrelated forms of chromosome change. Certain of these are intricately related to the speciation of that complex, whereas others are irrelevant to it. The *T. talpoides* complex involves extensive polytypism based on chromosome fusion, although certain pericentric inversions are also present. Chromosome numbers range from 2n = 40 to 60 (Thaeler, 1974) and the different chromosome races or subspecies are parapatrically distributed and chromosomally monomorphic. Hybrid zones of varying width have been recorded between the chromosome races, although on some occasions hybridization was absent (Thaeler, 1974). Patton and Sherwood (1982) analysed one of these subspecies with C-banding (*T. talpoides fossor*), and found that it was generally depauperate in heterochromatin, but for small paracentromeric blocks.

The *T. bottae* species complex displays a very different pattern of chromosomal evolution. Here, the highly polytypic and polymorphic populations all have 2n = 76. Adjacent populations may differ by up to 36 chromosome arms and can interbreed freely without deleterious side-effects. Patton and Sherwood's (1982) C-banding analysis revealed that biarmed elements had been produced by the addition of wholly or predominantly heterochromatic segments. *Thomomys umbrinus*, on the other hand, has characteristics of both the *T. talpoides* and *T.*

bottae complexes. That is, negatively heterotic differences distinguish certain of the karyomorphs, whereas others are highly polymorphic. Thus, the $2n = 78$ cytotype of *T. umbrinus* is very similar morphologically to the $2n = 76$ morph of *T. bottae*, with which they hybridize in southern Arizona (Patton, 1973). Patton and Smith (1981) have shown that the progeny of such crosses include sterile F1 males and partially fertile females. The formation of multivalents in hybrid meiosis also indicates that numerous reciprocal translocations distinguish these taxa.

There are two chromosome number morphs of *T. umbrinus*, a $2n = 76$ race which is restricted to the forested regions of the Sierra Madre Occidental, and the $2n = 78$ race which occupies deserts and forests from Arizona to the southern edge of the Mexican plateau. All but two of the $2n = 76$ populations have totally biarmed complements, the exceptions having 9 and 15 pairs of acrocentrics respectively (Patton and Feder, 1978). Northern populations of the $2n = 78$ race have tiny heterochromatic microchromosomes. Superimposed on this variation is a north to south gradient in the number of uniarmed chromosomes, with an increased number of biarmed elements in the south. Patton and Sherwood (1982) have shown that the northern $2n = 78$ karyomorph from the Patagonia mountains in Arizona has numerous interstitial heterochromatic blocks distributed in the larger biarmed elements of this generally acrocentric complement. This particular population was used in hybridization studies with *T. bottae*, and it appears that metacentricity has been attained by the addition of heterochromatic segments to the basic acrocentric complement (Hafner *et al.*, 1987).

The results of a series of electrophoretic analyses on the three species complexes are most illuminating. Nevo *et al.* (1974) analysed 31 loci in ten populations from six chromosome forms of *T. talpoides* representing five species ($2n = 40$, 44, 46, 60, 48W and 48NM). Coefficients of genic similarity between the forms (I) were particularly high and more often greater between reproductively isolated karyotypes than between geographically isolated populations. Thus, $I = 0.94$ between $2n = 44$ and $2n = 40$ races, whereas in geographically isolated populations within a karyotype of the $2n = 40$ race, $I = 0.84$. Genic similarity decreased with isolation distance and ranged from $I = 0.996$ for the $2n = 44$–46 dichotomy to $I = 0.858$ for the $2n = 60$–48NM dichotomy. A general tendency for monomorphism in geographically and reproductively isolated populations was present, and heterozygosities were particularly low ($H = 0.005$–0.067). Genetic differences between chromosome races were very small, thus supporting the primacy of chromosome race formation and suggesting a lineage of descent ($2n = 48$NM, $2n = 60$, $2n = 40$, $2n = 44$, $2n = 46$, $2n = 48$W).

An electrophoretic analysis of *T. bottae*, made by Patton and Yang (1977), examined 23 loci in 825 individuals collected from 50 populations. This sample included the total known range of chromosome variability for this highly polymorphic and polytypic species. Genic variability was particularly high between and within populations. Of the loci analysed, only two were monomorphic and these were fixed for the same allele in all sampled populations. The average individual was heterozygous at 9.3% of its loci and polymorphic at 33%. Genetic distances were very high between populations (mean $D = 0.144$ (range 0.008 to 0.348)) and the populations could be separated into two clusters, one very similar to *T. umbrinus*. Patton and Yang (1977) found a high degree of concordance between the allozyme pattern and distribution of chromosome morphs, this producing an array of regionally delineated units which could interbreed freely with adjacent units. The authors found a significant correlation between overall variability, population density and degree of geographic connectedness. They concluded that gene flow had played a major role in the population structuring of this complex. Clearly, these results show a completely contrasting genic pattern to *T. talpoides* and all other examples of chromosomally speciating complexes cited in this chapter, wherein very low genic distances occur between the most recently differentiated forms. *T. bottae* typifies a highly polymorphic/polytypic complex where the chromosomal rearrangements are either neutral, or positively heterotic. However, populations of this species appear to be highly subdivided, with significant disruptions to gene flow occurring over small geographic areas. Indeed, Smith and Patton (1988) were able to relate morphological characters to electrophoretically defined distributions.

In the case of *T. umbrinus*, two chromosome races are present, one of which is known to have had numerous translocations established within its complement, at least over a part of its range (2n = 78 northern), and also has a directionally distributed and highly polymorphic system of heterochromatic segments attached. Patton and Yang (1977) included three of the northern populations of this species in their electrophoretic analysis of *T. bottae*. As is often typical of chromosome races which are isolated by powerfully negatively heterotic rearrangements, they found a greater genic similarity between the reproductively isolated chromosome races than between geographic entities of the same race, indicating the potential for population isolation and genic differentiation.

Patton and Feder (1978) analysed 133 individuals from 18 populations of *T. umbrinus* from the northern and western portion of the species range and scored them at 23 electrophoretic loci. While the 2n = 78 chromosome form showed genic homogeneity ($S = 0.864$), it was clear that two distinct phenetic groupings within the 2n = 76 form were also present and each was equally homogeneous ($S = 0.930$ and $S = 0.930$). Comparisons between these three groups showed a similarity of only $S = 0.751$. Patton and Feder (1978) argued that this suggested reproductive isolation existed between these forms, but that the level of isolation may be a function of which pairs of adjacent populations are involved.

Hafner *et al.* (1987) analysed 26 populations of *T. umbrinus* which were distributed throughout the range of both the 2n = 76 and 2n = 78 chromosome races. The specimens were surveyed for 23 electrophoretic loci. Both chromosomal and electrophoretic data suggest that the 2n = 76 and 2n = 78 populations are specifically distinct, as are the northern desert and central plateau groups from the 2n = 78 distribution. Genetic similarities and distances suggest a pattern of extreme genic differentiation with similarities ranging from $S = 0.991$ to 0.610. Five electrophoretic subgroups could be identified and these delineate geographic groups defined by chromosomal characteristics. The 2n = 76 karyomorph, however, was divided into three electrophoretic groups. Numerous, fixed allelic differences suggest that gene flow between groups may have been heavily restricted in the past (Fig. 8.9).

At the intrapopulation level, heterozygosity and polymorphism frequencies were particularly low and Hafner *et al.* (1987) argued that their population genetic statistics support the view that founder events and past bottlenecks resulted in the loss of certain alleles and chance fixation of others within the population isolates. In fact, rather than solving problems, *T. umbrinus* has created more, for Hafner *et al.* (1987) believe that *T. umbrinus* includes several biological species that should be recognized as such.

It is possible that in *T. umbrinus*, two temporally superimposed forms of chromosome change have become established. That is, 2n = 76 and 2n = 78 chromosome races were formed by the fixation of translocations. A high level of chromosome segment polymorphism was present throughout the primordial species distribution before this racial subdivision. Subsequent changes to the habitat of these forms, including the isolation of populations by watercourses, tectonic or erosion activity, have resulted in a heavily dissected distribution in which numerous population isolates now occur. Genic and chromosomal differences established in these isolates have been established after the geographic isolation of these populations and are specific to them.

8.3.2 The *Peromyscus maniculatus* complex

The deer mice of the genus *Peromyscus* are a diverse assemblage of some 65 species distributed throughout north and central America. These animals are unusual in that some species are chromosomally monomorphic, whereas others are highly polymorphic. All species have a chromosome number of 2n = 48, which is generally acrocentric, and the primitive karyotype has metacentric pairs 1, 22 and 23 present. In this ancestral karyomorph, the C-positive heterochromatin is paracentromeric in its distribution. Both G and C-band chromosomal information

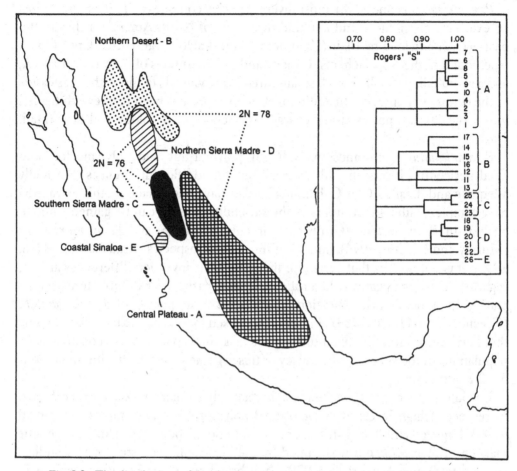

Fig. 8.9. The distributions of the chromosome races and electrophoretically distinct forms of *Thomomys umbrinus* are indicated by different shading patterns. The inset shows the phylogenetic relationships between populations from each of the electrophoretic forms. (Redrawn from Hafner *et al.*, 1987.)

is now available on 26 species of *Peromyscus*, and this largely stems from the work of Robbins and Baker (1981) and Rogers *et al.* (1984). Many species are unchanged from the ancestral 2n = 48 (1, 22 and 23) format; however, some are monomorphic for numerous additional pericentric inversion differences. For example *P. lepturus* (2n = 48, FN = 68 metacentric, 1, 2, 3, 5, 6, 7, 9, 10, 15, 22, 23) has established many fixed pericentric inversions. An additional form of chromosome change is superimposed over conserved ancestral and inversion-derived karyotypes. That is, many heterochromatic short arms may be present in a fixed or polymorphic state, and in some cases pericentric inversion polymorphisms may also be found (Gunn and Greenbaum, 1986).

P. maniculatus is one of the most diverse species complexes. This organism has an extensive distribution and is found over most of North America. It has in the past been subdivided by Blair (1950) into a long-tailed, large-eared, large-footed type, found in the Appalachians, Canada and the mountains of the western USA, and a short-tailed, small-footed, small-eared form which is found in the grasslands of the continental interior. In addition, *P. maniculatus* has been further subdivided into some 30 subspecies using characters such as coat colour and size (Hall, 1981).

P. maniculatus is chromosomally highly polymorphic and polytypic for both fixed pericentric inversion and heterochromatin addition differences. No really adequate and detailed C or G-banding studies have been made on this extraordinary complex, although an index of the variability present can be gained from an analysis of *P. m. austerus* (Gunn and Greenbaum, 1986) and *P. m. nebrascensis* (Murray and Kitchin, 1976), and 13 of the other subspecies (Bradshaw and Hsu, 1972). It is noteworthy that certain of the pericentric inversion differences are not regarded as being negatively heterotic, since heterozygotes exhibit heterosynaptic pairing in a pericentric inversion polymorphism in pair 6 of *P. m. gambelii* (Greenbaum and Reed, 1984). These authors used these data as a basis for claiming that pericentric inversion fixed differences lack the capacity to genetically isolate populations of the species. The fallacy of this argument was dealt with in Sections 5.1.2.2 and 7.1.

It would appear that *P. maniculatus* is relatively similar to *Thomomys umbrinus*, in respect of high levels of chromosomal polymorphism and polytypism spread over a large geographic distribution. An additional perspective on the genetic makeup of *P. maniculatus* was gained by the electrophoretic studies of Loudenslager (1978) and Avise *et al.* (1979). In a broad survey of 716 specimens of *P. maniculatus* taken from throughout its North American range, Avise *et al.* (1979) analysed allozyme variation at 23 loci. Genetic distances ranged from $D = 0.04$ to 0.126 between populations. They found that 75% of the loci are monomorphic and that the same alleles predominate in all samples. Of the six polymorphic loci, the same alleles often reoccur, but at different frequencies. Heterozygosities are very large (5.4–12.4%), and most of the differences result from frequency shifts at these polymorphic loci. Avise *et al.* were not able to establish geographic trends. Such a result was to be expected since population sampling was undoubtedly at too crude a level; one might only hope to find geographic variants in 30 subspecies by sampling 71 populations.

A different approach was taken by Loudenslager (1978) who analysed 190 specimens of *P. m. nebrascensis* from nine demes and scored them for 23 loci. Genetic identities ranged from $I = 0.95$ to $I = 0.98$ between populations. Heterozygosities ranged from 9 to 21%, and 41% of loci were polymorphic, thus indicating extensive genetic variation within the subspecies. Moreover, variation in the frequency of heterozygosities was significant between localities, and Loudenslager argued that adaptive divergence correlated with genetic heterozygosity, and that high heterozygosity reflected an adaptation to a variable environment. Indeed, this pattern of genic variation is most similar to that in the *Thomomys umbrinus* complex.

It is perhaps fortuitous, that *P. maniculatus* is one of the few groups in which there has been a detailed mtDNA analysis. Lansman *et al.* (1983) examined 135 specimens collected from localities throughout North America. The pattern and nature of the changes which have occurred in the evolution of the mtDNA molecule were made by mapping the restriction sites for eight endonucleases. A total of 61 clones were recognized and these could be further subdivided into major clonal assemblages. However, in some cases diversity between clones was often large within an assemblage. The conclusions reached by Lansman *et al.* are most significant, for they found that there was no relationship between the major morphological forms and mtDNA genotype. The phylogenetic subdivision of Blair (1950) was not supported. Second, of the animals sampled, the mtDNA data agreed with four of the defined subspecies (*P. m. blandus, P. m. rafinus, P. m. nubiterrae* and *P. m. gambelii*), but in most other cases relationships cut across the recognized taxonomic boundaries. Third, there was little disagreement with the chromosomal analyses, because these remain at a generally rudimentary level. Nevertheless, the study did suggest that some of the mtDNA defined clones may provide an interesting perspective on the evolution of the *P. maniculatus* complex. While the majority of allozyme studies which have been published have sampled different populations from those used by Lansman *et al.*, there was an overlap with the Wyoming populations investigated by Loudenslager (1978), and the Colorado populations investigated by Nadeau and Baccus (1981). Despite the fact that high levels of allozyme variation were reported between these populations, they were essentially identical in mtDNA sequence. This suggested to Lansman and his colleagues that allozyme variation may simply reflect differential selection acting over a short time on populations which initially possess the same genetic resources or, alternatively, that differences exist in the inheritance patterns of maternal and paternal phylogenies. Unfortunately, it is likely that neither of these explanations are correct, for, unless the same specimens are analysed for all techniques (as in a study by Nelson *et al.* (1987) on *P. leucopus*), it is difficult to know what subspecies or populations were being compared.

Populations of *P. maniculatus* which are characterized by both presumed negatively heterotic polytypism, and high levels of population polymorphism for heterochromatic addition and pericentric inversion, also display high levels of genic diversity reflected by allozyme electrophoresis of their nuclear genome, and restriction endonuclease mapping of their mitochondrial genome. Moreover, this genic diversity is correlated with highly subdivided geographic distributions in some cases, but not in others.

8.3.3 The *Sorex araneus* complex

Of the 61 species of shrews known throughout the world, those of the *Sorex araneus* complex are among the most diverse. Seven of the eight species have been karyotyped and have chromosome numbers of 2n = 20 (*S. coronatus*), 2n = 34 (*S. granarius*), 2n = 24 (*S. daphaenodon*), 2n = 22 (*S. caucasicus*), 2n = 34 (*S. sibiriensis*) and 2n = 26 (*S. arcticus*). These forms are distinguished by fixed paracentric and pericentric inversion differences, reciprocal translocations and fusions. Hausser *et al.* (1985) proposed that these species had speciated chromosomally, and in an electrophoretic analysis of 22 presumptive loci found that the mean genetic distance between *S. granarius, S. coronatus* and *S. araneus* was $D = 0.055$, and *S. coronatus* was the only species that had a fixed difference.

Sorex araneus reveals one of the most complex patterns of chromosomal evolution known in mammals and appears to have established a similar degree of variation to *Mus domesticus* and *Peromyscus maniculatus*. In this complex of chromosome races, chromosomal differentiation has occurred by Robertsonian fusion between acrocentric elements. To complicate matters, complex chromosomal polymorphisms occur within many of the chromosome races and broad hybrid zones exist between chromosome races (see Fig. 8.10). A total of 13 chromosome races are now recognized in western Europe, although Searle (1988b) suggests a possibility of 40 chromosome races having been described if the eastern European and Siberian assemblages are included. Chromosome races are now known from Sweden, Denmark, Britain, Germany, Switzerland, Poland, Czechoslovakia, Hungary, Austria, Yugoslavia and Finland in the western European assemblage (Hausser *et al.*, 1985; Wójcik, 1986; Halkka *et al.*, 1987; Searle, 1988b).

Some of the chromosome races have unusual distributions. For example, of the three British races (Oxford, Hermitage and Aberdeen), the Oxford race is also found in Denmark, whereas the Aberdeen race also occurs in Poland. Searle (1988) has interpreted the present-day distributions to be a product of colonizing radiations and retreats to refugia associated with climatic changes. The three broad hybrid zones in Britain reflect subsequent interactions between what are now relic populations, whereas the polymorphisms which overlie portions of this racial distribution are the result of more recent events (see Fig. 8.10).

A total of 36 different metacentric combinations have been encountered out of the 66 possible (Searle, 1988). All of these characteristics suggest that the vast majority of Robertsonian changes established in these populations are either

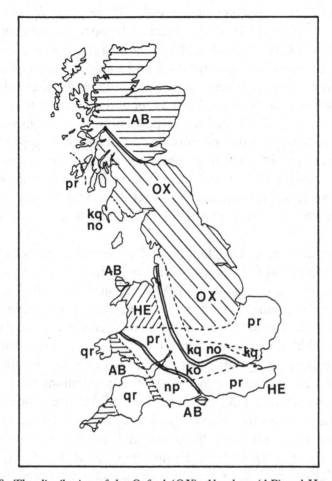

Fig. 8.10. The distribution of the Oxford (OX), Aberdeen (AB) and Hermitage (HE) chromosome races of the *Sorex araneus* complex in Great Britain. The chromosomally monomorphic areas are shaded, whereas the polymorphic areas are unshaded. The boundaries of interracial hybrid zones are marked by a double line. Lower-case letters refer to the particular fusion metacentric chromosomes present in animals at that site. (Redrawn from Searle, 1988b.)

positively heterotic or neutral. The wide hybrid zones indicate considerable chromosomal introgression between races and suggest that there are only very minor fertility and viability effects associated with structural hybridity. Indeed, it was suggested by Searle that the numerous unfused acrocentrics present in hybrid zones as fusion polymorphisms, ameliorated the effects of the less benign structural rearrangements. Searle (1984a, 1986) examined polymorphic animals from the Oxford race and found non-disjunction rates of from 1.0 to 2.5% per heterozygous arm combination and 1.5 to 3.7% per heterozygous parent in females, and

0.7% per heterozygous arm combination and 1.0% per heterozygous individual in males. In some of these cases monobrachial homologies were formed which segregated normally. Garagna *et al.* (1989) examined spermatogenesis in animals obtained from the same chromosome race and found a similar low impact on fertility. However, complex heterozygotes which formed quadrivalents, had a higher proportion of defective seminiferous tubules and lower testis weight and had reduced fitness when compared to independent fusions. Laboratory hybridization between the three British races was successful and there appeared to be no strong pre- or post-mating reproductive barriers between these races in terms of independent fusions (Searle, 1984b). However, the lower frequency of heterozygotes for multiple fusion differences recovered from both hybrid zones and the laboratory, suggested to Searle that there was a high level of selection against these.

In the majority of cases, electrophoretic analyses revealed that the chromosome races of *Sorex araneus* had failed to establish fixed genetic differences, once again indicating the primacy of chromosome race formation. This finding is not surprising if the low levels of non-disjunction and minor impact on fertility detected so far are a true indication of the impact of the many fixed chromosomal differences. However, even with such minor fertility effects, levels of genic isolation were detected between races, suggesting that they are a barrier to gene flow. Frykman *et al.* (1983) used 27 presumptive loci to examine populations from three of the Swedish chromosome races and found a frequency difference of one of the alleles at the *Mpi* locus between the northern and central races. Searle (1985) found similar differences in both allele frequency and heterozygosities between the Aberdeen race and both Oxford and Hermitage races at the *Mpi-1*, *Pgm-2* and *Pgm-3* loci. Halkka *et al.* (1987) referred to similar, although unpublished, differences between the Finnish races. However, Frykman and Bengtsson (1984) detected what they assumed to be gene flow between the northern and central chromosome races of *S. araneus* in Sweden using three genetically variable loci. Clines in alleles of the *Mpi* locus over the hybrid zone suggested that gene flow was occurring.

8.4 Concluding remarks

8.4.1 Genic expectations in chromosomally speciating populations

Founding populations, which are inextricably involved in the process of chromosomal speciation, have been characterized by presumptive changes in the degree and nature of genic differentiation established in their genomes. These are undoubtedly contentious grounds, yet some workers have been inclined to place the postulated degree of genic variation (as reflected by levels of heterozygosity and allelic polymorphism), in a predictive framework, despite the fact that there is a lack of concordance between predicted and real genic differences in founding populations.

More recently, this predictive approach has been extended by Sites and Moritz (1987), who have proposed that the pattern of genic variation in chromosomally speciating complexes could be used to determine the mode of speciation, since, they reason, different modes of speciation would have a characteristic population structure and predictable level of genic variation (average heterozygosity (H) and mean number of alleles per locus (A) being two of the parameters considered).

A number of significant problems are associated with such an approach and it is of value to consider some of these. First, it is most difficult to visualize how the levels of heterozygosity, or the mean number of alleles per locus, can be predicted for chromosomally speciating plants and animals when such an array of genetic systems are known to exist. Second, the predicted levels of genetic variability present in Mayrian founder populations (decreased variability) are markedly different from those advocated by Templeton (1980) and Carson (1990) (increased variability), (see Chapter 4). One might ask how Sites and Moritz (1987) can sustain their arguments for precise genetic expectations in such an environment. Third, the least satisfactory aspect of the Sites and Moritz (1987) approach is its logic. To make the step from a generalized model of speciation, to the construed expectations for levels of heterozygosity associated with that model, involves numerous precarious assumptions. To then reverse this order and argue that a mode of speciation is not operating because it fails to comply with these assumed criteria is simply unacceptable. Fourth, the relevance of the continued use of heterozygosity and polymorphism in electrophoretic analyses, as an index for the capacity of a species to produce isolated populations, is questionable. Present-day indications of genic variability may be irrelevant to the past characteristics of a species when it existed as a founding population. Fifth, recent studies on Hawaiian *Drosophila* comparing chromosomal, mtDNA and electrophoretic variability suggest that electrophoresis has failed to provide the resolution for changes which have occurred in founding populations (De Salle *et al.*, 1986a, b; Templeton, 1987; Craddock and Carson, 1989; see also Section 4.8).

Comparisons between species which have very high levels of genetic heterozygosity and polymorphism, such as *Peromyscus maniculatus, Thomomys bottae* and *T. umbrinus*, reveal enormous differences in genetic variability between isolated populations (Sections 8.3.1 and 8.3.2). However, they generally fail to show fixed allelic differences between isolates or chromosome races. Such species are characterized by high levels of chromosomal polymorphism as well as chromosomal polytypism. Indeed, in the case of *T. umbrinus* the two 'old' chromosome races (2n = 76 and 78), which are reproductively isolated by numerous translocations, are subdivided into population isolates with high levels of genetic heterogeneity and polymorphism distinguishing them. This highly subdivided population structure might suggest that the great level of genetic variability is locally adaptive. It could also suggest that this population structure is inappropriate for the fixation of negatively heterotic chromosomal rearrangements, since high levels of heterozygosity and polymorphism are not generally correlated with chromosome race formation, or chromosomal speciation, according to Patton and Yang (1977) and Sites and Moritz (1987). But if this is so, how did the initial 2n = 76 and 2n = 78 translocation races reach fixation if the population structure was unsuitable? Were the levels of heterozygosity and polymorphism different when these rearrangements reached fixation? If so, the study of present-day genic variability might be a waste of time. Claims by Sites and Moritz (1987) that chromosomal speciation can be discounted as operating in the *Sceloperus grammicus* complex, because heterozygosities are high and hence suggest a more panmictic population structure than would be predicted by them for chromosomal speciation, cannot be taken seriously. While there is little doubt that actively speciating complexes distinguished by fixed negatively heterotic chromosome differences do generally display great genetic similarity (Patton and Yang, 1977), they also have the capacity to establish population isolates with fixed chromosomal or allelic differences. Indeed, isolated populations of *S. grammicus* which have several fixed genic differences were detected by Sites and Greenbaum (1983).

The most realistic means of determining whether a species has the capacity to produce isolated populations is to survey the species distribution for electrophoretic anomalies. Thus, fixed allelic differences encountered within isolated populations, that differ from the existing species' electrophoretic phenotype, suggest that such populations have been isolated for a sufficient period of time to enable the fixation of these allelic differences. By the same token, the discovery of a population isolate which has differences in heterozygosity and polymorphism frequency to the remaining populations, suggests a degree of isolation at the present time, but nothing more.

8.4.2 The primacy of chromosome change in speciation

If we consider both chromosomal and non-chromosomal modes of speciation and compare chromosomal speciation to allopatric speciation by founding and non-founding means, a dichotomous basis for comparison exists. That is, in the allopatric modes of speciation, geographic isolation accompanied by the fixation of genetic differences and the attainment of reproductive isolation are key events. Alternatively, in chromosomally mediated speciation, geographic isolation accompanied by the fixation of negatively heterotic rearrangements which impose reproductive isolation are the key events. In both processes, isolation from gene flow permits the fixation of the primary means of differentiation, whether it is genic or chromosomal. However, isolation by itself is not responsible for species formation. Differentiation is the key to that process, and the type of differentiation determines the likelihood and nature of speciation.

By the same token, the period of isolation needed to accrue genic differences sufficient to cause reproductive isolation would generally be much longer than the period of time necessary to produce chromosomally mediated reproductive isolation. The former is thought to be due to the accumulation of multiple genic differences which may impinge on reproductive isolation in many ways, while the latter is a single step, the fixed rearrangement itself acting as a reproductive barrier.

If it can be demonstrated that the fixation of genic differences has occurred as a primary event, then speciation cannot be construed as being chromosomal. Alternatively, if it can be demonstrated that the fixation of negatively heterotic chromosomal rearrangements has occurred as a primary event then speciation is shown to be chromosomal and not genetic. This has a proviso, and that is that the chromosomal change fixed must also be capable of acting as a reproductive isolating mechanism.

Three major assumptions have been made in testing the numerous complexes of species described in this chapter. First, that the most recently differentiated species in a recognizable complex will have few differences established. Second, older forms which have been isolated for a longer period will have accrued additional morphological and genetic differences. That is, differences between isolated forms gradually accrue with time. Third, that clusters of species/chromosome races in an evolving complex reflect the sequence of chromosomal and genetic differentiation. These assumptions appear to be at the very basis of evolutionary and systematic studies and a vast body of evidence suggests that they are reasonable assumptions to make.

Thus, in all of the examples of species complexes cited in this chapter, the entities of which are distinguished by single or multiple fixed chromosomal differences, two features are apparent. First, increased morphological and genetic differentiation occurs with increased age of species in the complex. Second, the most recently differentiated chromosome races are distinguished from their direct ancestor by either minor genic differentiation, or no form of genic differentiation at all (see Fig. 8.11). That is, the fixation and establishing of a chromosome race has been the primary act of speciation in these speciating complexes. In the examples of class three and four speciating lineages described in this chapter and in Chapter seven, this same pattern of differentiation is apparent. Moreover, the chromosome races differ from each other by one or more fusion or fission event, and in those cases where multiple fusion changes have occurred, chromosomes

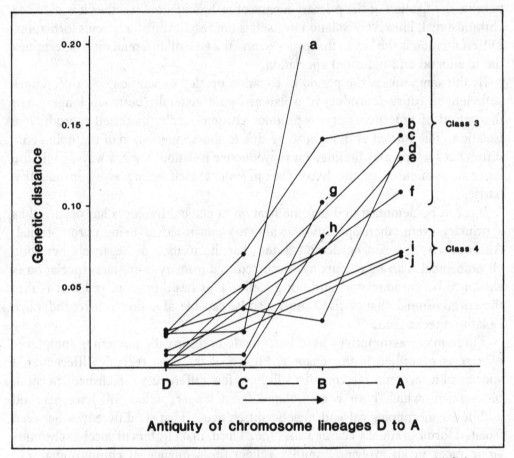

Fig. 8.11. Genetic distances are plotted against chromosome races or species, in their known order of descent within ten independent lineages. The entities within these lineages are shown as A, B, C and D. For the sake of simplicity, the genetic distance for any form has been plotted above that form rather than between that and its closest relative. Note that in every lineage the most recently derived chromosome race (D) has been established before any significant genetic differentiation has occurred between it and its director ancestor. These data emphasize the primacy of chromosome race differentiation and the secondary occurrence of genetic differentiation. The following complexes are considered: *Acomys cahirinus* (a); *Gehyra variegata–punctata* (b); *Thomomys talpoides* (c); *Proechimys guairae* (d); *Rattus rattus* (e); *Sceloporus grammicus* (f); *Gerbillus pyramidum* (g); *Rattus sordidus* (h); *Mus domesticus* (i); *Spalax ehrenbergi* (j). The genetic distance data are shown in Chapters seven or eight under the description of each of the complexes.

may share brachial homologies, which presumably enhance adverse fertility effects. The chromosome differences are all putatively negatively heterotic.

The salient features of the lineages described are very similar:

1 The most recently derived chromosome race has attained a defined geographic distribution, before any significant genic differentiation has been established to distinguish it from the parental form.

2 Most of the daughter races are morphologically indistinguishable from the parental forms.

3 Most of the daughter races are reproductively isolated from the parental form, but not all.

4 The ancestor/descendant chromosome forms are generally parapatrically distributed and in some cases narrow hybrid zones occur between them.

5 In some cases, lineages of species are geographically distributed in the order of descent, suggesting a pattern of colonization associated with speciation.

6 In certain daughter races, fixed genic differences are found between populations within that race, whereas these differences do not characterize that race.

7 Genetic discontinuities in terms of divergence in allele frequencies may occur within, or between, chromosome races, a feature which at least emphasizes the subdivided nature of populations within speciating forms.

8 The most recently derived chromosome races may be distinguished by one or several fixed chromosomal differences which may have a minor, moderate or profound impact on the fertility of hybrids.

A further test for the causative relationship of negatively heterotic chromosomal rearrangements and speciation can be made by comparing lineages which have different forms of chromosomal rearrangements established. That is, comparing what is happening at the genic level in lineages of chromosome races distinguished by negatively heterotic fusion changes, such as in *Thomomys talpoides*, with those populations which have multiple fixed or polymorphic heterochromatic blocks, such as *Thomomys bottae*, and those cytotypes and races which have both translocations and heterochromatic blocks such as *T. umbrinus* (Section 8.3.1), is highly informative. Thus, while genic differentiation is minimal in *T. talpoides*, as might be expected in a case of colonizing chromosomal speciation, in *T. bottae* where the heterochromatic blocks are probably neutral, or adaptive, relatively enormous genetic distances, heterozygosity and polymorphism frequencies occur between isolated populations. Similarly, in *Peromyscus maniculatus*, where both presumptively negatively heterotic inversion differences and high levels of heterochromatic addition polymorphism are present, genic variability between and within populations is most significant, and this can be detected within the nuclear genome and reflected in the mitochondrial genome. It can be argued that in these situations two fundamentally opposed genic systems are in operation, one resulting in speciation and the other resulting in local adaptation.

The argument has been raised that the reason why genic differentiation is not observed between the most recently evolved chromosome races in speciating complexes is because negatively heterotic chromosomal differences between these forms are an ineffectual barrier to gene flow which continues despite them (Futuyma and Mayer, 1980; Baker, 1981; Greenbaum 1981; Sites and Greenbaum, 1983; Thompson and Sites, 1986). This is basically a fatuous argument since electrophoretic comparisons between chromosome races assess whether differentiation between populations or races has occurred. These comparisons cannot provide any evidence for gene flow, since this character is not measured.

In a most thorough analysis of the gene flow problem, Larson et al. (1984) analysed the genetic structure of 22 species of plethodontid salamanders. They found that by calculating the Nm (the product of effective population number and rate of migration among populations) from protein polymorphisms, they were able to provide a much more accurate historical perspective on gene flow between populations. They found that most plethodontid species could not be viewed as units whose cohesion was maintained by continuing gene exchange. They also found that the maximum possible migration rate was less than the mutation rate, suggesting that most species analysed had populations that were completely isolated from genetic exchange. 'Species of plethodontids generally do not comprise units connected by gene flow' (Larson et al., 1984, p. 305).

If appropriate comparisons are made it can be shown that gene flow does occur to some degree over certain hybrid zones which are distinguished by weakly deleterious rearrangements. The case of Sorex araneus provides some evidence for this. Here, hybrid zones are very wide, and some of the fusion differences are relatively ineffectual in terms of fertility effects on hybrids. Frykmann and Bengtsson (1984) detected the introgression of some electrophoretic loci over hybrid zones between Swedish chromosome races of this species. Nevertheless, levels of genetic differentiation between chromosome races of S. araneus are also known (Frykmann et al., 1983; Searle 1985). Indeed, in some of the speciating complexes described in this chapter, which have only one or two fixed differences distinguishing them, fertility effects between recently derived chromosome races do not result in total reproductive isolation. However, hybrid zones are narrow and there is no evidence for chromosomal introgression. In most cases, experimentation has not been done to determine whether F2 or backcross hybrids are viable, or fertile.

Nevertheless, most chromosome races are distinguished by multiple chromosome differences which would have a significant additive fertility effect, or extreme effects due to monobrachial homology. Hybridization between many of the chromosome races is either unlikely, or restricted to a narrow hybrid zone through which there is no genic introgression possible. In some complexes, such as the *Rattus sordidus* and certain *Mus domesticus* races, the chromosomal barriers are so profound as to prevent the production of fertile hybrids (Baverstock *et al.*, 1983b; Britton-Davidian *et al.*, 1989). That is, these chromosomal differences between genically undifferentiated forms are impermeable barriers to gene flow.

There is a substantial body of evidence which argues against the ineffectuality of chromosomal barriers to gene flow and much evidence to support the absence of effective gene flow occurring in many of these species. When chromosome races are analysed electrophoretically, there are often significant regional discontinuities found between populations within a race, and between that race and its direct ancestor (see Nevo *et al.*, 1974; Sites and Greenbaum, 1983; Baverstock *et al.*, 1986; Nelson *et al.*, 1987; Sites and Moritz, 1987). These may involve either fixed differences between populations, or variation in the frequency of polymorphic loci or levels of heterozygosity between populations. Clearly, such evidence argues against gene flow occurring between the populations, let alone across chromosome race boundaries. There is no shortage of evidence for discontinuities in H and P between chromosome races in the literature, and the above are a minor selection.

Indeed, if hybrid zones between chromosome races were an ineffective barrier to gene flow, then the distribution of these races would simply represent area-effects within the distribution of a single panmictic species. That is, chromosome races would have been established within a species distribution. If Futuyma and Mayer (1980) were correct in this assumption, they are providing tacit support for the most unlikely aspect of stasipatric speciation. The primary act of this mode of speciation is the fixation of negatively heterotic rearrangements within the distribution of an existing species (see Section 9.1.2). It would appear that those who have argued against the primacy of chromosome change in speciation and have invoked the 'no barrier to gene flow' argument, have not thought through the ramifications of this position, and its associated implications.

In all of the examples described in this chapter, the evidence points to chromosomal differentiation involving negatively heterotic rearrangements acting as a primary and causative agent for speciation, permitting secondary, genic and morphological differentiation.

9
Chromosomal speciation

There was a widespread belief among early cytogeneticists that chromosomal rearrangement was the essential step in speciation. Proposed as an alternative to geographic speciation, the chromosomal speciation hypothesis is not valid.

(Mayr, 1963, p. 439)

and

The fact of chromosomal speciation poses a problem.

(Mayr, 1982b, p. 13)

When Michael White wrote his volume *Modes of Speciation* in 1978, he included four major models for chromosomal speciation, the triad hypothesis of Wallace (1953, 1959), the saltatory model of Lewis (1962, 1966), quantum speciation (Grant, 1971) and his own stasipatric concept (White, 1968). In the years since 1978 a spate of models defining chromosomal speciation have been published and many of these have centred on founding situations.

Nevertheless, the traditional views on allopatric speciation by isolation remain widely accepted and in some cases justifiably so. There are numerous examples of species which have speciated without any significant change in chromosome morphology. In most of these instances, the criteria for what we accept to be allopatric speciation have been met. Chromosome change is not a *sine qua non* for speciation, a fact that is well demonstrated with the homosequential polytene chromosomes of certain Hawaiian *Drosophila* (Carson *et al.*, 1967). Concomitantly, even those cases where species, or lineages of species, do have differences in chromosome morphology, many of the changes which have occurred may have had nothing to do with past cladogenic events. Thus, chromosomal polymorphism, chromosomal additions, or fixed chromosomal differences which are neutral, or relics of ancient speciation events which now characterize lineages of otherwise chromosomally identical species, testify to this.

In most of those instances where chromosome change is believed to be associated with speciation, these fixed negatively heterotic, or potentially negatively heterotic, structural rearrangements have been implicated. Such rearrangements reach fixation in a daughter population before any recognizable genic or morphological differences have been fixed which would distinguish it from the parental population. When chromosomally distinguishable parental and daughter races or species are in contact, hybridization between the forms is either non-existent, or restricted to a narrow hybrid zone. There is therefore little or no genic or chromosomal introgression of one form into the distribution of the other. The many species complexes which have these generalized characteristics lead us to suspect that the primary mechanism for speciation in these forms has been by the fixation of negatively heterotic chromosome changes and the population isolation which they ensure. Morphological and genic changes which reach fixation and eventually distinguish species are subsequent events. However, some of the models which have been proposed, suggest that the chromosomal differences involved in speciation may be either neutral or, in one case (the triad hypothesis), polymorphic.

Despite the enormous body of data now available, many biologists still only accept the possibility of speciation by allopatric means. They argue that chromosomal changes are irrelevant to cladogenic processes and, at best, that chromosomal rearrangements have no specific role, other than the vague possibility that they might reinforce the isolation of already differentiated species (Futuyma and Mayer, 1980). Many of these views can be attributed to an absence of discrimination between those changes which can be implicated in speciation and those which cannot.

The onus of proof placed on those who advocate a chromosomal involvement in speciation is less than even-handedly applied, when compared to the proof required for allopatric speciation. In many respects this may be attributed to allopatric speciation being an easy way out. It is universally accepted as a means of speciation, yet the data used to support the concept are no more profound than those for chromosomal speciation. Thus, allopatry is the accepted and often unquestioned dogma (see Chapter 3). Indeed, Bush (1982) made the point that it is no longer acceptable for the advocates of allopatric speciation to dismiss non-allopatric models on the grounds that they lack convincing evidence, when the counter arguments they use are tenuous and based on equivocal evidence and speculation.

The fact remains that the geographic distributions of species, species complexes, populations, or even demes, are crucial to our understanding of how new species have been generated. Models for chromosomal speciation can be subdivided into two broad areas delineated by where the chromosomally derived daughter species were established. If the new form arose within the confines of the parental species distribution, it is regarded as having an internal, or sympatric origin. Alternatively, if the daughter species was established in a founding population, or peripheral isolate, apart from the distribution of the parental species, it must be regarded as having an external, or allopatric origin. In this chapter, the numerous models which have been proposed to explain chromosomal speciation are considered under these broad headings. This step has been taken even though the distinction between the categories is often less than precise.

9.1 Internal modes of chromosomal speciation

It is implicit in the definition of internal modes of chromosomal speciation, that speciation has occurred within the distribution of a species while populations are in contact, or at least not isolated from each other in terms of gene flow. This point is of considerable importance, for if a population is isolated within the distribution of a species from other populations (internal allopatry), there is little difference between this and a population on the periphery of a species distribution also being isolated (external allopatry) (see Fig. 9.1 and Key, 1981).

It is also valid to make a distinction between internal allopatry and internal models for speciation, for the latter suggest that speciation had occurred in a

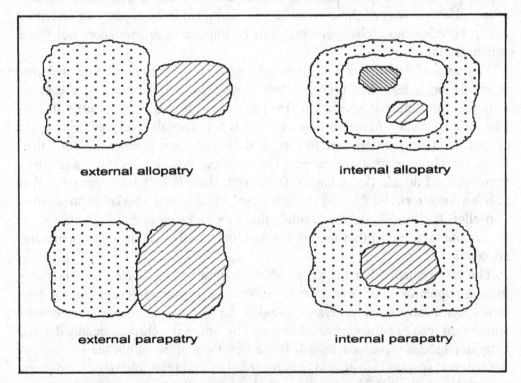

external allopatry   internal allopatry

external parapatry   internal parapatry

Fig. 9.1. A schematic representation of four of the distributional patterns found in plant and animal populations and used in the text.

sympatric situation. That is, chromosomally mediated speciation was occurring while populations were in contact and not isolated from gene flow. If any form of isolation was present, such as internal allopatry, it is not sympatric speciation, but chromosomally mediated allopatric speciation. This distinction is important to our understanding of the models for chromosomal speciation.

9.1.1 The triad hypothesis (Wallace, 1953)

A vast literature exists on the structural evolution and interaction of polytene chromosomes in the Diptera, in general, and *Drosophila* in particular (see White, 1973). In most cases, species of *Drosophila* are distinguished by chromosomal fusion differences (Patterson and Stone, 1952). High levels of chromosomal reorganization are present within populations of most species and these often involve paracentric inversion polymorphisms. Pericentric inversion polymorphisms also occur, but are much less frequent. Many of the side-effects of inversions are deleterious and recombination within inversion loops can lead to the formation of duplications, deficiencies and aneuploidy. Nevertheless, selection appears to compensate for these disadvantages, because of the high adaptive value of heterozygotes. Dobzhansky (1950) termed this adaptive value the co-adaptation of the gene complexes in chromosomes of one rearrangement with those of another.

When heterozygotes for pericentric inversions are present, chiasmata are moved to a terminal position, thus avoiding the deleterious meiotic effects of the inversions. The co-adapted gene sequences of inverted regions are effectively 'locked up', because of this reduction in recombination when one or two overlapping inversions are present. However, when three overlapping inversions occur in a population (triads), the chromosomal segments and inverted blocks of genes are not separated from one another (Fig. 9.2). Thus, in the former the co-adapted portions of the genome remain intact, whereas in the latter, crossing-over destroys the integrity of the genetic material within the inverted segments and co-adapted combinations are destroyed (Wallace, 1953).

A mechanism for the origin of *D. persimilis* from *D. pseudoobscura* was proposed by Wallace (1953). This was based on triads, the maintenance of co-adapted systems and the distribution of the two major complexes in California. The following criteria were applied:

1 Two inversion mechanisms for establishing co-adaptation were present in defined geographically separated areas.
2 If the two inversions extended their range so that local populations possessing them came into contact, the addition of another inversion

sequence into the interbreeding populations and resultant triad would cause the disruption and breakdown of both co-adapted systems.

3 Following the formation of the triad, local semi-isolated populations which had genetic systems with a high adaptive value, could have developed in what would have been a zone of low adaptive value.

4 If the adaptive value of the population was higher than surrounding populations, a locally co-adapted system would result, despite migration.

5 Sexual isolation would be automatic if genic variability was present,

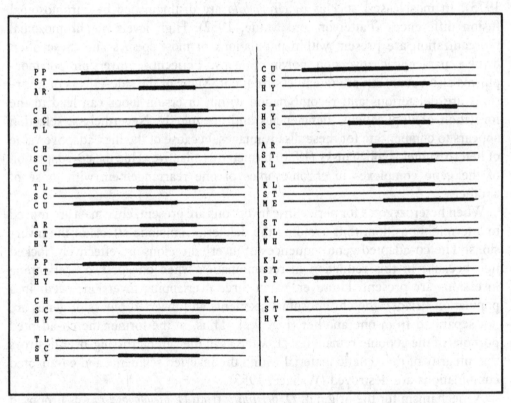

Fig. 9.2. The relative sizes and position of different chromosome three inversions in triads of *Drosophila pseudoobscura* and *D. persimilis*. The centre arrangement of each triad was regarded as uninverted. The standard (ST) arrangement is found in both species, as is hypothetical (HY), whereas Whitney (WH), Klamath (KL), and Mendocina (ME) occur in *D. persimilis*. The remaining arrangements are found in *D. pseudoobscura*: Pykes Peak (PP); Arrowhead (AR); Chiricahua (CH); Treeline (TL); Santa Cruz (SC); and Cuernevaca (CU). Of the triads illustrated, ten are *D. pseudoobscura*, four are *D. persimilis*, and in one (AR-ST-KL), inversions of both species coexist in the one geographic area. (Redrawn from Wallace, 1953.)

since surviving individuals would be the product of intrapopulation matings.

In a subsequent article, Wallace (1959) suggested that numerous *Drosophila* species and race hybrids were distinguished by overlapping inversions, and he reasoned that this may be a significant mechanism by which species evolved and extended their geographic range. Wallace also drew an analogy between triad inversion systems and chromosomal translocation polymorphisms in cross-fertilizing plants such as *Oenothera*, where chromosomal interchanges were inherited as single units. That is, any chromosomal rearrangement that prevents crossing-over would permit the isolation of gene sequences and protect them from recombination. For example, if a translocation changed the combination of chromosomes 1.2, 3.4 into 1.3, 2.4, there were segments on each inherited as a unit. However, if a third interchange, 5.6, was introduced, a triad resulted. If any two interchanges in the triad coexisted in a population, they would result in a population polymorphic for blocks of genes. If the three were present, neither 1.2, 3.4 nor 3.4, 5.6 could be held intact and segregation would be disrupted.

The existence of interchange triads, like inversion triads, could disrupt selection for co-adapted gene blocks. The triads cannot coexist sympatrically. Because the constraints of co-adaptation were removed, one member of the triad could expand its range into a new territory. This may be done by displacing one of the parental forms, or by moving into a new external area.

Wallace (1959) pointed out several of the weaknesses of the triad hypothesis. These included the fact that inversion polymorphisms have varying frequencies over their geographic range; that geographic distributions of triad members may be overlapping, but that they are dissimilar and that other combinations of gene arrangements might lead to the same disadvantageous recombination as do triads. If the triad hypothesis has had a role in speciation, it is undoubtedly an uncommon one, although it has been implicated in the speciation of the *Anopheles gambiae* complex (Coluzzi *et al.*, 1985).

9.1.2 Stasipatric speciation (White *et al.*, 1967; White, 1968, 1978a, b)

White proposed that this mode of speciation could generally be applied to low vagility insects and also to vertebrates with similar lifestyle characteristics. The model was designed to account for those cases where low vagility organisms found in species complexes have visibly different karyotypes and where the parental species had an unreasonably small distribution when compared to the derived forms. Although the stasipatric model was initially preferred as a mechanism for the speciation of grasshoppers, it has a much wider application.

The stasipatric model (adapted from White, 1978a, b) included the following criteria:

1 A widespread species generated within its range a daughter species which was characterized by one or more negatively heterotic chromosomal changes. These may include translocations, fusions or pericentric inversions.

2 These rearrangements are primary in the speciation process, diminishing both fecundity and viability when heterozygous and are adaptive when homozygous.

3 The initial rearrangement arose within the distribution of the parental species despite the probability that it would be eliminated by selection against heterozygotes. Both genetic drift in small inbred populations and meiotic drive, particularly in females where the first meiotic division is asymmetrical, assisted in the fixation of the rearrangement.

4 The daughter species extended its range, displacing the parental species because of the greater fitness of the new homozygote.

5 A narrow hybrid zone existed between parental and daughter species. The narrowness of the zone was due to the low vagility of the species and selection against the homozygote for the derived rearrangement in the parental distribution.

6 Hybridization occurred between daughter and parent species within this contact zone leading to the production of genetically inferior individuals with meiotic irregularities.

7 It is axiomatic that one should be able to distinguish between the derived and parental forms. The derived taxa occupy central areas and the ancestral taxa peripheral, or external areas (White, 1978b).

8 Only those incipient daughter species that are genetically sufficiently different from the original parental population would be capable of displacing it from previously occupied territory (White, 1982). Disruptive selection would be implicated following the initial dichotomy.

9 Regulatory gene changes could be implicated in subsequent divergence (White, 1982).

The two greatest assaults on the stasipatric model came from Key (1968, 1981) and Futuyma and Mayer (1980). While the former placed emphasis on problems associated with White's logic, expression and improbabilities of the concept, the arguments advanced by Futuyma and Mayer (1980) were a trenchant attack on chromosomal speciation in general.

The major criticisms raised by Key (1981) centred around the inconsistency of White's presentation of stasipatric speciation. Subtle differences in the wording and parameters associated with the model were found in several publications (see White, 1968, 1978a, b, 1979). The criticisms are relevant and should be considered.

First, White was inconsistent in his statements on the degree of isolation experienced by the daughter species within the parental population. This is variously reported as being internally isolated and in continuous contact. Key points to the fact that structural chromosome changes can only be realistically fixed in isolation. Thus, if the population is isolated, the situation of stasipatry is in reality internal allopatry (see Fig. 9.1), a point also made by Futuyma and Mayer (1980).

Second, it was unclear to Key whether the chromosomal rearrangements became fixed as homozygotes in the derived population, or remained as polymorphisms. White's definition has undoubtedly varied from paper to paper in its clarity. In the 1978b publication, the daughter population was distinguished by a fixed homozygous difference isolated by a tension zone. In 1979, the central part of the colony was occupied by homozygotes, whereas the periphery was surrounded by a polymorphic population in which heterozygotes were selected against. In this regard, Key was also unclear as to whether the spread of the chromosomal rearrangements of the daughter species, into the parental species distribution, was due to these changes filtering in to this distribution, or progressive displacement of the hybrid zone to the disadvantage of the parental species. White's (1978b) definition leaves little doubt that it is the latter. Nevertheless, Key could not accept that a rearrangement fixed in a central isolate, should acquire a greater fitness than the parental form which was clearly adapted to the area.

Third, Key also attacked both the role of chromosome change as a post-mating isolating mechanism and the existence of the process of meiotic drive. He considered that there was a lack of evidence for both processes. Futuyma and Mayer (1980) also made significant attacks on both of these mechanisms. Consequently, Key devoted much of his time to showing compelling distributional evidence that many of the speciating complexes which White cited as having a stasipatric origin, were more readily explained by allopatric models of speciation, whether they were chromosomal or genic. Indeed, in 1968 Key reinterpreted the chromosomal evolution of the *Vandiemenella* complex. Similarly, Futuyma and Mayer (1980) reinterpreted the chromosomal evolution of *Didymuria*, whereas Bickham and Baker (1980) reassessed that of *Mus domesticus*. In all cases, these authors favoured an allopatric origin.

Indeed, one of the great flaws in White's approach was his continued attempt to force the stasipatric concept onto species complexes which had been analysed chromosomally and were at least arguably of allopatric origin. This extended to the marine isopod *Jaera albifrons*, the beetles of the genus *Chilocerus*, the rodent genera *Spalax, Thomomys, Ctenomys* and *Perognathus*, and the mole crickets of the genus *Gryllotalpa* (see White, 1978a).

Many of the remaining criticisms made by Futuyma and Mayer (1980) on non-allopatric modes of speciation, were also covered by Key (1981). Unfortunately, the criticisms made were on some occasions arithmetically inappropriate (see John *et al.*, 1983), others were factually incorrect, and much was simply uncorroborated opinion (see introductory quotation for Chapter 6). Indeed, some of the 'evidence' cited by Futuyma and Mayer to support their case against a role for chromosome change in speciation, fell into the classic trap of including chromosome changes which were either polymorphic, or not negatively heterotic (chrommosome addition), and thus not implicated in speciation. For example, they argued that structural chromosomal rearrangements did not decrease karyotype fitness and cited as an example the chromosome forms of *Thomomys bottae*, which have a highly polymorphic karyotype modified by heterochromatic C-block additions (see Section 8.3.1). Other cases that they cited are at least open to question. For example, in *T. talpoides* hybrid zones occurred between the chromosome races, they were narrow and there was very limited chromosomal introgression; in others there were significant levels of reproductive isolation. Heterozygote fitness was decreased by the chromosomal rearrangements which distinguish the races (see Section 8.3.1). It is indeed unfortunate that those who cite Futuyma and Mayer (1980) as evidence against chromosomal speciation have not critically examined the data which these authors have used to support their case.

9.1.3 Chain processes in speciation (White, 1978b)

'The roles of chromosomal rearrangements in speciation should be seen as varied rather than conforming to a single stereotyped model' (White, 1978b, p. 296). Thus, White utilized the extensive cytogenetic database available for *Mus domesticus* in Italy, as the frame for a mechanistic approach to explaining stasipatric speciation. In reality, a completely different and unorthodox model for chromosomal speciation resulted, the characteristics of which appear to be contrary to the stasipatric principles.

The numerous chromosome fusions which distinguish low vagility species such as *Mus domesticus*, were thought to have been accrued by a process of multiple succeeding mutations (Capanna *et al.*, 1977). White utilized this approach to argue that if conditions facilitated the fixation of a single fusion, the same conditions would favour the fixation of additional fusions in a chain process.

White (1978b) was keen to emphasize that his model 'does not invoke founder effects, local extinctions of populations, or invasions of occupied or unoccupied territory by individuals' (p. 295). It was based on a population composed of a large number of demes, inbred to varying degrees, but without permanent barriers to gene flow between them. Within this framework a basic level of genic differentiation was present and this occurred as an area-effect. The chain processes model had the following characteristics (see Fig. 9.3):

1 Chromosomal fusions have an adaptive role in that they reach fixation to protect the integrity of co-adapted gene complexes, which are endangered by introgression from neighbouring populations.

2 The 'area-effects' detected in low vagility species by allozyme electrophoresis reflected the co-adapted gene complexes.

3 A derived chromosomal fusion would spread until it was co-extensive with the area-effect. The fertility effects of this structural rearrangement prevented gene flow from the adjacent population. Subsequent rearrangements would only be established if the first was insufficient to arrest introgression. These would only penetrate that population to the boundary zone of the area-effect. A sufficient number of fusions would be accrued to prevent introgression.

4 When additional fusions were established they would do so within the boundaries of the derived race and extend throughout that race. Alternatively, the additional rearrangement could be established in isolation, in a second population, and alternate hybrid zones would pass through each other until congruence with the area-effect was achieved. White thought that this alternative was an unlikely mechanism.

5 The effect of the structural rearrangement in isolating a local population would be through group selection, although individual selection on structural heterozygotes and homozygotes would operate within the boundary zone.

A number of criticisms were made of the chain processes concept by Capanna *et al.* (1985). First, because of the unusual distribution of *Mus* and their commensal association with man in the valleys of northern Italy, the distribution of chromosomal variants was considered to be peripheral and not central as White suggested. Second, in no way could the distribution of mice be considered as being continuous. The species is distributed in discontinuous demes. The studies by Capanna and his group suggested that chromosomal variants were established in allopatry and that many of the contact zones observed were secondary contacts in which hybridization occurred without any danger of extinction of the new homokaryotype.

There are several other criticisms which may be levelled at the chain processes model. Thus, White's suggestion that additional fusions could only be established

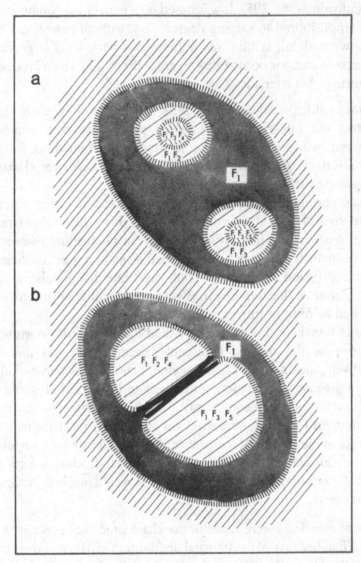

Fig. 9.3. The Italian populations of *Mus domesticus* were used as a basis for the chain processes model advocated by White (1978b).

a. An initial fusion (F1) was followed by two later fusions (F2 and F3) which arose independently within the parental race range. Later, F4 is established within F2, and F5 is established within F3.

b. The different chromosome fusions coalesce within each of the parental cytotypes which are now F1 F2 F4 and F1 F3 F5, and are more strongly genetically isolated from each other than they are from the F1 populations which surround them.

(Redrawn from White, 1978b.)

within the area-effect population 'if the first one is insufficient to completely arrest introgression' (p. 295) is quite unusual. It implies that a population is in control of its evolutionary destiny, being able to manipulate the number of fusions which can be induced and then established. Bickham and Baker (1980) also made this point in their critique of the chain processes model.

The basis of White's chain processes concept, is that the chromosomally derived race which occupies the area-effect zone, is immutable in its distribution within the parental species distribution. The essence of stasipatric speciation was that the 'tension zone', or hybrid zone, was effectively mobile and that the derived form would displace the parental species from that distribution. It would therefore appear that this model and stasipatric speciation were incompatible.

The chain processes model also supposed that genically detectable 'area-effects' characterized the new chromosome race distribution and that attaining chromosomal differentiation was secondary to genic differentiation. The chain process concept was, therefore, not a primary cladogenic process, but a secondary reinforcement of genically distinct populations. Data from a variety of speciating complexes suggest that the most recently differentiated chromosome races are genically identical to their direct ancestor. Indeed, this applies to many chromosome races of *Mus* (see Section 7.3.1.5).

Bickham and Baker (1980) were also critical of White's 'chain processes model'. They attacked it on a number of grounds, some of which are worth reiterating.

First, they suggested that White advocated group selection, which they regard as being unparsimonius and unlikely. This process required extinction to occur, and White specified that this did not occur. Second, White argued that evolution of chromosome morphology in *Mus* was based on karyotypic orthoselection. Bickham and Baker claimed that this process was adaptive and White previously argued that the changes established were non-adaptive. Third, the essence of White's argument was that the chromosome fusions established reduce fertility in heterozygotes. Bickham and Baker cited populations at the Val Bregagli, Grisons and eastern Dolomites which were polymorphic for fusions and they suggested that these were an example of heterozygote fertility not being reduced. This is an interesting point, for in the next sentence they explained that these populations were hybrids. It would appear that Bickham and Baker equate balanced polymorphisms with hybrid zones. Errors of this type are common and are responsible for continued confusion as to the nature of chromosomal rearrangements involved in speciation (see King, 1987, and Sections 5.1.2.2 and 7.1). Fourth, Bickham and Baker also questioned what force made the fusion which was occupying the area-effect, actually stop at the boundary of this zone. This is undoubtedly a valid criticism. Fifth, Bickham and Baker regarded the events observed in the chromosomal evolution of *Mus* in the Italian and Swiss Alps as being rare occurrences, reasoning that *Mus* was basically chromosomally conservative. They argued that the radiation and speciation of *Mus* in these areas was a special case and that it was more plausible that chromosomal evolution in this group was adaptive, in line with their canalization model. However, we now know that chromosomal repatterning in *Mus* is widespread throughout Europe and the Middle East (see Section 7.3.1 for details and Fig. 9.4). There is no evidence that any of the rearrangements that distinguish chromosome races of *Mus* are adaptive.

While many of the points of criticism against the triad hypothesis, stasipatric speciation and chain processes model are arguable, they undoubtedly point to the fact that examples of these internal models for speciation are not only rare, but are in most cases unlikely.

9.2 External models for chromosomal speciation

The models for chromosomal speciation discussed in the previous section, only included those involving speciation within the borders of a pre-existing species, where contact between the populations was maintained. It is important to make the distinction between these and the numerous external models that have been introduced, presumably as a product of the large number of population cytogenetic studies which associate chromosome change with speciation.

The traditionalist view of chromosomal reorganization occurring as a secondary event after the fact of speciation by isolation and genic differentiation has undoubtedly been challenged by White (1978a) and his co-workers. The realization that geographic isolation *per se* does not result in speciation has, in many respects, been the key to changes in our understanding of speciation. Geographically isolated and undifferentiated populations are that and no more. Speciation is a product of attained reproductive isolation between the allopatrically isolated daughter population and its parental population, due to the fixation of differences whether they are chromosomal, or genic. Prevention of gene flow between daughter and parent population is critical, which ever way this is obtained. The issues are: First, did speciation of an allopatric daughter from a parental species result from gradually accrued genetic differences, which were responsible for the eventual reproductive incompatibility and isolation from the parental species when they secondarily contacted? Alternatively, was reproductive isolation achieved by the fixation of a structural chromosomal rearrangement in the daughter population which, because of its deleterious fertility effects when crossed with the parental species, resulted in a barrier to gene flow between the forms? That is, in chromosomal speciation the role of physical isolation was restricted in time to that period

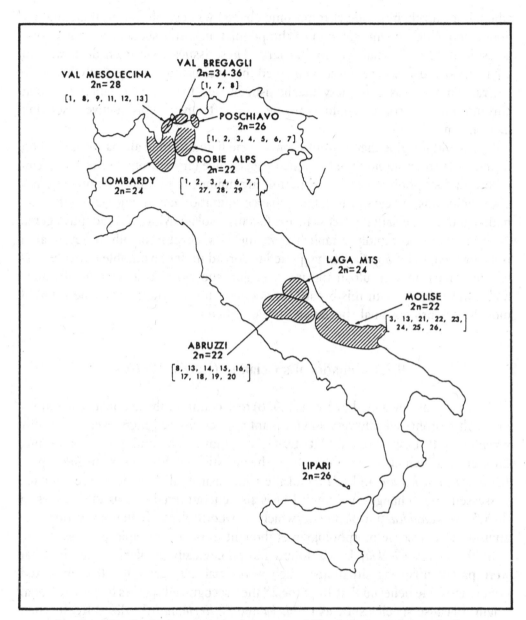

Fig. 9.4. The distribution of chromosome races of *Mus domesticus* in Italy and adjacent areas of Switzerland. The numbers 1 to 29 designate individual fusions found in each of the shaded areas. The surrounding areas were believed to be made up of $2n = 40$ *Mus domesticus* populations. (Redrawn from White, 1978b.)

when a negatively heterotic chromosome change was established in the daughter population. Subsequent isolation of the populations after secondary contact was a result of chromosomal fertility barriers. These issues have been dealt with in Chapters 6 and 7, where it was confirmed that the fixation of negatively heterotic rearrangements was a primary mechanism in this form of speciation and that chromosomal barriers could provide an effective reproductive isolating mechanism.

One might imagine that a chromosomally mediated form of allopatric speciation would be far more rapid than by genic means, since chromosomal differences can enforce a disruption to gene flow, permitting unimpeded differentiation, whereas after thousands of years of isolation, chance migration could halt genic differentiation and the speciation process in an allopatric isolate. Moreover, adaptive genic changes could be rapidly established in the new chromosomally differentiated isolate which, if the population happened to spread under favourable environmental conditions, may result in the invasion and successful colonization of a new niche and habitat. With this broad distinction in mind, a number of the external models for chromosomal speciation are described.

9.2.1 Saltational speciation (Lewis, 1966)

This model was developed by Lewis (1966) to account for those unusual situations in which morphologically very similar plant species showed high levels of sterility when crossed. Lewis argued that sterility was induced by multiple chromosomal differences which distinguish the species he investigated. Species from *Holocarpha*, *Lasthenia*, *Allophyllum* and *Clarkia* of the Compositae, all of which were annuals, possessed these changes, although Lewis also encountered similar differences in the *Chamaelaucoideae* and *Boronieae*, which are woody plants. Saltational reorganization of chromosome morphology was thought to be a very rapid process.

In the genus *Clarkia*, Lewis found morphologically similar species in what were parent/offspring situations which were unable to grow together in mixed populations. He believed that 14 of the 28 then recognized species had a saltational origin. Because species such as *C. biloba* and *C. lingulata* hybridized freely when placed in mixed populations, Lewis reasoned that the elimination of one form would occur, since all progeny of any individual plant would be of hybrid origin and have very low fertility. Thus, the mutual exclusion of individuals of one species in the other species' range would result. A narrow hybrid zone some 3 metres broad occurs between the forms.

In other situations, species such as *C. williamsi* and *C. speciosa* readily hybridized if plants from different areas were crossed, but if individuals from a sympatric area were crossed, hybridization failed. This suggested that a secondary genic response to selection had been established.

Lewis defined the following criteria when detailing the saltational model for chromosomal speciation:

1 Most species of *Clarkia* were distinguished by multiple fixed chromosome differences which have profound effects on fertility. These include both reciprocal translocations and pericentric inversions. Because of the conceptual difficulty of establishing each of these changes in a population in sequential order, Lewis proposed that they were established simultaneously.

2 One or more individuals heterozygous for these multiple changes could have been isolated in a small population where they would not be at a selective disadvantage. If the population did not become extinct, heterozygous hybrids with reduced fertility would be eliminated. If not, and if hybrids which had formed with the parental species had low fertility, the inbred derived population would be able to maintain itself against encroachment from the parental species, even though it was less vigorous and less fecund.

3 Spatial isolation of one, or a very few individuals, was necessary and had to be maintained until the derivative population could eliminate migrants through hybridization, or grow without hybridizing. This isolation could be achieved by dispersal into an unoccupied site, or by survival after others had been eliminated.

4 Catastrophic selection was most likely to lead to saltational speciation. That is, selection operated on individuals which survived in environmental extremes at the limits of that species range. In *Clarkia*, which lives in great aridity with highly seasonal rainfall, all derivative species have a shorter growing season than their direct ancestors. Unusually short growing seasons limited by water have catastrophic affects on populations (Lewis, 1962).

5 To account for multiple simultaneous rearrangements Lewis argued that enforced inbreeding in the daughter population could induce chromosome breakage. This is something which might be expected in founder populations, or survivor populations.

6 Chromosomal reorganization could preserve the adaptive gene complexes emerging from intense inbreeding by preventing recombination.

7 The chromosomal barrier to gene exchange resulting from saltation could be an accidental byproduct of inbreeding, resulting in an immediate barrier to gene flow. If hybridization was frequent, secondary bar-

riers preventing hybridization would eventually develop by selection.

8 New genetic combinations could be established in the course of reconstructing an adaptive genetic system. These could result in rapid divergence from the parental populations. The relationship between saltational species was parent and offspring. Species complexes were geographically arranged in ancestor/descendant lineages and on some occasions the daughter species were self-fertilizing (see Fig. 9.5).

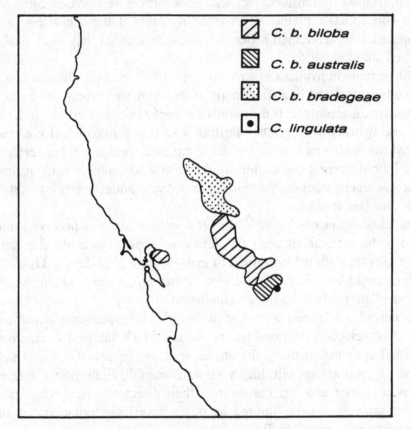

Fig. 9.5. The distribution of members of the *Clarkia biloba* complex of subspecies and the colony of *C. lingulata*. Note the linear distribution of ancestor/descendant related forms. (Redrawn from Lewis and Roberts, 1956.)

9.2.2 Quantum speciation (Grant, 1971)

Grant (1971) defined this 'as the budding off of a new and very different daughter species from a semi-isolated peripheral population of the ancestral species in a cross fertilizing organism'. This model of chromosomal speciation was based on the founder effect (*sensu* Mayr, 1963), and was a rapid and radical means of speciation in terms of both its phenotypic and genotypic effects. The following criteria were defined by Grant, who also thought that *Clarkia* was an example of this form of speciation:

1 An ancestral population was assumed to be polymorphic for translocations, inversions and other rearrangements present in a balanced or transitory form.

2 A daughter colony founded by a few immigrants was spatially isolated and peripherally distributed.

3 The founding individuals were a non-random sample of the gene pool of the ancestral population.

4 Inbreeding and genetic drift resulted in individuals with drastic genotypic and phenotypic effects, most of which were unsuccessful. New chromosomal segmental homozygotes were formed, differing from one another and the ancestral population, and separated by chromosomal sterility barriers.

5 One of the homozygous gene combinations could have a superior adaptive value in the new habitat. Homozygous segregants for the favoured gene combination could have colonized the new habitat. The daughter population differed genetically and structurally from the parental population and adaptive genetic differences were associated with the chromosomal repatterning.

6 The homozygous chromosomal differences provided a sterility barrier to interbreeding with the parental population.

Grant (1985) proposed that the species-specific karyotypes in plants and animals may have an adaptive role, a possibility also suggested by Bickham and Baker (1979). Either the species-specific segmental arrangement determined a set of developmental and physiological responses as a result of gene arrangements. That is, numerous position effects added up to a 'pattern effect'. Or, alternatively, chromosomal rearrangements brought together the linkage of adaptive gene combinations in the ancestral and daughter species, preventing them from being disrupted by crossing-over. White (1978a) suggested that this could explain the relationship between reduced chromosome number and reduced chiasma frequency. White's attitude was that chromosomal rearrangements must not have too great an effect on the fecundity of heterozygotes, or they would not reach fixation. Thus, changes in chiasma localization could favourably modify the reduction in fertility.

One of the difficulties with Grant's model, and indeed with part of White's explanation, is once again the confusion of the role of balanced polymorphism and negatively heterotic rearrangements which act as sterility barriers. One might ask how chromosomal polymorphisms were transformed from balanced rearrangements, without any fertility effects in the parental population, into negatively heterotic rearrangements providing sterility barriers in the daughter population? Nothing could have changed in the pattern of segregation of the rearrangement to allow this transformation to have occurred, for it was based on the 'simple' process of cellular/chromosome movement mechanics. This basic premise poses a severe constraint on the acceptability of Grant's parental population structure.

9.2.3 Parapatric speciation and model 1B: speciation by the founder effect (Bush, 1975)

The two models of speciation described by Bush (1975) are distinguished on the basis of population characteristics, but they are both essentially external models of speciation, one of which (type 1B), need not involve chromosome change. In this respect, it is very similar to quantum speciation and will be dealt with relatively briefly. Interestingly enough, Bush regarded parapatric speciation as being most similar to stasipatric speciation. However, in reality this is an invasive model of speciation involving 'the penetration and exploitation of a new habitat' (p. 348), whereas stasipatric speciation is internal.

9.2.3.1 Speciation by the founder effect: type 1B

1 This model involved the establishment of a new colony by a number of founders, among which there was some degree of inbreeding.
2 The new colony was the product of a population 'flush' (*sensu* Carson, 1975), where selection was rewarded during a rapid population increase.
3 The population was peripherally located and under an intense R-selection regime (high reproductive rate, early sexual maturity, large number of offspring, short life span and low competitive ability).
4 The peripheral population became permanently isolated from the parental species and was close to an unexploited area suitable for invasion.
5 Selection for genic homozygosity occurred.
6 Speciation occurred rapidly.

Bush pointed to the Hawaiian *Drosophila* as an example of this process, where no genic revolution has occurred, or where chromosomal changes had played an insignificant role, with some species being homosequential. However, he then discussed a number of vertebrate groups where chromosomal evolution could be important in speciation by the founder effect. That is 'species of mammals of moderately low vagility that are subsocial or social frequently form small cohesive closed bands or small family groups (primates), harems, herds (some ungulates), or family groups with permanent pair bonds (e.g. foxes). Others are solitary, flightless, or cave-dwelling with homing behaviour similar to that of bats' (Bush, 1975, p. 347).

In populations which became temporarily isolated and in which a dominant member had a major adaptive chromosome change, this rearrangement could be rapidly fixed in the homozygous condition. Rearrangement such as fusions, fissions, whole-arm translocations, or pericentric inversions, provided a rapid means of reorganizing regulatory mechanisms and thus had an enormous impact on developmental changes. They permitted a homozygous population to penetrate and exploit an unsuitable habitat at the periphery of a species range. Bush (1975) equated these rearrangements with Goldschmidt's (1940) 'macromutations'.

Significantly, Bush argued that reproductive isolation arose after the founders had established themselves in a new territory and that chromosomal rearrangements were involved. Some post-mating reproductive isolation occurred due to the deleterious meiotic effects of the rearrangements. However, because hybrids were never produced between parent and daughter populations, there was no direct selection for intrinsic barriers to gene flow. Bush suggested that the resultant species may eventually become sympatric with the parental form.

9.2.3.2 Parapatric speciation

This mode of speciation is distinguished from the type 1B founder model on three criteria. First, no spatial isolation is required during speciation. Second, the organisms involved have exceptionally low vagility. Third, reproductive isolation arises by selection at the same time as penetration of a new habitat. The following characteristics of the processes were described by Bush (1975), some of which appear contradictory:

1 The species involved were R-selected and were found in small semi-isolated peripheral populations where inbreeding occurred. They were low vagility forms and the populations were never strictly isolated from each other.

2 Chromosome rearrangements were frequently responsible for initiating speciation in population isolates and a major effect of these was the alteration of regulatory gene pathways. The release of genetic constraints permitted the new chromosome race to expand into unoccupied territory, for which it was previously poorly adapted.

3 Intrinsic factors that reduced gene flow between parapatric races, such

as chromosomal rearrangements, were strongly selected for. These permitted the population to shift into a new niche and erect a strong
post-mating barrier to gene exchange. Divisive rearrangements such as
fusions and fissions, which had adaptively superior homozygotes and
negatively heterotic heterozygotes, were involved.

4 Although gene flow occurred at the borders between adjacent populations, introgressive genes would penetrate no more than a few feet
and the effects on differentiation were small. Individuals in the parental
and daughter populations were under strong directed selection in the
different habitat.

5 Pre- and post-mating isolation mechanisms arose by selection as individuals penetrated a new habitat.

6 Bush argued that little or no genic differentiation occurred between
populations even after speciation. Despite this general statement, he
later proposed that rapid genic differentiation over short distances could
be associated with parapatric speciation and that these intrinsic factors
could reduce gene flow and thus enhance the fixation of chromosomal
rearrangements.

7 No long-range dispersal took place and diverging populations remained
in constant contact.

Organisms which Bush (1975) suggested had evolved by parapatric speciation
included morabine grasshoppers, rodents such as *Spalax*, *Peromyscus*, pocket-
gophers and shrews, *Sceloporus* lizards, mole crickets, stick insects and *Clarkia*.
That is, most of the species complexes White (1978a) cited as having evolved by
stasipatric speciation!

The key to distinguishing this mode of speciation from allopatric models, at
least in Bush's view, is whether the population in which the new rearrangement
is established was isolated from, or was in constant contact with, the parental
population (parapatry). Nevertheless, there appears to be no means by which a
derived chromosomal rearrangement can reach fixation in a population if it is
other than isolated. Bush recognized this fact in one significant statement: 'It
seems realistic to envision a situation in which fixation of the new arrangement
in only a few generations could occur in a small colony established by a single
fertile female bearing a major chromosome rearrangement located only a few
hundred meters from other members of her population' (p. 350). That is, the
colony is in isolation (internal allopatry), and in reality this model is basically
allopatric. The distinction Bush made between the type 1B founder model of
speciation and parapatric speciation, is more imaginary than real. The major
problem appears to be Bush's understanding of the term 'vagility' and his equating

this with 'mobility'. 'Vagility' is the distance between where an adult specimen of a species reproduces and where the progeny of that individual reproduces. Highly mobile species such as bats can have very low vagility if they all breed in one cave, even though they may fly for hundreds of miles around it. Equally, horses and other ungulates may have similarly low vagility characteristics, but these may be produced by a very different lifestyle. Species do not have to be snails or burrowing rodents of low mobility to be of low vagility.

The point I am making is that in both models the population characteristics are much the same. Whereas the type 1B model catered for larger animals, the parapatric concept catered for small terrestrial organisms. Nevertheless, both groups involve animals which, because of some attribute of their lifestyle, have the capacity to form population isolates which are removed from the gene flow of their parental population, and which can inbreed for prolonged periods, establish novel negatively heterotic chromosomal rearrangements, speciate and colonize new territory, or a new habitat. The distinctions which Bush made between what are both founder-effect-mediated models of speciation are unnecessary and probably unreal.

9.2.4 Alloparapatric speciation (Key, 1968, 1974, 1981)

This model was originally developed by Key (1968, 1974) to describe speciation in the morabine grasshoppers and was subsequently named alloparapatric speciation by Endler (1977). Key (1981) made a meticulous re-examination of this process which undoubtedly has a wider application than that which he gave it. This is essentially a two-stage model of allopatric raciation followed by completion of speciation in hybridization parapatry. That is, a race is established in allopatry in which at least one fixed genic or chromosomal difference distinguishes it from the parental population. The attainment of reproductive isolation at a parapatric tension zone is the second part of this speciation process.

Key pointed out that fixation of a negatively heterotic chromosomal rearrangement could only work if long-term extrinsic isolation occurred, followed by fixation by drift, expansion of the new monomorphic population and renewed contact with the parental population on a front. The difficulty of understanding the means by which negatively heterotic chromosomal rearrangements could reach fixation in populations has plagued evolutionary biologists. Whereas White (1978a) argued for rearrangements with small fitness reductions, Key pointed to the possibility that rearrangements with severe heterozygous inferiority could reach fixation by acute population bottlenecking followed by population 'flushes' (*sensu* Carson, 1975). Key did not believe that meiotic drive was a possibility for forcing these changes to fixation (but see Section 6.4.3), nor was Key convinced of the efficacy of fertility effects which had been attributed to chromosomal rearrangements. He argued that because of the small reduction in fertility of even multiple chromosomal heterozygotes, there has recently been a tendency to postulate the accumulation of a number of genic differences in the allopatric stage of alloparapatric speciation, rather than rely entirely upon subsequent genic differentiation in a chromosomal tension zone. He emphasized that any form of differentiation in allopatry (chromosomal or otherwise), that did not give rise to full reproductive isolation before renewal of contact with the parental population, would have to pass through a parapatric stage if it was to speciate, otherwise the populations would simply fuse together. That is, it would need to incorporate a fixed chromosomal difference with heterozygote inferiority, producing a tension zone behind which further genetic differences could accrue. Nevertheless, Key conceded that a single chromosomal rearrangement had the potential to produce a level of heterozygote inferiority that would rarely result from a single gene mutation. However, he felt that the role of meiotic irregularities had been exaggerated and cited studies by Barton (1980), Futuyma and Mayer (1980), and Moran (1981), to support this view.

It is of some interest that these three studies all reach questionable conclusions. Thus, in *Podisma pedestris* Barton (1980) described a hybrid zone between chromosome races distinguished by an X–autosome fusion, in which F1 hybrids had substantial embryonic inviability and survival problems. These effects were not attributed to the impact of the sex chromosome fusion (which can have a profound fertility effect, see Section 5.1.1.3), but to 'inviability loci'. These loci were not detected by a comprehensive allozyme analysis (Halliday *et al.*, 1983). More recently, Hewitt *et al.* (1989), by using selective matings, were able to demonstrate sperm precedence, homogamy and a biased sex ratio favouring the survival of females. All of these characteristics are suggestive of a perturbation of the sex-determining mechanisms and are reminiscent of similar effects in *Drosophila* and rodent hybrids. Equally, the study by Moran (1981) on F1 hybrids of *Caledia captiva* in which no fertility effects were encountered has been superseded by subsequent studies by Shaw *et al.* (1986) which show F2 inviability (also see p. 216).

Key postulated the following events in the newly created tension zone:

1 A race fixed in allopatry for a chromosomal rearrangement, or one or more major gene mutations, resumed contact with the parent population and formed with it a front along which unrestricted mating between the two populations occurred.

2 If the hybrid progeny were infertile, or if no progeny were produced, reproductive isolation was complete. In this case, the tension zone was

a genetic sink through which neither race could penetrate. If F1 hybrids had reduced fertility, a tension zone would contain hybrids, parental forms and their progeny. The zone would be wider and the width would be related to the vagility of the insects and the degree of fertility depression. Deep introgression would be prevented.

3 Introgression of neutral or positively heterotic chromosome changes or genes would occur, but one could only detect partial introgression not complete introgression.

4 Additional chromosomal rearrangements, or gene mutations, would be arrested at the tension zone, further reducing the fertility of hybrids and differentiating the races.

5 Movement of the zone would occur in the direction of the less adapted race, but this would be slow because of the elimination of new carriers, and in many respects the tension zone would be a 'fossil'.

6 The piling up of genic and chromosomal differences could lead to full reproductive isolation completing the process of alloparapatric speciation, but with the species in parapatric confrontation at the tension zone.

7 Evolution of pre-mating isolating mechanisms would be unlikely because of the gene flow into the hybrid sink. There would be no selective advantage in attaining pre-mating isolating mechanisms.

Undoubtedly, the alloparapatric concept is the most realistic of the models dealt with so far, in terms of the events which lead to the speciation process. It is most similar to Lewis' saltational model, although that author undoubtedly placed a greater significance on the apparent fertility effects of multiple structural rearrangements than did Key (1981).

9.2.5 Chromosomal transilience (Templeton, 1981)

Chromosomal transilience as defined by Templeton (1981) is a model which suffers from many inherent drawbacks; but, then, chromosomal transilience was a 'straw man', which had little chance of being an effective mode of speciation. Perhaps its inclusion here is unnecessary, although it does provide a good example of how information can be presented.

1 Templeton (1981) suggested that under this mode of speciation a chromosomal rearrangement arose that substantially lowered the fitness of the carrier, but had no other major impact on the phenotype of fitness.

2 Drift and inbreeding overcame the normal selection against such rearrangements allowing fixation in an isolated population.

3 That population would acquire a substantial isolation barrier with respect to the ancestral karyotype which could be reinforced to yield speciation.

4 Small population size and inbreeding were necessary to fix the rearrangement.

After so defining chromosomal transilience, Templeton attacked his own concept on the grounds that reinforcement could not operate in a population in which a rearrangement had no significant fitness effect except negative heterosis. This enabled him to conclude (p. 34):

> Thus, even if fixed in a local deme, extinction of one karyotypic form, most like the new and rare one, will be the ultimate outcome, not speciation. Consequently, population genetics theory implies this mode is impossible.

Clearly, if you make a straw man without legs, it can never get off the ground. Templeton (1981) went to some length to argue that while chromosomal change may be associated with speciation and contribute to reproductive isolation, it is not as a primary agent in that process. 'Chromosomally induced negative heterosis is a most unlikely primer of speciation' (p. 36).

9.2.6 Primary chromosomal allopatry (King, 1981, 1984)

This model was originally produced to account for the situation encountered in colonizing radiations of gekkonid lizards of the genera *Phyllodactylus* and *Gehyra* (see Sections 8.2.7, 8.2.9). Each of the ancestor/descendant array of species or chromosome races were distinguished from the next by one or more fixed chromosomal differences. The taxa had a series of linearly arrayed and abutting geographic distributions. The most recently differentiated chromosome race was found at one end of the lineage, whereas the older and morphologically distinct ancestral species occurred at the other end of the lineage. There was no evidence of chromosomal introgression of one chromosome race into the next and each of the chromosomally distinguishable forms had established one or more unequal chromosome fusion, or tandem fusion differences. Similarly aligned complexes have been found in the rodents *Perognathus goldmani* (Patton, 1969) and *Spalax ehrenbergi* (Wahrman *et al.*, 1969a, b) and a large number of other species complexes (see Chapters 7 and 8 for examples).

The name of the model was chosen to make a distinction between this active mode of allopatric chromosomal speciation, where the chromosomal change was primary and the derived chromosome race colonized new territory, from those situations where chromosomal changes were secondary and unrelated to speciation.

The following characteristics of primary chromosomal allopatry were drawn from King (1981, 1984):

1 The species involved in this model were low vagility forms which had the capacity to form isolated demes.

2 The isolated population was genically and morphologically undifferentiated from the parental form.

3 The isolated population underwent the extreme selective gradients common to founder regimes, during which chromosomal mutants were produced.

4 One or more negatively heterotic chromosomal fusion, fission or tandem fusions, adaptive in the homozygous state, reached fixation in this peripheral isolate from the parental species, because of intensive inbreeding.

5 The derived population expanded and secondary contact was made with the parental species. If a hybrid zone was formed chromosomal introgression would be prevented because of the chromosomal fertility effects on hybrids.

6 During favourable environmental periods the chromosomally derived population could expand its distribution into previous uninhabited territory as a colonizing chromosome race. Genetic and morphological differences may be established during this radiation and adapt the speciating form to a new and different habitat type. Alternatively, such changes could reach fixation within population isolates of the derived chromosome race and spread throughout the distribution of this race during periods when numbers were high and all populations were in contact.

7 The hybrid zone acted as a chromosomal barrier to gene flow, preventing introgression of parental genes into the daughter population, thus enforcing isolation and permitting genetic divergence. Equally, genetic differences established in the daughter chromosome race would be prevented from introgressing into the range of the parental species by the hybrid zone, thus enforcing its isolation and permitting its differentiation and eventual speciation.

This relatively rapid means of chromosomal speciation was based on a revolutionary situation experienced during a colonizing radiation. The chromosomally derived daughter species could occupy an extensive geographic distribution adjacent to the parental species. Subsequently, new changes could reach fixation in peripheral populations of the daughter species, which may also take part in colonizing radiations resulting in a phylogenetically related linear series of chromosome races and species. With the passage of time, new genetic and morphological differences are established within the chromosome races and species. The linear sequence of chromosome races or species has morphologically divergent forms at the ancestral end of the distribution, and less divergent and/or distinguishable forms at the recently derived end of the distribution.

In cases where the ancestral forms of such a lineage are quite ancient, or indeed if the lineage itself is ancient, those species which may have long ago attained very large distributions under favourable climatic regimes may occur today as a series of relics in isolated populations. *Gehyra nana*, which is found in north Queensland, in northwestern Australia and in central Australia in completely isolated populations which are chromosomally and electrophoretically indistinguishable (King, 1984; Moritz, 1986), falls into this category. Equally, isolated populations within such distributions may themselves speciate by allopatric differentiation (see King, 1984 and Section 8.2.7).

9.2.7 Speciation by multiple centric fusions which share monobrachial homologies

These models of chromosomal speciation are distinguished by a particular type of chromosomal rearrangement which most of the previously described models included under their broad view of multiple rearrangements. Capanna (1982) introduced a model of chromosomal speciation which was designed to cope with the complex of chromosomal races encountered throughout the distribution of *Mus domesticus*. Here, numerous fusion races are present within the distribution of the species and these range from the ancestral acrocentric $2n = 40$ karyotype to races with $2n = 26$.

It is believed that commensalism with man together with peculiar population characteristics of the mouse, shattered populations into microdemes which were an essential premise to the onset of chromosomal diversification. Capanna suggested that chromosome fusions were accumulated sequentially by multiple succeeding mutations (*sensu* Capanna *et al.*, 1977). The fixation of these rearrangements in isolated demes led to reproductive isolation enforced by these changes. Subsequent gene mutations led to ethological differentiation of populations and eventual fixation of pre-mating isolating mechanisms. Anatomical and physiological selection resulted in the adaptation of chromosome fusion races to separate niches. Genic differentiation was then established in parallel and eventually full speciation was achieved.

Most significantly, Capanna argued that the accumulation of metacentric fusion products reinforced isolation between populations, but that definitive isolation was only reached when metacentric elements with monobrachial homologies appeared, since the impairment such rearrangements have on fertility is so profound that these constitute a basis for divergence. The rearrangements thus isolate the subpopulation possessing it from the parental population from which it arose. Capanna emphasized that the free flow of metacentric chromosomes across a fusion system would meet a barrier where monobrachial homologies were formed. Capanna *et al.* (1985) proposed that alternative fusions could be established in different demes. When these came into contact, double heterozygotes sharing monobrachial homologies were produced.

Baker and Bickham (1986) termed this model 'chromosomal speciation by monobrachial centric fusion' and in an unorthodox statement suggested that Capanna's concept was too intricately associated with the population characteristics of *Mus* to be universal. This model proposed the following sequence of events:

1 Extrinsically isolated populations such as peripatric (*sensu* Mayr, 1982) founder populations were necessary for this mode of speciation.
2 Centric fusions were the only rearrangement involved and these would be established in two different subpopulations. The rearrangements were to have little or no loss of fertility when heterozygous.
3 The isolated subpopulations had to independently become fixed for different centric fusions in which the same acrocentric had to be fused, but in different combinations of arms in each of the subpopulations.
4 If the two subpopulations hybridized, because of the shared combinations of chromosome arms, the interpopulation hybrids would be sterile producing chain multiples at meiosis.
5 However, while each of the subpopulations would be sterile when crossed, because of the monobrachial homologies, Bickham and Baker claimed that the subpopulations would still produce fertile hybrids when crossed to the ancestral acrocentric population.
6 Significantly, and as Capanna (1982) pointed out, balanced gametes made by monobrachial hybrids either reconstruct the hybrid state or rebound to the parental state (see Fig. 9.6).

Capanna's (1982) model for speciation by monobrachial homology and Baker and Bickham's (1986) modification of that model, both provided a mechanism where one of the more powerful negatively heterotic rearrangements could reach fixation in daughter populations. There are numerous examples of species which have diverged from each other by multiple fusion differences and many of these may have monobrachial homologies with all of the sterility side-effects implicated in this phenomenon. Baker and Bickham (1986), made a number of statements and assumptions in their model which are of interest and are undoubtedly open to question.

For example, Baker and Bickham regard speciation by monobrachial homologies as a special case and argue that it may be widespread among mammals. They insist that independent centric fusions in natural populations have low levels of meiotic malsegregation due to the ability of trivalents to segregate normally and are therefore an ineffective barrier to gene flow and are not involved in speciation.

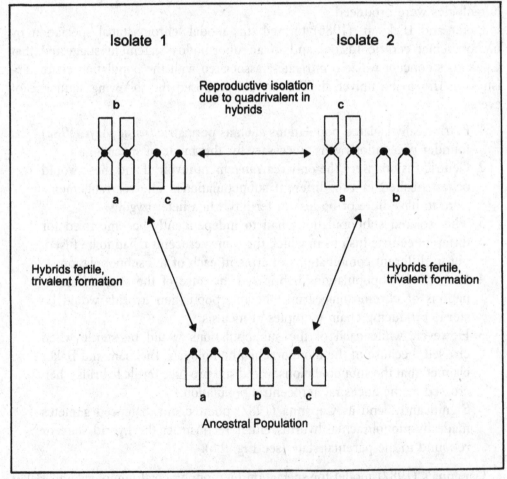

Fig. 9.6. A model proposed by Baker and Bickham (1986) depicting how reproductive isolation could be obtained between geographically isolated populations which had established chromosome fusions involving the same chromosome arms in different combinations. Hybridization would result in the formation of multivalents at meiosis which may have significant fertility or viability effects. (Redrawn from Baker and Bickham, 1986.)

To support this, Baker and Bickham quote Capanna's (1982) figures of 5–18% loss of fertility and argue that even three independent fusions produce less than 25% aneuploidy. These authors appeared to ignore contrary data produced by Gropp *et al.* (1982) which showed a non-disjunction rate of up to 61% in females and up to 28% in males. They failed to consider species which have more than one pair of centric fusions which do not form monobrachial homologies and which have additive fertility effects, just as they failed to consider the impact of X–autosome association. Clearly, many species and chromosome races are distinguished by six to eight independent centric fusions, where additive fertility effects would be very high, even though only trivalents are formed at meiosis. Gropp and Winking (1981) and Capanna *et al.* (1985) pointed out that hybrids heterozygous for multiple independent fusions in *Mus* have fertility impaired by 51 to 77%. In a similar vein, the high incidence of centric fusion polymorphism in some organisms (*Oryzomys*) was cited by Baker and Bickham as evidence that fixed fusion differences have little effect on fertility. Once again, these authors fell into the trap of confusing neutral rearrangements which occur as balanced polymorphisms, with negatively heterotic rearrangements which are implicated in speciation.

Baker and Bickham (1986) claimed that 'One of the more interesting aspects of speciation by monobrachial centric fusions is that reproductive isolation can be achieved between two or more populations that possess derived karyotypes while these incipient species maintain reproductive compatibility with populations possessing the primitive acrocentric karyotypes' (p. 8247). Such a claim is difficult to comprehend. If there were no reproductive constraints on the fusion population interbreeding with the ancestral form, there is no way that otherwise undifferentiated populations could maintain their chromosomal integrity. That is, inbreeding of the derived with the ancestral population would result in the formation of chromosomal polymorphism throughout the distribution of all chromosome races. It is a most unusual situation where morphologically and genically undifferentiated chromosome races, which are distinguished from the direct ancestor by one or more centric fusions, hybridize freely with that ancestor. In most cases, the chromosome races have discrete distributions and, if a contact zone is present between the ancestral and derived state, hybridization is either absent, or present but with very limited chromosomal introgression.

The models of both Capanna, and Baker and Bickham, provide a clear-cut opportunity for chromosomal speciation, if the massive fertility effects caused by malsegregation of the chain multiples are realized. The extensive hybridization studies of Gropp *et al.* (1982) and Capanna *et al.* (1985) indicate that this is so. However, it is worth reflecting on a most similar situation in the evening primroses of the genus *Oenothera* (see Cleland, 1962, for a review). In this case, numerous species and populations are distinguished by whole-arm translocations. Here, because the metacentric chromosome morphology is so uniform, translocation heterozygotes produce zig-zag arrangements in metaphase and so suffer little in the way of non-disjunction or sterility. Cleland (1957) proposed that segmental diversity between populations of *Oenothera* had been attained by extensive hybridization with other forms. This resulted in the formation of modified chromosomes and the transfer of particular chromosomes from one species complex to another, producing marked changes in segmental arrangement. Balanced lethal and self pollination led to the preservation of heterozygosity in the complement with minimal loss of fertility. Thus, balanced lethals have transformed segregating hybrids into true-breeding strains. This most unusual situation is a product of the genetic system and the reproductive mode of the plant involved, and has certain similarities to the model advocated for *Mus domesticus* by Baker and Bickham (1986).

9.2.8 A dual-level model for speciation by pericentric inversion (King, 1991)

A considerable body of evidence suggests that the deleterious meiotic effects of pericentric inversions in F1 hybrids can be overcome by changes in chiasma location and various means of non-homologous pairing (see Section 5.1.1.4). Such overriding mechanisms may render pericentric inversions benign and increase the likelihood of their fixation in population isolates. It has in the past been argued that overriding mechanisms of this type negate the involvement of pericentric inversions as reproductive isolating mechanisms in speciation. However, it is suggested that the involvement of pericentric inversions in speciation should be considered on two levels. First, that by reducing meiotic effects in F1 hybrids, overriding mechanisms facilitate the fixation of pericentric inversions. Second, when contact hybridization occurs between the chromosomally derived and parental populations, second-level effects may be encountered. That is, the recombinational effects of pericentric inversion differences on co-adapted gene complexes (*sensu* Brncic, 1954; Shaw and Coates, 1983), enforce profound sterility barriers in F2 and backcross matings.

The implications of these findings on the possibility of pericentric inversions acting as post-mating isolating mechanisms are significant in four respects. First, in a derived population distinguished by multiple pericentric inversions, the question arises of how such changes could ever reach fixation when they provide second-level sterility barriers in later crosses (i.e. F2 and backcrosses). The most reasonable answer is that the individual inversions must have been established sequentially, so that their effects on the fertility of the population were minimized. That is, it is assumed that the second-level effects of fertility are less profound with single inversions. If the inversions were established simultaneously rather than sequentially, it is probable that the second-level effects would achieve the same level of sterility encountered in F2 and backcrosses and consequently none could reach fixation. Nevertheless, the possibility of multiple inversions simultaneously reaching fixation in populations undergoing particular dynamic changes cannot be ruled out (Rouhani and Barton, 1987). Second, it is probable that the second-level recombination effects described by Shaw *et al.* (1986) in their hybridization studies of *Caledia captiva* are a product of the disruption of numerous pericentric inversions acting in concert. Presumably, genomes with one or a few pericentric inversions established would provide less profound second-level fertility effects in hybrids. Third, the fact that pericentric inversions appear to have no impact on fertility in F1 hybrids and profound effects in later crosses, suggests that those studies which have discounted chromosomal fertility on the basis of F1 results alone are suspect. The implication is that an analogous second-level effect may apply to other types of rearrangements. Fourth, this dual-level situation provides a new perspective on the role played by pericentric inversions as post-mating isolating mechanisms in speciation. The following general model is immediately apparent:

1   If level one overriding mechanisms are present, pericentric inversions established in an isolated population are not selected against because of their heterozygosity and can thus reach fixation due to stochastic processes (*sensu* Lande, 1984).

2   Pericentric inversions established in a concerted fashion could reach fixation sequentially in the daughter population. Presumably, the second-level fertility effects are individually small, when one or two changes are established at a time.

3   When contact is made between the parental and daughter populations, which are distinguished by numerous fixed inversion differences, viable F1 hybrids could form in the contact zone due to the overriding of first-level mechanisms.

4   Further inbreeding and backcrossing would lead to profound sterility barriers, due to the second-level effects; the result would be a narrow hybrid zone with an absence of gene flow across it. The magnitude of second-level effects would be proportional to the number of fixed inversion differences established between parental and daughter populations.

Thus, the overriding of first-level effects facilitated the fixation of neutral pericentric inversions, whereas second-level effects allowed these to operate as major post-mating isolating mechanisms. That is, the overriding mechanisms which rendered pericentric inversion heterozygotes as benign in F1 hybrids, contributed to the increased likelihood of the inversions' eventual fixation. It was only when secondary contact was made between populations with multiple fixed inversions that hybrids which were formed in the contact zone could display the profound second-level fertility effects.

For this model to be general, two basic assumptions are necessary. First, that the second-level fertility effects described by Vetukhiv (1953) and Brncic (1954) in *Drosophila pseudoobscura* and by Shaw and Coates (1983) in *Caledia captiva* are the same, are also present in other organisms, and are a product of pericentric inversion heterozygosity. Second, that the second-level effects of individual inversions are less deleterious than those associated with multiple inversions, and that they may be additive. Clearly, this model for chromosomal speciation is not universal since the existence of balanced polymorphisms involving multiple pericentric inversions (see John and King, 1977a, b, 1980, for examples) suggests that some inversions lack second-level effects. Nevertheless, those species in which second-level inversion effects are present have a profound post-mating isolating mechanism which could be implicated in chromosomal speciation.

9.3 Speciation by hybridization

9.3.1 Hybrid recombination (Templeton, 1981)

This unusual mode of speciation utilizes the revolutionary reorganization of the genome in hybrid individuals and the potential of that reorganization to produce evolutionary novelty and result in speciation. Templeton (1981) proposed the following sequence of events:

1 Hybridization between species occurred, and this was followed by inbreeding and hybrid breakdown due to genetic or structural incompatibility.
2 Selection favoured those surviving F2 and later individuals with the highest viability and fertility.
3 A new recombinant phenotype may be stabilized as a product of selection if it was reproductively isolated from parental species, otherwise it would be overcome by gene flow.
4 a. Reproductive isolation could be enhanced in the recombinant by karyotypic evolution. That is, structural differences between distinct

parental karyotypes could be further modified by recombination, or additional rearrangements may be induced. Inbreeding allowed for the fixation of such rearrangements.

b. Reproductive isolation could also be produced by mutator activity induced by the action of hybridization (hybrid dysgenesis), resulting in high levels of chromosome breakage and genic divergence (see Sections 6.1 to 6.3 and 10.2.2 for detail).

5 If a subdivided population structure was present, hybrid dysgenic events could lead to karyotypic divergence and speciation.

6 Once a stabilized recombinant form was produced, it would either coexist with the parental species, or expand into a new area creating a separate distribution.

Templeton (1981) then proposed that hybrid recombination can occur through pre-mating barriers alone and that post-mating barriers between recombinants and the parents, or between parents, are not necessary. This conclusion is a little unusual, for the model is based on structural or genetic isolation.

There is now a substantial amount of evidence which suggests that hybrid dysgenic-like effects can occur in interspecific hybrids and thus increase the incidence of genic variation (Barton *et al.*, 1983) and chromosome breakage and reorganization (Sections 6.1 to 6.3 and 10.2.2). Clearly, the difficulty of establishing this variation in an isolated recombinant population in a hybrid zone is a major obstacle to Templeton's model. There is also the difficulty of making a distinction between speciation involving negatively herotic rearrangements in isolated recombinant populations in hybrid zones and those in peripherally isolated founding populations. Both forms of derived population can colonize new territory, or displace the parental species distribution, and they can give all the appearance of being the same.

9.3.2 Polyploidy, parthenogenesis and hybridogenesis

The most clear-cut examples of speciation involving chromosomal mechanisms are the processes of polyploidy, parthenogenesis and hybridogenesis. All may result from the act of hybridization and are products of meiotic misadventure, the isolation of parental genomes and the duplication of genomes. The reason why such processes are readily accepted as modes of speciation is because the act of speciation can often be recapitulated by artificially hybridizing the parental species. Thus, the classical experiments of Karpechenko (1927) determined that the fully fertile tetraploid *Raphanobrassica* was an allopolyploid species produced by the unreduced diploid gametes of a sterile diploid hybrid between the radish and cabbage. Sterile triploid hybrids resulted when it was crossed back to either of the parental species. Equally, one of the great coups in White's analysis of the parthenogenetic grasshoppers of the *Warramaba virgo* complex, was the formation of 'synthetic' diploid *virgo* embryos by hybridizing the parental species P169 and P196 in the laboratory (White *et al.*, 1977). Moreover, by crossing these parthenogenetic females to one of the parental species male triploid hybrids could be produced.

Polyploidy, parthenogenesis and hybridogenesis have had a significant evolutionary impact, and despite the fact that they are beyond the scope of this book it is worth making note of the contribution that each of these processes has made. The reader is referred to White (1973, 1978a) and Bell (1982) for a broader perspective on these topics.

Polyploidy has been the dominant mode of speciation in plants and it is estimated that between 70 and 80% of all angiosperms have had a polyploid origin. Polyploidy arises when the diploid chromosome set is doubled and this can result in the formation of tetraploids. Ploidy levels can be further enhanced by duplication of these diploid genomes (hexaploid, octaploid), or by the acquisition of haploid genomes (triploid, pentaploid). A distinction is made between the mode of origin of polyploids. If both chromosome sets in a tetraploid have been derived from a single species, they are referred to as autopolyploids, whereas if they have originated from two or more species and have a hybrid origin they are allopolyploid. It should be pointed out that this is a substantial oversimplification and distinctions can be far more complex. The most important aspect of polyploid species survival is meiotic segregation. Thus, allopolyploids tend to form bivalents at meiosis between elements of the diploid complement of each parental genome and segregate normally. However, autopolyploids form multivalents at meiosis, since there are multiple copies of the same chromosome from the one genome. Because of segregational difficulties, autopolyploids face considerably impaired fertility. However, autopolyploid genomes tend to diploidize over time, a factor which often confuses our appreciation of the origin of polyploids.

Polyploidy appears to be a highly adaptive means of speciation, and alterations in cell size, form, structure and breeding system are associated with changes in ploidy. Thus, diploid plants are generally outcrossing, whereas polyploids are self-fertilizing. Moreover, selfing is commonly associated with allopolyploidy, whereas autopolyploids (a rare plant group) are outcrossing species. The evolutionary success of polyploidy in plants has variously been attributed to gene duplication, fixed heterozygosity and the reduced effects of inbreeding depression (Barrett, 1989).

In animals, polyploidy has played a far less significant role as an evolutionary mechanism and this may be attributed to the disruptive effects of change in ploidy on the chromosomal sex-determining mechanisms. Polyploidy has been commonly encountered in amphibians and the great majority appear to be allopoly-ploids (see King, 1990, for a review). These generally lack recognizable sex chromosomes. Many invertebrate groups are polyploid (White, 1973, 1978a). Both parthenogenesis and hybridogenetic speciation often display polyploidy as a product of their differentiation. These unusual modes of speciation are restricted to plants, invertebrates or lower vertebrates such as fish or reptiles.

Hybridogenetic speciation is known from fish of the genus *Poeciliopsis* and has been responsible for speciation of *Rana esculenta*, the European edible frog, which results from the hybridization of *R. lessonae* and *R. ridibunda*. A pre-meiotic exclusion of the *R. lessonae* genome in males and females is followed by endored-uplication of the *R. ridibunda* genome, which is then followed by normal meiosis. That is, hybrids produce gametes of one parental type and breed with members of the other parental type. Every generation of frogs produced by such a mechanism is hybrid in origin. A series of undescribed members of this complex have been found in the Balkans, Spain and Italy. These have also originated by hybrid-ogenesis, although the population in the Balkans fails to reproduce (Bullini, 1985).

The most common form of parthenogenesis, at least in vertebrates, involves hybridization followed by meiotic changes which facilitate the transmission of the hybrid genome. In automictic thelytoky, females give rise to female progeny with-out fertilization, due to changes in the meiotic cycle which allow the reconstitution of the diploid genome. Several forms of reconstitution have been encountered; these include a pre-meiotic doubling of the complement followed by normal meiosis, or normal meiosis followed by fusion of the egg nuclei in pairs to reconsti-tute diploidy, or meiosis which lacks the second meiotic division.

Polyploidy is common in parthenogenesis and this may result from the hybridiz-ation of a diploid parthenogen with one of the parental males to produce a triploid. In the case of apomictic thelytoky, meiosis is suppressed altogether and eggs go through simple mitosis. Thus, all barriers to polyploidy are removed.

9.4 Concluding remarks

In this chapter I have presented a variety of models for chromosomal speciation along with many of the problems which have been raised about them. These models have been proposed to variously account for chromosomal speciation within the parental species distribution, within isolated populations and within hybrid zones. This is not intended to be a comprehensive description of all models produced, but rather a description of major genres. A number of models have been omitted because they simply duplicate existing concepts. Thus, White's

(1982) invasive speciation is most similar to primary chromosomal allopatry (King, 1981), whereas Mayr's (1982b) peripatric speciation is similar to quantum speciation (Grant, 1963).

A number of the models are particularly specific, involving certain types of rearrangements in unusual circumstances (Wallace, 1966; Baker and Bickham, 1986). This is not surprising when the variety of rearrangements and complexity of organismal lifestyles are considered. Nevertheless, all of the models invoke the primacy of chromosome change in speciation. The internal models have little to support them and this can be seen from the onerous criticisms which have been made. The great body of evidence supports the external modes of chromosomal speciation. Most of these models argue that negatively heterotic chromosome differences have been fixed in founding populations which subsequently make contact with the parental species forming a hybrid zone which is an impermeable barrier to gene flow. Some of the models suggest that the enforced isolation from gene flow permits the rapid genetic and morphological differentiation of the derived form. Thus, when this new species colonizes new territory during times of population expansion, it can readily adapt to the new environment without the constraint of gene flow emanating from the parental species.

10
Molecular mechanisms and modes of speciation

Molecular drive is not an alternative to the evolutionary processes of natural selection and genetic drift in the fixation of mutational variants of single copy genes (. . .); it constitutes a third mode of evolution that would of necessity be subject to natural selection and genetic drift in an interesting variety of real biological situations

(Dover, 1982, p. 116).

The enormous drive toward molecular research which has occurred over the last 20 years, has provided an array of new processes and raised new possibilities for mechanisms involved in speciation. The power of the techniques lies with the incredible resolution which they provide. However, the fact that sequences and variation associated with these can be catalogued, does not provide an answer to the evolutionary significance of this variation. It is critical to our successful use of the molecular approach to be able to integrate these findings with what we already know about the structure and function of chromatin, and what we think we know about the genetic basis for speciation. There is little value in producing molecular mechanisms as models for speciation which are simply unworkable from a chromosomal or genetical perspective, despite what they may contribute to our understanding of the evolution of the genome.

Two questions have been addressed in this chapter; first, can molecular turn-over mechanisms provide processes by which morphologically, chromosomally and genetically indistinguishable allopatric populations establish post-mating isolating mechanisms which are sufficiently powerful to enable speciation to occur? Second, have molecular mechanisms been detected which can enhance the formation and fixation of chromosomal rearrangements, or genetic divergence, thus providing additional support for existing modes of speciation?

To attempt to answer these questions, those molecular processes which may act as significant evolutionary mechanisms are described, in conjunction with possible modes of speciation which have been derived from them.

10.1 Concerted evolution: the pattern

The great majority of coding genes occur as unique DNA sequences. However, multiple copies of particular DNA sequence units, or gene families, also occur and these may have a complex genomic distribution. This applies equally to repetitive sequence satellite DNA which is often visualized as heterochromatin, or to major coding loci, such as the site of 18s and 28s rDNA in the nucleolus organizing region, or regions (NOR).

Multigene families may be distributed in a single region where all sequences are arrayed in a functional unit. This is commonly seen in the NOR of many vertebrate species. Indeed, many hundreds of amphibians have a single and often heavily amplified NOR on one chromosome pair, although this may vary in position between higher taxonomic groups (King, 1990). However, in some plethodontid salamanders numerous sites are present, with *Plethodon glutinosus*, *P. vehiculum* and *P. vandykei* having three sites, and *P. elongatus* four sites (Macgregor and Sherwood, 1979). Such dispersed groups of repeated gene families may be found on many chromosomes, or different sites on the one chromosome. The location of these sequences may be in similar positions within the chromosome arms of some or all of the chromosomes involved, although this is not necessarily so. Moreover, in some instances the degree of amplification of a sequence may be similar at all sites, but not at others. The overall picture is of a unified and concerted evolutionary pattern. This is particularly well demonstrated in the distribution of heterochromatin, where particular sequences may have been amplified to a similar degree, and may be found in a similar position, on most if not all chromosomes of the complement.

Concerted evolutionary patterns have been commonly encountered in the analysis of heterochromatin distribution in amphibians (see King, 1990, 1991). These patterns have been revealed by the use of base-pair-specific fluorochromes which bind to the heterochromatin. Particular bands of this heterochromatin have been detected in similar sites, sharing a similar degree of amplification on all chromosomes in the complement. On some occasions, groups of chromosomes which are a certain size within the karyotype may also share a common type of heterochromatin. The instances where molecular analyses have been made on amphibians support these conclusions. Thus, the AT-rich sites which fluoresce intensively with quinacrine in *Xenopus muelleri* are also the sites to which an isolated DNA satellite was located with *in situ* hybridization by Pardue (1974). Similarly, Macgregor and Kezer (1971) isolated a highly repeated GC-rich satellite which was localized at all centromeres in *Plethodon cinereus*. Indeed, Mizuno and Macgregor (1974) found that this same satellite was common to other *Plethodon* species and that it clustered on blocks of C-positive heterochromatin.

Salamanders from the genus *Triturus* display a most similar concerted evolutionary pattern. Barsacchi-Pilone *et al.* (1986) isolated a DNA satellite (Sat G) from *Triturus vulgaris meridionalis* and hybridized this AT-rich fraction of highly repeated DNA to *T.v. meridionalis*, *T. italicus*, *T. apuanus*, *T. carnifex* and *T. marmoratus*. The sequences hybridized to specific terminal and subterminal sites as well as to the multiple bands of paracentromeric heterochromatin, most of which share a similar degree of amplification, and which are found on all chromosomes in the complement. Indeed, of the five satellite families found in *Triturus*, three of these are located at the centromeric or adjacent paracentromeric C-bands, and are present in all species examined (Baldwin and Macgregor, 1985; Batistoni *et al.*, 1986).

Extensive studies on the *Atractomorpha similis* complex of short-horned grasshoppers (John and King, 1983), indicate that the six recognizable cytotypes differ by complex C-block polymorphisms for multiple variants on every chromosome in the complement. Certain cytotypes have a much greater level of block amplification than do others. John *et al.* (1986) cloned a 537 bp *Taq*I restriction fragment from the satellite 1 DNA of *A. similis* and *in situ* hybridized tritium-labelled RNA copies of this probe, or a related *Sau3A* fragment obtained from the same source, on to the genome. Both probes were uniformly represented throughout all distal C-bands in individuals from three cytotypes which were compared. The probes did not hybridize to the paracentromeric C-bands on any chromosomes. The complex heterochromatin polymorphism involving distal C-bands can be accounted for by the amplification of Sat 1 DNA and its concerted spread to similar positions on all chromosomes in the complement (see Fig. 10.1). That this is a derived condition can be judged from the absence of similar C-bands in the closest relative of this species and in most other pyrgomorphine grasshoppers studied (King and John, 1980; John and King, 1983).

A quite different form of concerted evolution is observed in some species and this may be found in the same genome where the more defined pattern described above is in evidence. That is, particular sequences may be uniformally dispersed over all chromosomes in the complement. For example, in the grasshopper *Caledia captiva* Arnold and Shaw (1985) *in situ* hybridized ^{3}H-RNA derived from 168 base-pair highly repeated DNA sequences isolated from the southeastern Australian representatives of this taxon. The satellite is present in interstitial and telomeric constitutive heterochromatin blocks in this and the Moreton taxon. It is only found on a few C-bands in the Torresian taxon. A second 144 bp sequence family which is AT rich, occurs on procentric blocks of pair 2 to 7, 9 and 10 of this taxon. In contrast, Arnold (1986) found an additional highly repeated 185 bp sequence family, which had a dispersed distribution throughout both the euchromatic and heterochromatic areas of the genome. This sequence family showed a

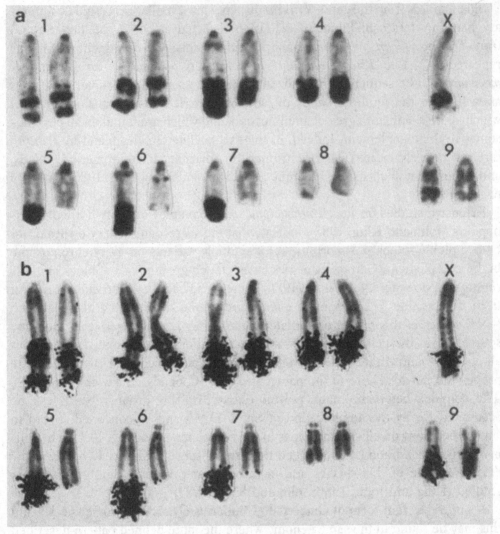

Fig. 10.1. a. A C-banded male mitotic cell from the grasshopper *Atractomorpha similis* (Dee Why cytotype).

b. An *in situ* hybridized male mitotic cell from the same species and population showing the accumulation of a satellite 1 DNA probe in all of the terminal and most interstitial C-bands. The probe was a 537 bp *Taq*I restriction fragment, which had been cloned from the satellite 1 DNA. This provides an excellent example of a concerted pattern of chromosomal evolution, where the one sequence is distributed in specific sites throughout the genome. (From John *et al.*, 1986.)

greater level of sequence variation, as well as a more randomized distribution than the 168 and 144 bp sequence families. This might be regarded as an extreme form of concerted evolution involving the homogenization of this sequence family throughout the genome.

The plethodontid salamanders also provide an example of this homogenized pattern of concerted evolution. IC DNA content has been estimated to range from 18.2 picograms per nucleus (pc/n) for *Plethodon nettingi shenandoah* to 69.3 pc/n for *P. vandykei* (Mizuno and Macgregor, 1974). Accompanying this great variation in DNA content, is a corresponding change in chromosome size. However, chromosome morphology in terms of arm ratios (a reflection of centromere position), and percentage size of each element in the complement, has remained the same (Mizuno and Macgregor, 1974). C-banding of these chromosomes revealed that there has not been any change in the distribution of heterochromatin in the genome. However, molecular analyses detected major changes in the proportions of middle repeated sequence DNA between species. Mizuno and Macgregor (1974) proposed that multiple amplifications of these sequences, followed by the dispersion of the amplified sequences throughout the genome, had led to the retention of the basic karyomorph.

10.1.1 Molecular drive: the process

Molecular drive is not regarded as an alternative to the evolutionary process of natural selection and genetic drift in the fixation of mutational variants of single copy genes. Rather, it constitutes a third mode of evolution which is subject to both natural selection and genetic drift. It is a mechanistic synthesis resulting from the splurge of molecular research which has dominated the 1970s and 1980s and which will continue to dominate our understanding of evolutionary processes. Molecular drive has been defined as the fixation of variants in a population as a consequence of stochastic and directional processes of family turnover (Dover, 1982).

The nuclear genome of eukaryotes is subjected to continued turnover by unequal chromosome exchange, gene conversion and DNA transposition. This was the view held by Dover (1982) and it is of some interest that the number of turnover mechanisms was increased by the inclusion of replication slippage and RNA-mediated transfer of genetic information (Dover, 1988), whereas John and Miklos (1988) also included amplification. Amplification, like replication slippage, can increase the number of copies of a given sequence and in this way enhance turnover, but such processes cannot play a role in the transposition of these sequences from one chromosome to another and both are thus subsidiary mechanisms.

The basic evidence for genome turnover comes from the abundance and distribution of multiple copy gene families and of non-coding DNA sequences within the genome. The major processes which provide for genome turnover (see below) and the redistribution of multigene families form the basic mechanism of molecular drive. The directional component of molecular drive is thought to have arisen from persistent non-random exchanges of sequence information within and between chromosomes by duplicative transposition and the biased direction of gene conversion. Molecular drive remains as an independent process to both genetic drift and natural selection for it is the product of numerous sequence exchanges between and within chromosomes resulting in non-Mendelian patterns of inheritance (Dover, 1982, 1988).

10.1.1.1 Unequal chromatid exchange

Since crossing-over can occur between two chromatids (as it can between two chromosomes), unequal chromosome/chromatid exchange (between chromosomes of different arm lengths) results in one element gaining material and the other losing it (see Fig. 10.2). Continued crossing-over events may lead to the accumulation of tandem arrays of sequences. Unequal crossing-over between chromosomes ensures that a mutant copy can spread to a number of chromosomes in the complement. After sexual reproduction, these chromosomes are given to different individuals, thus ensuring that homogenization continues. Unequal crossing-over is one of the turnover mechanisms that underpin molecular drive (Dover, 1988).

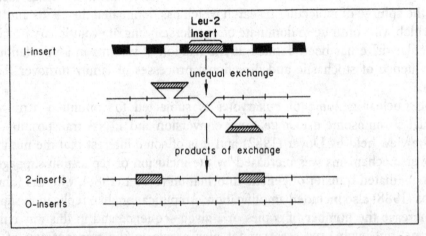

Fig. 10.2. The consequences of unequal exchange following the insertion of a *Leu-2* locus from chromosome 3 of yeast into the yeast rDNA locus on chromosome 12 (Szostak and Wu, 1980). (Redrawn from John and Miklos, 1988.)

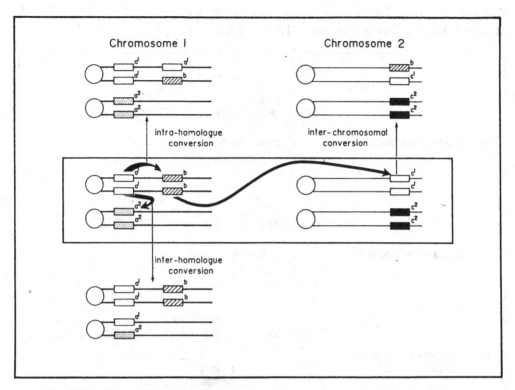

Fig. 10.3. The possible modes of gene conversion between and within chromosomes. (Redrawn from John and Miklos, 1988.)

10.1.1.2 Gene conversion

This process involves the non-reciprocal transfer of a sequence between copies of a gene. It is the product of a mismatch repair in a heteroduplex DNA region resulting in the conversion of function of a particular sequence and leads to the production of two identical DNA copies (Holliday, 1964). This process does not alter the copy number of a sequence family, but results in the homogeneity of pairs of closely linked genes. Gene conversion can occur within chromosomes, between chromosomes which are homologous and between non-homologous chromosomes (see Fig. 10.3). Biased gene conversion in one direction, together with the duplication and transposition of sequences, provides the directional component to molecular drive and results in the persistent non-random exchange of sequences within and between chromosomes (Dover, 1982, 1988).

10.1.1.3 Transposition

This process involves the movement of lengths of DNA from one position to another in the genome. Mobile elements are generally responsible for transpositions (Fig. 10.4). Duplicative transpositions involve the duplication of a length of DNA followed by its insertion (Dover, 1988).

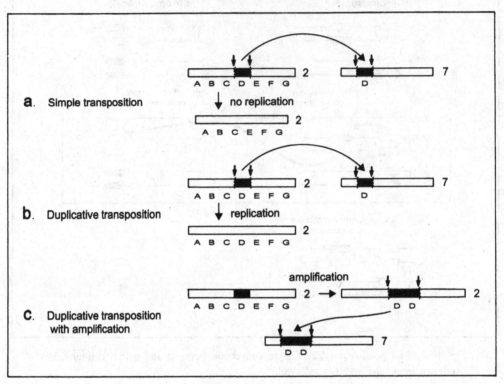

Fig. 10.4. The mechanisms of genome transposition.

10.1.1.4 Replication slippage
It is thought that this process involves the gain, or loss, of small segments of DNA produced by the repair process which occurs after the two strands of the DNA helix slip and mispair, creating a gap and corresponding loop. Repair processes may excise the loop. This process only occurs within a chromatid double helix and cannot lead to the spread of the addition. Like amplification, this can only be regarded as a subsidiary mechanism for it does not induce movement of sequences between chromosomes (Fig. 10.5).

10.1.1.5 RNA-mediated transfer of genetic information
Dover (1988) suggested that the presence of reverse transcriptase, transcribes RNA into its complementary DNA, and this cDNA is reinserted into the genome at a series of different loci. This mechanism can lead to a rapid accumulation of repetitive elements and thus greatly increase differences in copy number for a repetitive family even between closely related species (Fig. 10.6).

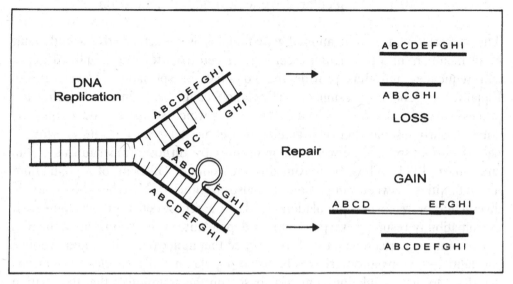

Fig. 10.5. The mechanisms of replication slippage.

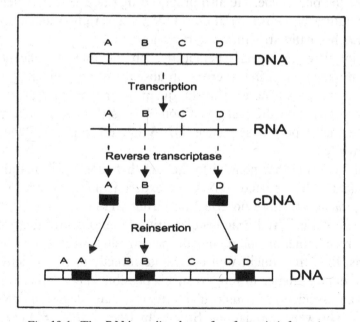

Fig. 10.6. The RNA-mediated transfer of genetic information.

10.1.2 Molecular drive as a mode of speciation

The concerted pattern of fixation of gene families by molecular drive would result in all members of a population receiving the redistributed and amplified genes and sequences and thus provide the potential for speciation (Dover, 1982). Equally, the genetic cohesion of a population could be maintained during the process of molecular drive because the rate at which sexual reproduction randomizes chromosomes between generations is greater than the rate at which a new mutation spreads between chromosomes by turnover mechanisms (Ohta and Dover, 1984). Thus, by maximizing the genomic cohesion of a population, discontinuities between populations are also maximized. Whether this genomic divergence between two populations leads to a reproductive or developmental incompatibility resulting in speciation, is dependent on the nature of sequences undergoing concerted changes. Dover argued that molecular drive could produce substantial phenotypic differences between populations of a species without any ill-effects to that population. This was based on the assumption that the fixation of variants in multigene families for histones, DNA, globins or immunoglobins would effect the phenotype. He also proposed that non-genic sequence families may in the future be shown to affect rDNA transcription and transcript processing, and chromatin structure and behaviour.

Two major flaws in Dover's perspective on the role of molecular drive in speciation are apparent. First, there is an absence of any evidence to support a relationship between an increase in the quantity, or type, of repeated sequence DNA and an effect on the pairing of homologues in meiosis. Second, there is no evidence that repeated sequence DNA has an impact on the fertility of structural hybrids.

John and Miklos (1988) pointed to the fact that substantial quantitative alterations in the highly repeated DNAs of experimentally constructed stocks of *Drosophila melanogaster* have no effect on fertility. However, they suggested that if the mechanisms involved in molecular drive were to contribute to a failure of pairing, or recombination, or if meiotic pairing sites were subject to turnover mechanisms, then they might well contribute directly to reproductive isolation. There is as yet no evidence to support such a possibility, nor is there any evidence to support an association of molecular drive with any degree of morphological differentiation between populations. Studies by Rees *et al.* (1982) examined pairing and synaptonemal complex formation in hybrids between *Lolium* and *Festuca* species and demonstrated that there was no encumbrance to pairing, or synaptonemal complex formation, associated with huge differences in chromosome size and DNA content between the species hybridized.

As with most forms of genetical or molecular differentiation, a distinction cannot be made between those sequence changes which had been established before the speciation process and those which occurred subsequent to this as phyletic evolution. Thus, the question remains as to whether those differences attributed to the action of molecular drive are the primary cause of speciation, or whether they are but a consequence of reproductive isolation (Rose and Doolittle, 1983).

10.2 Genomic transposition

While this process is a subcategory of the general genome turnover mechanisms, its implications for our understanding of chromosomal evolution and speciation are profound. The phenotypic effects of mobile elements and their activation by genomic stress, together with their impact on structural rearrangements and fertility, provide a new perspective on speciation.

10.2.1 Selfish DNA

The great majority of genes found within the eukaryote genome and which encode for proteins are single copies of unique DNA sequences. In certain situations where high levels of production of structural RNA occur, such as the ribosomal RNA loci, multiple copies of this unique sequence DNA are present.

The remainder of the genome is made up of repetitive DNA sequences which appear to have little or no known function and may be irrelevant to developmental or evolutionary processes. This broad group includes short segments of non-coding repetitive DNA found within introns, around genes, or as the spacer sequences between genes. Tandem arrays of highly repeated sequences are also present, but by far the largest component are the middle-repetitive DNA sequences which occur in dispersed or clustered form and may make up 30% of the total genomic DNA (Doolittle and Sapienza, 1980). Repetitive DNA is generally non-transcribed and is often located in heterochromatin.

All of these repeated sequence DNAs are included by Orgel and Crick (1980) under their broad definition of selfish DNA. That is, DNA which makes no specific contribution to the phenotype and which arises when a DNA sequence spreads by forming additional copies of itself within the genome. At the same time, and in the same volume of *Nature*, Doolittle and Sapienza (1980) also provided a perspective on selfish DNA. It is of some interest that the main consideration of these authors was also the middle-repetitive DNA. This type of DNA is considerably more conservative in sequence homology than unique sequence DNA. Doolittle and Sapienza suggested that middle-repetitive DNA comprised too large a fraction of most eukaryote genomes to be kept homogeneous by Darwinian selection on the organismal phenotype and thought it was likely that it retained its conserved sequence integrity by transposition. Consequently, middle-repetitive DNAs in general could be regarded as transposable elements, or degenerate descendants of such elements which had lost their transposability. Doolittle and Sapienza (1980) reasoned that if this was so, the observed range of sequence divergence within families and changes in middle-repetitive DNA family sequence and abundance, were the result of non-phenotype selection within genomes. Thus, no cellular function at all was required to explain either the behaviour, or the persistence, of middle-repetitive sequences as a class of DNA.

Once established within a genome, useless DNA sequences which were perpetuated by transposability, or any other genomic turnover mechanisms, would be difficult to remove from the genome and would appear to have a long life expectancy. Orgel and Crick (1980) suggested that a kind of molecular struggle for existence within the DNA of the chromosomes would occur, based on the intragenomic spread of selfish sequences and phenotypic selection against an excessive amount of DNA. Indeed, the existence of selfish DNA was possible because DNA is a readily replicated molecule and it occurs in an environment in which DNA replication is a necessity: 'It thus has the opportunity of subverting these essential mechanisms to its own purpose' (p. 606).

In a subsequent reappraisal of the concept of selfish DNA, Orgel et al. (1980) made a distinction between those sequences which were repeated in tandem arrays and those which were dispersed throughout the genome and occurred as a few copies in any one place. They suggested that dispersed sequences, if they had any specific function at all, were likely to be involved in the control of gene expression, whereas the tandemly repeated type were likely to influence chromosome mechanics. Indeed, this view is in line with Orgel and Crick's (1980) proposal that the 'host organism' occasionally found some use for particular selfish DNA sequences, especially if they are widely distributed over the genome. In this regard, Davidson and Britten (1979) had proposed that middle-repetitive DNAs had a significant regulatory function and that quantitative and qualitative changes in this DNA could provide an adaptive advantage to a species. There appears to be no substantive evidence to support this view.

10.2.2 Hybrid dysgenesis

Hybrid dysgenesis refers to a most unusual series of phenomena which are observed when particular strains of *Drosophila melanogaster* are crossed in the laboratory. When hybrid dysgenesis was first discovered it was attributed to physical mechanisms such as the spatial organization of chromosomes (Sved, 1976). It was only when Kidwell (1979) was able to distinguish strains of reactivity that an understanding of the mechanisms involved appeared possible. In the *I–R* system (inducer–reactor), dysgenic events such as X-chromosome non-disjunction and enhanced mutation were found in the germ line of hybrid females, but not males. In the *P–M* system (paternal–maternal), dysgenic effects are more profound and were observed in both sexes. Temperature-dependent sterility in the form of gonadal dysgenesis, where the gonads of interstrain hybrids atrophy, is a major effect. Enhanced female recombination is also associated with this system. Male recombination commonly occurs on chromosomes 2 and 3 and is thought to be a pre-meiotic effect. When chromosomes carrying *P* elements are placed in an *M*-cytotype, chromosome breakage, distortion of transmission ratios, and high frequencies of mutation predominate. Numerous unstable genic mutations are also produced (Bregliano and Kidwell, 1983).

In the *I–R* system, the *I* factor is now known to be a transposable element which may be linked to any of the four chromosomes in the inducer strain. Transposition occurs by crossing *I* and *R* strains, but only in the direction $(R)\female \times (I)\male$. This results in transitory male sterility. Transposition can only occur if *R* cytoplasm is present. This reactor cytoplasm is genotypically determined and maternally inherited, but is extrachromosomal (Bucheton *et al.*, 1984). The structure of the *I* mobile elements is quite distinct from other elements found in *Drosophila*. They are 5.4-kb-long sequence units and lack any homology with *P* elements, nor are they *Copia*-like, or similar to foldback (*FB*) elements (John and Miklos, 1988).

In the *P–M* system, the *P* factors are also transposable elements which occur as multiple copies in some strains (*P* strains), but not *M* strains. They are stable in *P* strains and are induced to operate in an *M*-cytotype. *P* factors are a heterogeneous family of elements which display low levels of transpositional activity, but which may be switched to a highly active state resulting in hybrid dysgenesis. Activity results when a male of the *P* strain from the wild is mated to a female *M* strain, but not when the reciprocal cross occurs. The *P* elements are 2.9 kb sequences with terminal inverted repeats of 31 bp. The *P–M* system, like the *I–R* system, involves a combination of cytoplasmic and chromosomal inheritance (John and Miklos, 1988).

The presence of an *MR* (male recombinant) element which maps to chromosome 3 in *D. melanogaster* and not only increases the frequency of mitotic recombination, but transposes and induces complete *P*-insertion mutations at specific loci, was demonstrated by Green (1986). Green argued that the transpositional and genetic mapping data demonstrate that mutational and recombinational components of hybrid dysgenesis are caused by discrete MR elements located in the genomes of wild flies.

It is now believed that transposable elements in *Drosophila* had a very recent origin. Most wild populations in the world prior to 1930 were R–M in constitution, lacking both I and P factors. Between 1930 and 1960, I factors appeared in natural populations and spread throughout the world. Thus, R strains found in laboratories represent flies which were collected from the wild and established as lines before the spread of I factors occurred. It is also probable that the P factors arose in about 1950 and continue to spread through populations of *D. melanogaster* throughout the world today (Bregliano and Kidwell, 1983).

10.2.3 Do transposable elements induce speciation?

The significance of hybrid dysgenesis determinants can be viewed in two ways. When I and P factors are in a destabilized state they have an enormous potential for developing genetic diversity by mutation. Such diversity may be developed at the DNA sequence level, or may produce structural changes in chromosomes. However, if a population has the ability to rapidly evolve suppressors for transposition activity, transposable elements may remain inactive for considerable periods (Bregliano and Kidwell, 1983). The mobilization of transposable elements through the act of hybridization in *Drosophila melanogaster* can produce a spate of mutator activity. Equally, genomic stress, such as that instituted by the breakage–fusion–bridge cycle in *Zea mays*, may also release dormant transposable elements resulting in mutator activity (McClintock, 1984).

The possibility that transposable elements may be able to act as post-mating isolating mechanisms by reducing the viability of hybrids between populations which carry them, thus suggesting that they were directly implicated in speciation processes, was advocated by Bregliano and Kidwell (1983). Rose and Doolittle (1983) took a slightly different position, although still supporting the involvement of these elements in speciation. They proposed that if an isolated population had either lost, or failed to acquire, transposable elements that had spread throughout the remaining populations of the species, then it may lack immunity to the elements. Hybridization between that population and neighbouring populations could lead to abnormalities produced by the proliferation of these elements, resulting in F1 or F2 hybrid sterility. The attaining of reproductive isolation by this means was termed by Rose and Doolittle as the 'genomic disease model' of speciation.

Bingham *et al.* (1982) and Ginzburg *et al.* (1984) produced a subtly different model of speciation induced by transposable elements. They proposed, first, that the genomes of two geographically isolated populations were independently contaminated by different transposable elements, second, that these 'infections' spread in their respective populations in spite of the loss of fitness shown by individuals carrying the transposons, and, third, that effective reproductive isolation of the two populations followed even if they became sympatric. This reproductive isolation would be due to the cumulative effects of a combination of unrelated transposable elements.

Rose and Doolittle (1983) listed four criteria which would have to be satisfied by a molecular mechanism for speciation and they reasoned that the evidence derived from the activity of mobile elements in *D. melanogaster* fulfilled these requirements.

1 A known biological effect would lead to incompatibility. In this regard, hybrid dysgenesis had a major impact on germ line dysfunction including high mutability, frequent chromosomal rearrangements, male recombination, failure of embryonic development and sterility.

2 The molecular mechanism must be known. The fact that the *D. melanogaster* genome has numerous types of mobile elements (*P–M* and *I–R*), provides an adequate molecular mechanism for speciation.

3 The coupling of molecular mechanisms (mobile elements), with a direct biological effect (hybrid dysgenesis), must occur.

4 There should be parallels with biological effects and speciation. Rose and Doolittle (1983) pointed to the striking effects seen in *D. melanogaster* × *D. simulans* hybrids and their similarity to those of hybrid dysgenesis. Equally, Naveira and Fontdevila (1985) were able to release a high level of mutator activity in hybrids of *D. serrido* × *D. buzzatii* caused by the introgression of a chromosomal segment of *D. serrido* into the *D. buzzatii* genome.

The impact of hybrid dysgenesis on speciation is weakened by the absence of examples where hybrid dysgenesis actually precludes gene flow between populations. Consequently, dysgenic effects cannot be regarded as effective post-mating isolating mechanisms. Moreover, the fact that certain of the fertility effects induced by mobile elements are temperature dependent, and in the absence of the appropriate temperature no impairment to fertility occurs, suggests that such dysgenic responses are totally ineffective barriers to gene flow. It was for this very reason that Ginzburg *et al.* (1984) proposed a model which had as many operating mobile elements as possible: that is, to attempt to maximize the impact on fertility.

10.2.4 The induction of chromosomal rearrangements by transposable elements

The controlling elements of maize and the numerous transposable elements of *Drosophila melanogaster* are characterized by the capacity of these elements to initiate multiple chromosome breaks and thus induce novel large-scale chromosomal rearrangements. If these rearrangements are of the type which would enable them to act as post-mating isolating mechanisms, the likelihood exists that these transposable elements could play a significant role in initiating chromosomal divergence and thus speciation.

It is now known that in *D. melanogaster*, *P* factors (Engels and Preston, 1981, 1984), *L* factors (Lim, 1981) and *FB* transposable elements (Collins and Rubin, 1984) can all produce major chromosomal rearrangements in destabilized genomes. In *Zea mays*, the action of the breakage–fusion–bridge cycle and the production of the resultant ruptured chromosome ends on pair 9, induces a spate of disruptive structural reorganization leading to a complex array of major and minor rearrangements. Most similar mutator activity has been encountered within certain interracial and interspecific hybrids of Chironomidae (Hägele, 1984), *Drosophila* (Naviera and Fontdevila, 1985) and grasshoppers (Shaw *et al.*, 1983). In all cases, the authors suggested the likelihood that transposable elements were responsible for the elevated and often directed mutation rate (these examples were discussed in greater detail in Sections 6.2 and 6.3). In both genomically destabilized species and hybrids, the potential for the formation of major chromosomal mutations is apparent (Fig. 10.7).

It is appropriate to examine the nature of the effects of transposable elements as chromosome mutators, for the type of rearrangements induced are critical to the possible role of transposable elements in speciation (Fig. 10.8).

Foldback elements are dispersed, middle-repetitive DNA sequences with long inverted repeats at their termini. Collins and Rubin (1984) used the *white–crimson* mutation in *Drosophila* to screen for phenotypically detectable *FB*-element-induced rearrangements. This unstable mutation is a recessive allele of the X-linked *white* eye colour locus. The *white–ivory* eye colour is a phenotypic revertant resulting from the duplication of 2.9 kb of DNA within the *white* locus, whereas *white–crimson* results from the insertion of a 10 kb *FB* element into this duplication. By examining the sequence structure of 10 *white*-eyed derivatives of *white–crimson*, the nature of chromosomal rearrangements induced by these changes could be ascertained. Six of these rearrangements were deletions which arose by recombination between closely linked *FB* elements. Remnants of *FB* DNA remained at the deletion junctions. Recombination between foldback elements appeared to be a preferred means of deletion formation. Transposition

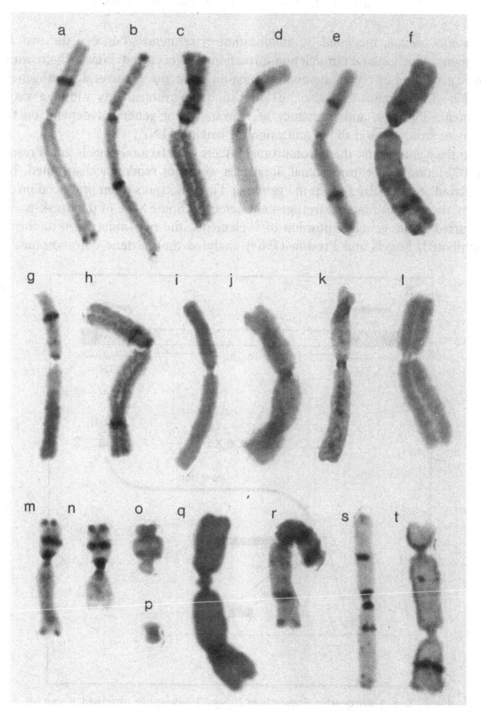

Fig. 10.7. These most unusual chromosomal rearrangements were identified among viable backcross progeny of the grasshopper *Caledia captiva* and may represent a release of mutator activity, analogous to hybrid dysgenesis, in interracial hybrids. Compare the individual chromosomes with the cytotypes illustrated in Fig. 7.10. (From Shaw *et al.*, 1983.)

of *white–crimson*, mediated by mobilization of sequences between the two *FB* elements which interact in deletion formation, also occurred. New *FB* structures were generated by recombination between the elements. Structural heterogeneity within *FB* elements can also be induced by rearrangements within a single element. The type and frequency of rearrangement generated depend on the element structure and the organization of flanking DNA.

In their analysis on the chromosomal effects of *P* factors, Engels and Preston (1981) found that chromosomal breakages were not randomly distributed, but occurred at particular sites in the genome. These hotspots often produced inversions, the most common rearrangement detected. Some 85% of the break-points occurred at the genomic position of *P* elements, the remainder were uniformly distributed. Engels and Preston (1984) analysed the polytene chromosomes of

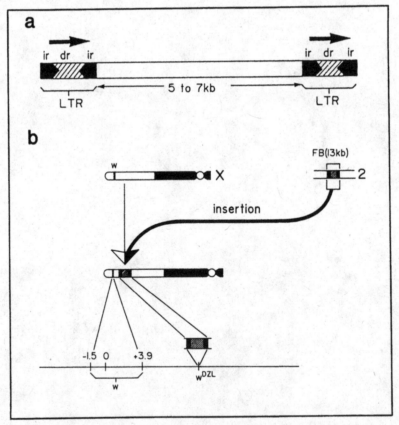

Fig. 10.8. a. A transposable element from *Drosophila melanogaster* comprised of long terminal repeat (LTR), inverted repeat (ir), and direct repeat (dr) units.
b. The action of foldback (FB) mobile elements in *Drosophila melanogaster*. Here a chromosome 2 segment has been introduced into the X chromosome by two foldback elements. (Redrawn from John and Miklos, 1988.)

746 rearrangements recovered from the screening of the progeny of dysgenic flies which had mutations at the *hold-up-wings* locus on the X chromosome. This locus is the site of a *P* factor, so breaks were associated with this and other sites. Most breaks were the product of simple two-break events and were inversions. This was due to a bias of the scoring procedure and a result of the position of *P* factors. The fact that most X–autosome translocations are male sterile greatly reduced the number of these mutations scored. Complex multibreaks with random rejoining were also encountered. Most break-points were either at, or within a few hundred nucleotide pairs of, the site of existing *P* elements. Some *P* elements were lost during the rearrangement, others were gained. In some cases, inversions were followed by reversion to the original sequence which could be detected by the loss of small bands, or other irregularities, in the banding pattern. Engels and Preston (1984) suggested that many of the rearrangements occurred prior to meiosis, and that breaks would occur at the terminus, or both termini, of a *P* element, thus explaining their loss during rearrangement formation. Highly active *P* sites, in terms of large number of breaks, were related to whether the *P* factor was itself complete. Complete elements were most active.

McClintock (1978, 1984) proposed that mobile elements in *Zea mays* had the potential for rapidly reorganizing the genome by simultaneous structural rearrangements. Destabilization of the maize genome by the breakage–fusion–bridge cycle produced rearrangements including reciprocal translocations, inversions, deficiencies, duplications and the formation of short mobile segments which included transposable gene control systems. The restructuring was non-random and resulted in numerous centromere to centromere, centromere to knob, or knob to knob attachments. The structural rearrangements not only involved chromosome 9, the element involved in the breakage–fusion–bridge cycle, but extended to other chromosomes in the complement.

The three studies indicate that major chromosomal rearrangements most similar to those encountered in chromosomally derived species and populations can be induced in destabilized genomes by the action of mobile elements. The point at issue is whether such rearrangements can singularly, or simultaneously, reach fixation in isolated populations and, because of their deleterious effects as structural heterozygotes, act as post-mating isolating mechanisms. The possibility that destabilization of the genome could be induced by the vigorous selective regime of 'bottlenecks' in numbers of animals present in founding populations (McClintock, 1984; Carson, 1990), adds considerable impetus to the value of mobile elements as initiators of chromosomal divergence in isolated populations. While there is evidence to support this relationship (Chapters 4 and 6), John and Miklos (1988) had a very different opinion: 'As we have commented earlier, there is no evidence that any of the chromosome rearrangements which distinguish different species of *Drosophila* function in reproductive isolation, so there is no reason to believe that rearrangements induced by hybrid dysgenesis would lead to speciation' (p. 276).

10.3 Does repetitive DNA have a regulatory function?

A very different approach to the genomic role of non-coding repetitive DNA was advocated by Britten and Davidson (1971) and Davidson and Britten (1979). These authors suggested that genes could be regulated in their activity by specific interactions occurring at repetitive sequences in the DNA genome. That is, it was assumed that repetitive DNA was functional and that its function was major. Short repeated sequences would act as regulator modules to a group of structural genes with which they were associated. Davidson and Britten (1979) suggested that nuclear RNA contains continuously synthesized RNA copies of the structural gene regions of the genome, and that regulatory interactions occur between these copies and complementary repetitive sequence transcripts by the formation of RNA–RNA duplexes. The pattern of gene expression would be established by the transcription of regulatory DNA regions into control RNAs.

Clearly, the regulatory elements would have to be in a position and of a sequence type which would influence developmental pathways in a complex and coordinated fashion. Rapid divergence between isolated populations would be expected since these pathways would be under substantial selective pressures. For speciation to occur, one family of dispersed regulators would be replaced by another. Indeed, even if the pathways were neutral they could lead to pre-zygotic isolation through coincidental divergence of mating behaviour, or coincidental post-zygotic incompatibility in the germ line, or somatic cell development, or function, in hybrids (Rose and Doolittle, 1983).

Based on their analysis of the molecular organization of primate DNA, Gillespie *et al.* (1982) proposed that portions of the animal genome could be reorganized, or 'reset', periodically. Genome reorganization meant an alteration of the long-range organization of DNA sequences rather than structural rearrangements. The evolutionary consequence of genome resetting was speciation. Genome resetting would be mediated by the amplification of repeated sequence DNA. Thus, changes would appear in the form of reorganization of the distribution of repeated and unique sequence DNA and structural differences in the karyotype associated with amplification. Gillespie *et al.* (1982) argued that reorganization established new genetic programmes in development and differentiation which defined the morphological and physiological characteristics of new species.

One of the chief sources of evidence for this model was that major amplification of satellite DNA in Old World primates preceded speciation. That is, in the baboon and guenon groups, all species tested had the same *alpha* satellite and each had a similar quantity of this in their 2n = 42 complement. Gillespie *et al.* proposed that the baboon group *alpha* satellite was amplified 6 to 8 MYBP during the period after baboons and guenons diverged, but before the 20 or more species of baboons, macaques and mangabeys were formed. This amplification before speciation was assumed to be causally related to speciation and the convergence of a karyologically coherent phylogenetic group. Unfortunately, Gillespie *et al.* reached this conclusion without any evidence to support a causal relationship. Moreover, there is no evidence to suggest that repeated sequence DNA is involved in any form of regulatory mechanism of this type.

10.4 Concluding remarks

From a cladogenic perspective, the two most significant evolutionary developments which have resulted from comparative molecular analyses of genome structure are, first, the pattern of concerted evolution, together with the associated genomic turnover mechanisms unified under the process of molecular drive. Second, the possibility that segments of the genome may be mobilized by the action of mobile elements and transposed to other areas of the genome. Both of these processes have had a major impact on evolutionary thought and both are intimately associated with possible mechanisms of speciation.

At the start of this chapter, two questions were posed, both of which attempted to define the impact of molecular mechanisms on modes of speciation. The answers to these questions are summarized below.

10.4.1 Can molecular turnover mechanisms provide a means of establishing post-mating isolating mechanisms in undifferentiated populations powerful enough to enable speciation?

Genome turnover mechanisms appear to be responsible for changes in the quantity and distribution of repeated sequence DNA in some organisms. The possible impact of such mechanisms on the formation of post-mating isolating mechanisms would arise from fertility effects associated with the presence of blocks of this material in proximal or terminal positions, or on interspersed segments. There is no evidence that the presence of repeated sequence heterochromatic blocks has an impact on interracial or interspecific hybrids distinguished by these sequences alone.

The remaining possibilities are that repeated sequence DNA disrupts pairing in interspecific hybrids at pachytene, thus providing a direct fertility effect or, alternatively, that the presence of heterochromatic segments may induce the formation of other types of structural rearrangements which may be able to act as post-mating isolating mechanisms. There is no evidence to support an impact of blocks of repeated sequence DNA on pairing, or on the formation of additional rearrangements (these aspects of heterochromatin effect were dealt with in more detail in Section 5.1.2.1).

There appears to be little doubt that genomic turnover mechanisms could lead to the genetic divergence of allopatric populations, if these processes resulted in the redistribution, duplication or deletion of meaningful DNA sequences. As it stands, these processes have not been implicated in this form of differentiation in association with the speciation process. As usual, the difficulty of establishing changes which have occurred subsequent to speciation and those which occurred as a primary role remains.

More significantly, there are difficulties associated with the generality of concerted evolutionary patterns and molecular drive processes. There appears to be little doubt that there are restrictions on the nature of changes to chromosome structure which can be tolerated. This gives the appearance that some organisms have experienced a degree of immunity to genomic turnover processes. For example, most species analysed chromosomally with C-banding show that heterochromatin is present at the centromeres and telomeres and is absent from interstitial regions (see, for example, King and John, 1980, in Orthoptera and King, 1990, in Amphibia). If this is a reasonable indication of the genomic distribution of blocks of middle-repeated DNA, which it should be, there seems to be little evidence for the homogenization of DNA sequences throughout the genome.

Indeed, there appear to be degrees to which genomic structure may be changed and this can be seen in many amphibian species. In some frogs (see Section 10.1), a concerted pattern in the distribution of repeated DNA sequences is associated with paracentromeric areas. While this suggests that turnover mechanisms may have been responsible for the distribution of these sequences within the paracentromeric heterochromatin, they are confined to tightly defined areas and are not distributed throughout the genome. Yet, in the same species, particular sequences may be distributed over all chromosomes in the complement. In the same species, all highly repeated 18s and 28s sequences may be confined to a single NOR found on the same chromosome in all species of the family (King, 1990, 1991). These observations suggest that genomic turnover mechanisms are highly constrained and in some groups, where there is no form of chromosomal differentiation between any species in a whole family, it may be that turnover processes are inoperable.

As is often the case, examples of extraordinary variation in sequence diversity, substantial variation in heterochromatin polymorphism or multiple genomic sites for NORs are published because of their interest and unusual form. In reality, such examples may be exceptional situations in groups which generally show great uniformity, thus giving a false impression of the generality of concerted evolutionary patterns.

Indeed, a series of questions can readily be asked, for while there is no doubt that concerted evolutionary patterns are present, there are difficulties associated with explaining why the differences in pattern exist. First, if concerted patterns of sequence evolution are occurring, why are the sequences often situated in what appear to be identical positions on each chromosome arm in every chromosome of the complement, often regardless of chromosome size in some species? Second, why do sequence amplifications often involve small clusters of chromosomes in the same size category rather than the entire genome? Third, why do particular chromosomes, or chromosome structures, such as the NOR, retain their sequence integrity and site on one chromosome, and why are they not distributed throughout the genome? Fourth, why is there a distinction between the pattern of concerted evolution involving satellites which are distributed throughout the genome, but at particular specific localities such as C-bands, and those other sequences which are spread throughout the genome in a uniformly homogenized pattern on all chromosome arms?

10.4.2 Can molecular mechanisms enhance the formation and fixation of chromosomal rearrangements or genetic divergence and thus support existing mechanisms for speciation?

There seems to be little doubt that genomic turnover mechanisms in general can have a direct impact on forming chromosomal rearrangements, although these are most commonly the amplification of heterochromatic segments, and there is no evidence that these have any impact on hybrid fertility. The question is, are any of the mechanisms encountered able to initiate the formation and fixation of negatively heterotic rearrangements which can possibly play a role in speciation? The answer appears to be that they do. The activation of transposable elements can result in the formation of translocations, inversions, deletions and fusions, all of which may have the potential to act as post-mating isolating mechanisms.

However, the formation of chromosome rearrangements as a hybrid dysgenic side-effect between strains of *Drosophila*, or genomically stressed maize, may not be an indication that comparable changes can be produced, or reach fixation, in population isolates of a species.

There is a body of evidence which indicates that the act of hybridization between individuals can at times result in the release of exceptional mutator activity analogous to that occurring in hybrid dysgensis. This can involve hybridization between relatively closely related forms: chromosome races of *Caledia*, subspecies of *Chironomus*, species of *Drosophila* (see Sections 6.2 and 6.3 for details). These studies provide direct evidence for the formation of chromosomally divergent individuals with completely novel complements in hybrid situations. Whether such individuals are simply bizarre mutants, or on rare occasions could go on to form viable species, remains unknown. However, it is highly likely that the mutator activity released was a direct result of the activity of transposable elements. Indeed, in the case of the *Chironomus* hybrids examined by Hägele (1984), possible transposable elements were detected (see Section 6.2).

It is now well documented that genomic stress, such as that induced in interstrain, or interracial hybrids, or that produced by the formation of ruptured chromosomes, can provide the impetus for the activation of mobile elements (McClintock, 1984). However, the act of hybridization in such situations occurs between forms which have differentiated to some degree either chromosomally or genically. The question is, can the same type of mutator activity be released in undifferentiated population isolates, thus enhancing the likelihood of the formation of chromosomal rearrangements by substantially elevating the mutation rate? This remains unanswered, but for a possibility raised by McClintock (1984) and Carson (1990) who suggested that the severe selective regimes associated with population bottlenecking in founding populations could initiate mutator activity and thus lead to the chromosomal or genetic divergence of stressed populations.

11
Conclusions and perspectives

The remarkable fact that such blatantly inadequate data have been so widely accepted as convincing proof of gradualism only reinforces our claim that gradualism has always rested on prior prejudice rather than paleontological data.

(Gould and Eldredge, 1977, p. 122)

and

I do not think that the attempt to explain morphological evolution by species selection can survive in the face of results of this kind. But there never was much sense in the idea anyway.

(Maynard Smith, 1987, p. 516)

11.1 An overview

When writing this book my aim was to set out in the simplest fashion the evidence which related to the association of chromosomal change with speciation. Comparison of evolutionary phenomena between disparate organisms is the only real basis for establishing unified evolutionary theory. With this in mind, simple comparisons have been made between distantly related species complexes to establish a commonality of evolutionary function, and between closely related entities within species complexes to reveal the nature of evolutionary change. In this regard, the fertility and viability effects associated with hybridization between closely and distantly related members of an evolutionary lineage display the impact of chromosomal and genetic differences accumulated since the time of genetic uniformity and initial chromosomal divergence (see Chapter 7). The impact of structural rearrangements on hybrid fertility and viability can be formidable and the effects cumulative. Similarly, Chapter 8 examines genetic divergence in diverse species complexes representing a range of organisms, all of which are linked by the commonality of chromosomal speciation, but which are sometimes distinguished by the fact that some have speciated recently, whereas others are much older complexes. In such a broad classification it has been possible to assess the relationship between the timing of chromosomal and genetic divergence. That is, the critical question as to whether genic, or chromosomal, divergence is the primary motivator of speciation can be answered.

For comparisons of this type to be valid, certain assumptions must be made. Prime among these is the assumption that from the moment a species is formed and until that species becomes extinct, it will accrue genetic and morphological differences linearly with time. This is not a difficult assumption to make for it forms the very basis of genetic gradualism. It is also supported by a significant body of biochemical evidence. For example, Thorpe (1983), in a substantial comparative study, revealed a linear increase in electrophoretic divergence associated with time (see also Section 3.2).

The second assumption which has been made in the analysis of species complexes, is that differences found between chromosome races, subspecies and species in evolving lineages, reflect the stages of differentiation which have occurred in the speciation of that complex. This can only apply to class 3 and 4 complexes (*sensu* Section 8.1), where the very earliest stages of cladogenic differentiation have occurred (i.e. where chromosome races have first been formed). Class 1 and 2 species complexes may reflect anagenic differentiation; subsequent to speciation.

Comparisons of this type reveal the primacy of chromosome race formation, before any form of genetic or morphological divergence has occurred. This has significant ramifications on a series of issues and it is worth considering each of these in turn.

The first issue is that the biological species concept is based on the view that reproductive isolation is the key character to species distinctiveness. In those models which have been proposed to replace the BSC, such as Paterson's (1985) recognition species concept, gamete recognition has been used as a criterion for distinguishing species. That is, if gametes recognize each other and form a zygote, the species are regarded as being conspecific. Clearly, the demonstration that powerfully negatively heterotic rearrangements reach fixation as a primary form of differentiation in the speciation process, indicates that in chromosomal modes of speciation the recognition concept is inoperative. That is, despite the fact that gamete recognition has occurred between genetically indistinguishable chromosome races and the basis for conspecificity is fulfilled under the recognition concept, the chromosomal isolating mechanisms can render the zygote inviable, or infertile. Total sterility, or profound infertility, have been demonstrated between genetically indistinguishable chromosome races of *Mus domesticus* (Capanna, 1982; Gropp *et al.*, 1982; Britton-Davidian *et al.*, 1989), and the *Rattus sordidus* complex (Baverstock *et al.*, 1983b, 1986). These effects have been attributed to as few as two Robertsonian fusions with monobrachial homologies, and sex chromosome fusion multivalent associations (Forejt, 1979, see Sections 7.3 and 8.2). Thus, the recognition concept is demonstrably false in cases of chromosomal speciation, for the chromosomal reproductive isolating mechanism overrules gamete recognition. Consequently, the BSC which is based on reproductive isolation, is the only valid biological species concept.

A second area which can be reassessed in the light of these findings is the concept of the founder effect. Given the fact that deleterious negatively heterotic rearrangements can reach fixation in isolated populations which have then expanded to form chromosome races, a reappraisal of certain stochastic processes can be made. First, the fact that such rearrangements have reached fixation provides an insight as to how this has occurred. Theoretical modellers indicate that negatively heterotic rearrangements can only reach fixation in small isolated demes in which there are high levels of inbreeding, and sib mating, powerful selective gradients, substantial changes in population numbers and where the population may be reduced down to a few individuals and is bottlenecked (Lande, 1979, 1984; Hedrick and Levin, 1984). If meiotic drive is also occurring, the rearrangement can reach fixation under less stringent conditions, although the same basic population dynamics still exist (Hedrick, 1981; Walsh, 1982).

These findings suggest that the best opportunity for the fixation of profoundly deleterious rearrangements is in the genetic revolution of a Mayrian founder effect. There seems to be little doubt that the genetic gradualists' relatively bland concept of a founding population, as being an unexceptional isolated population (see introductory quotation for Chapter 4, Carson and Templeton, 1984), could not establish negatively heterotic rearrangements unless these changes were meiotically driven to fixation. Undoubtedly, Templeton (1980) would not regard it necessary for founding populations to establish deleterious rearrangements, for he does not support the possibility of chromosomal speciation. However, Mayr's (1963, 1982b) founder effect does meet these requirements. Given the fact of chromosomal speciation, it would appear that Templeton's view of the founder effect is untenable. Similarly, although the models for the founder–flush, and flush–crash–founder cycle advocated by Carson (1968, 1975, 1982a) provide for the fixation of deleterious changes, they remain without evidence to support them, as they did when attacked by White in 1978. Carson's view of founding populations was also designed around the requirements for genetic changes rather than chromosomal speciation, a concept which Carson (1985) fervently opposed (see introductory quotation Chapter 1). Thus, Mayr's (1963) concept of a genetic revolution in a founding population appears to be the most reasonable structure for the fixation of deleterious and undriven chromosomal rearrangements.

In its simplest form, the genetic gradualists' view of speciation may be summarized as the differentiation of a population in isolation, to the point where reproductive differences are accumulated, which may act as a reproductive barrier to hybridization when the parental and daughter species are once again brought into contact. Alternatively, behavioural differences may be accrued which result in pre-mating isolation and similarly impair hybridization.

Problems associated with this most common view of speciation became apparent when species were hybridized in the laboratory. In some situations, morphologically and genetically divergent allopatric species fail to show any reproductive incompatibility when hybridized. That is, despite displaying considerable divergence, reproductive isolating mechanisms had not been established. In the great majority of cases analysed in *Drosophila*, pre-mating isolating mechanisms were present, and in those instances where reproductive isolating mechanisms were absent, pre-mating isolating mechanisms were the sole means of species separation (Bock, 1984).

As apparent confirmation of the assumptions made in this volume, Coyne and Orr (1989a) found that greater genetic distances were present between species of *Drosophila* which produced inviable hybrids of both sexes, when compared to those of hybrids in which only sterile, or inviable, males were produced and females were fertile, or had impaired fertility. Coyne and Orr regarded this as a demonstration of the fixation of reproductive isolating mechanisms, where there was a stepwise accumulation of alleles causing male sterility (visualized as Haldane's rule), followed by the acquisition of female sterility. The mutation of loci associated with any of the multigene complexes responsible for fertility could conceivably result in sterility. While there is little doubt that such a theory is viable, there are major exceptions to it, and in the case of Coyne and Orr's analysis, the impact of chromosomal rearrangements was not partitioned out.

The uncertainty as to how this type of allopatric speciation occurs, is contrasted to the almost instantaneous reproductive isolation and sympatric speciation in some flowering plants. Here, changes in mating system associated with recombination and polyploidization can lead to reproductive isolation in a single step (Barrett, 1989, Section 3.4.6).

The largest section of this book involves a discussion of the types of chromosomal rearrangements which play a role in speciation, the methods of their formation and fixation, and their impact on fertility and viability (Chapters 5, 6 and 7). There seems to be little doubt that the only valid role that chromosomal change can have in speciation is that of a reproductive isolating mechanism, and this is the position taken in this book. There is no evidence to suggest that chromosome change is directly responsible for morphological change in speciation. This was clearly demonstrated by John and Miklos (1988).

Many of the studies attacking the role of chromosomal rearrangements in speciation, have incorrectly used as examples chromosomal polymorphism, heterochromatin addition, or other forms of change, which do not produce profound fertility effects in hybrids. Nevertheless, the very large amount of information produced by hybridization studies indicates the major classes of rearrangements associated with speciation. These include single Robertsonian fusions, multiple independent Robertsonian fusions, fusions which share monobrachial homologies involving any number of chromosomes, tandem fusions, reciprocal translocations, sex chromosome translocations (Robertsonian and reciprocal), fixed paracentric and pericentric inversions, multiple inversions of either class, or deletions. The status of chromosomal inversions is uncertain because of the possible avoidance mechanisms which are now known, and the impact of recombination on co-adapted gene complexes.

One of the major grounds for attacking chromosomal speciation has been the fact that many chromosome races are distinguished by only one or two independent chromosome fusions. Since these may have very minor malsegregation effects, they are often demonstrably ineffective as reproductive barriers. However, one of the features that has arisen from the comprehensive hybridization studies is that, in some cases, one or two Robertsonian fusions can produce total sterility, or inviability, in structural hybrids if the association of trivalents with the sex chromosomes occurs during meiosis (Luciani et al., 1984; Rosenmann et al., 1985). One of the characteristics of sex chromosome/multivalent association is the disproportionate increase in male sterility giving a typical Haldane effect. Although it is uncertain how these associations affect fertility, it is likely that the X-inactivation cycle is interfered with or, alternatively, that inactivation is transferred from the sex chromosome to the associated autosomes. Sex chromosome effects of this type have been demonstrated in organisms as diverse as Drosophila and man (see Section 7.3). It would appear that single fusions can no longer be ignored when considering chromosomal speciation.

Nevertheless, there are many instances where single chromosome fusions which distinguish species or chromosome races have been demonstrated to have a negligible impact on fertility, and yet these have still been a primary form of differentiation in chromosome races. Indeed, on some occasions effective hybrid zones appear to have been established between forms possessing such presumably ineffectual rearrangements. How this can occur remains a mystery.

Among the many advances made in the last few years is the appreciation that a number of the previous assumptions on mutation rate, the sequential or simultaneous origin of chromosome changes and the repetitious origin of particular types of chromosome change, have been eclipsed by new technologies. The impact of molecular drive mechanisms on speciation has revealed a significant underestimation of mutation rate, fidelity to the site of chromosome breakage and the types of change which can be established. One of the chief findings is the release of mutator activity in destabilized genomes. While most of the effects of transposable elements have been demonstrated within species, a growing body of data is now available which indicates that hybrids between species display evidence for major mutator activity. This may take the form of large numbers of rare electrophoretic variants being found in hybrid zones (Barton *et al.*, 1983; Nelson *et al.*, 1987), or interspecific hybrids which have totally restructured karyotypes associated with chromosome breakage (Shaw *et al.*, 1983; Hägele, 1984; Naveira and Fontdevila, 1985). It is most likely that these are a direct effect of mobile element activation, and may turn out to be one of the most significant contributions to our understanding of the role of chromosomes in speciation. However, at this stage there is no evidence that the events which occur in hybrid zones lead to speciation in those areas, although this has been proposed (Templeton, 1981). McClintock (1984) suggested that transposable elements might be activated by genomic stress. The possibility that founding populations under severe selective regimes might undergo sufficient stress to release mutator activity was suggested, but remains untested.

At the moment, the evolutionary significance of the remaining molecular drive or genomic turnover mechanisms is a matter of debate, for most involve the accumulation, or redisposition, of repetitive sequence DNA. The fact that much of this material is demonstrably inert, has in the past suggested that it has little impact on fertility or viability, and that it cannot play a role as a reproductive isolating mechanism. This may still be so, for despite the fact that controlling or responder genes associated with the meiotic drive mechanisms, male fertility and rDNA, are associated with, or are found in, heterochromatic areas (King, 1991, and Sections 5.1.2.1 and 10.3), this does not mean that the heterochromatin itself has those genetic functions, but rather it could mean that the heterochromatin was secondarily added to that site to protect that function from the impact of recombination (King, 1991).

Indeed, while the potential for chromosome change due to these molecular mechanisms is so apparent, the means by which profoundly negatively heterotic rearrangements can reach fixation in populations, has attracted considerable attention. White (1968, 1978a) was ridiculed for his advocacy of meiotic drive as a mechanism which could force such rearrangements to fixation. There was simply very little evidence at the time to suggest that this may have been possible. There is now no doubt that meiotic drive is a fundamental and substantial evolutionary mechanism which has been thoroughly documented in numerous plants and animals. One of the possible means for promoting the fixation of deleterious rearrangements is the likelihood that derived homozygotes possess adaptive characteristics. There is very little evidence to support this concept other than the case of the blue fox (see Section 6.4.3.5), where homozygotes for a meiotically driven Robertsonian fusion have larger litter sizes than the other chromosome morphs. These results were repeatable (Christensen and Pedersen, 1982; Mäkinen and Lohi, 1987).

The many models for chromosomal speciation, a number of which were described in Chapter 9, were often designed around the constraints imposed by the organisms that they were proposed for. There appears to be little evidence to support the possibility that internal modes of chromosomal speciation (stasipatric, chain processes, or the triad hypothesis), have played a significant evolutionary role. The intuitive difficulty of a chromosomal rearrangement reaching fixation within a population, while at the same time being exposed to gene flow, is a major drawback. White (1978b) argued that a weakly negatively heterotic rearrangement could form a tension zone which would advance and displace the parental race distribution. How this could occur without a profound barrier to gene flow being present is difficult to explain. Many criticisms of the stasipatric concept have been made, and with some justification. A number of examples of stasipatric speciation could simply be regarded as cases of internal allopatry. The rock wallabies of the *Petrogale assimilis* complex can be considered in this context, for their enormous species distributions are, in reality, made up of a series of isolated rock outcrops, or mountain ranges. It is quite conceivable that one of these internally distributed founder populations could readily establish a chromosomally derived population in complete isolation. Indeed, most situations where chromosomal changes are associated with speciation can be readily explained by one of the external allopatric modes. Of these, alloparapatric speciation and primary chromosomal allopatry (Sections 9.2.4 and 9.2.6) seem to me to best cover the two most significant aspects of chromosomal speciation. That is, the formation of the chromosomally distinct population in isolation, followed by the secondary contact of that species to the parental form during which the chromosomal barrier to gene flow comes in to play. This enforced genetic isolation permits the accumulation of advantageous mutations during the colonization of new territory by the derived race and thus assists in the formation of a new species. Subsequent peripheral speciation events may then be established, leading to a geographical array of ancestor/descendant related species established during colonizing radiations.

It is of some interest that recent models for speciation are based on particular types of chromosomal rearrangements, such as speciation by monobrachial homology, or a dual-level model for speciation by pericentric inversion (Sections 9.2.7 and 9.2.8). One's dedication to a particular model for speciation is a matter of personal choice and accommodation with the available data. Nevertheless, the fact of chromosomal speciation can no longer be validly questioned.

11.2 A jaundiced view

A number of arguments which have been used to attack, or support, the concept of chromosomal speciation have been shown in this volume, either directly or by inference, to be either logically unsound, or without evidence to support them. Four of the most commonly encountered issues are considered here.

11.2.1 Chromosomal hybrid zones and gene flow

In an attempt to negate the suggestion that the primacy of chromosome change is proven by the fact that parental and daughter chromosome races are genetically undifferentiated but chromosomally distinct, the argument has been used that the absence of genetic differentiation between chromosome races separated by a chromosomal hybrid zone, demonstrates that this type of hybrid zone is an ineffective barrier to gene flow. This supposes that genes in one chromosome race introgress freely with those of the other race and this accounts for the absence of genetic differentiation between the races or species. Presumably, if the hybrid zone was a barrier to gene flow, genic differences would have accrued on either side of it. This argument is regularly handed out by the genetic gradualists and those who disagree with chromosomal speciation to attack that concept (Futuyma and Mayer, 1980; Baker, 1981; Sites and Greenbaum, 1983; Sites and Moritz, 1987). The validity of this argument can be formally rejected on a number of grounds:

1 Analyses used for revealing fixed or polymorphic genetic differences between populations or species are limited in resolution to the status of static populations. The tests cannot assess whether gene flow is, or is not, occurring and to do so would require estimates of migration and other population dynamics. If no genetic differences are detected between chromosome races, or species, one can only conclude that no differences were established. It is logically unacceptable to extend these data and reach a conclusion as to why differences were not established, i.e. due to the presence of gene flow through the hybrid zone between the races. If certain populations do have polymorphic or fixed differences established in them, it is not unreasonable to conclude that these populations may have been isolated (presumably from gene flow), to enable differences caused by mutation to reach fixation. However, the reciprocal argument cannot be used (i.e. the absence of differentiation is due to the presence of gene flow), for it is equally likely that genetic uniformity may be due to the absence of mutation.

2   When appropriate analyses (including the assessment of migration rate) have been applied to species which display different levels of genetic differentiation, such as the plethodontid salamanders examined by Larson *et al.* (1984), it was found that species do not generally comprise units connected by gene flow. Rather, they are characterized by populations displaying considerable isolation. The generality of these conclusions may be questioned in highly mobile species such as birds or insects, or in species which have small distributions, but they may well apply to many of the low mobility or low vagility forms.

3   The only way that one can validly prove that a hybrid zone is an ineffectual barrier to gene flow, is to analyse populations on either side of it for biochemical or molecular markers. The markers characteristic of one chromosome race may have crossed the hybrid zone and introgressed into the sister chromosome race, thus demonstrating gene flow. Once again, these findings are based on positive results, not on the absence of differentiation.

In reality, the absence of any genetic differentiation between parental and daughter chromosome races or species indicates that genetic changes have simply not been established and that the chromosome races were formed before genic differentiation occurred.

11.2.2 Are present-day geographic distributions a valid tool for determining past evolutionary events?

Sites and Moritz (1987) proposed that present-day distributions of chromosome races could not be used to discriminate between different modes of chromosomal speciation. The arguments used were, first, that present-day distributions do not reliably reflect past distributional patterns. Second, that many of the models for chromosomal speciation have similar patterns of distribution and would not be easily differentiated. Third, the rearrangement might not have been associated with speciation, but could have been established as a neutral rearrangement which spread with a colonizing radiation.

The first of these arguments is quite remarkable and its logical extension requires that no form of speciation could be ascertained from examining present-day distributions, because they do not reflect past distributional patterns. The biogeographers among us would be pleased. It is satisfying to see that neither Sites nor Moritz take their argument too seriously. Both authors use present-day geographical distributions in their own research to infer the mode of speciation in *Heteronotia binoei* (Moritz, 1983) and demonstrate the related array of chromosome races and mode of speciation in *Sceloporus grammicus* (Sites and Davis, 1989, see also Section 8.2.8).

On a more serious note, there is little likelihood that all present-day species would have precisely the same species boundaries that they had in the past, since changes in environmental conditions have undoubtedly been responsible for substantial range expansions and contractions. Indeed, the clearing of forests by man may have greatly influenced many species distributions leaving some species isolated in a series of relic populations. Nevertheless, the fact that many species still retain broad distributions (particularly low vagility forms), even if these are broken up into relic populations, still provides an indication of the region which that species historically occupied. The historical occupation of an area can often be tested by biochemical or chromosomal means, where lineages of related species (A–B–C–D) of known phylogenetic sequence can be plotted geographically. There are many examples of lineages which show geographic distributions which reveal their evolutionary sequence of origin with ancestors directly abutting against daughter species' distributions (see Chapters 7, 8 and 9). One cannot reasonably attribute these patterns to pure coincidence.

Second, it is no doubt true to suggest that many of the models for chromosomal speciation do predict similar distributional patterns. Nevertheless, there does not seem to be a problem in discriminating between internal and external modes of chromosomal speciation. In the former, the parental distributions are displaced by the chromosomally derived form. In the latter, the derived form invades new territory. Thus, a sequential array of cytotypes characterized by shared derived chromosome markers, reflects a lineage of descent and at the same time a colonizing radiation. In this regard, it is less important to discriminate between the different external forms of allopatric and colonizing chromosomal speciation than to discriminate between these and the internal, or stasipatric forms.

Third, there is little doubt that certain rearrangements are not associated with speciation and could be the product of a neutral rearrangement which has been spread by a colonizing radiation. However, it would be most unlikely that such a form would be present as an entity in a linear series of races distinguished by shared derived characters and parapatric boundaries. In fact, a rearrangement of this type could only be established in total isolation and maintained after genetically induced reproductive isolation had occurred. Otherwise, if the derived chromosome race made contact and hybridized with the parental form, the neutral rearrangement would simply introgress through the parent and daughter species distributions as a polymorphism.

11.2.3 Chromosomal or genetic reproductive isolation?

It is commonly argued that chromosomal differences that characterize species are not responsible for the profound fertility and/or viability effects present in interspecific hybrids. Rather, it is claimed that fertility/viability effects are due to genetic perturbations.

This old chestnut has been used by those who argue against chromosomal speciation to undermine that concept (see Charlesworth *et al.*, 1982; John and Miklos, 1988; also see introductory quotations to Chapter 7). In reality, genetic perturbations are a red herring and one might suspect that this ploy would be more suited to politics than the scientific approach. If the comparative studies presented in this volume have demonstrated anything (Chapters 7 and 8), it is that there is precious little evidence for genetic reproductive isolating mechanisms between species or chromosome races, which are distinguished by powerful chromosomal reproductive isolating mechanisms and characterized by genetic uniformity when compared with their direct ancestor. There is no good reason why genetic reproductive isolating mechanisms could not be established between genetically differentiated species or chromosome races. Nevertheless, there is no reason to suspect, and no evidence to show, that these mechanisms should be present between genetically indistinguishable species.

Charlesworth *et al.* (1982) attempted to use this form of argument to discredit chromosomal speciation. This was apparently considered necessary by these authors, because chromosomal speciation was being used as evidence to support the concept of macroevolution and punctuated equilibria. Charlesworth *et al.* were arguing for genetic gradualism and against punctuated equilibria. They argued 'But caution should be exercised in interpreting chromosomal differences as primary agents in causing hybrid sterility, in the absence of genetic investigations. Although there are relatively few cases in which such analyses have been carried out, it is evident that hybrid sterility may often be genic rather than chromosomal in origin' (Dobzhansky, 1951, Ch. 8, p. 485). Apart from demonstrating a classic slide in argument '... relatively few cases ...' to '... sterility may often be genic', the source reference for this statement was derived from Chapter 9 of Dobzhansky's 1951 revision of a book written in 1937. There is something reassuring in using outdated data; you can always find something to suit your argument. Charlesworth *et al.* may also consider the following quotation from the same volume and compare this to the data in Chapter 7 of this volume. 'One must admit, I think, that apart from the production of abortive eggs, ovules, and pollen by certain translocation heterozygotes of *Drosophila*, maize, peas, and other forms, the sterility of no hybrid has as yet been definitely proven to be of the chromosomal type.' (Dobzhansky, 1937, p. 288).

11.2.4 The relationship between chromosomal change and morphological change

Of the two possible arguments supporting a role for chromosome change in speciation, that favoured in this volume is the role of chromosome change as a reproductive isolating mechanism. The second possible role for chromosome change is a direct responsibility for morphological evolution and thus speciation. A considerable amount of impetus was given to this view by the comparative studies of Wilson *et al.* (1974, 1975), who found that rates of chromosomal evolution were directly correlated with rates of speciation and morphological evolution when different taxa were compared. Not only did Wilson *et al.* detect a relationship between chromosomal change, speciation and rapid morphological evolution, but they proposed that chromosomal homozygotes could act as post-mating isolating mechanisms, the key being that morphological change associated with chromosomal repatterning was a product of the impact of chromosomal rearrangements on regulatory genes. This hypothesis has been widely quoted and has also been discredited, both from the lack of morphological or molecular evidence to support it (John and Miklos, 1988) and, on more prosaic grounds, on the total unreliability of the database the information was derived from (King, 1981, 1987). The fact of the matter is that the authors failed to discriminate between those chromosomal rearrangements that could be involved in speciation (negatively heterotic changes), and those which could not (neutral or positively heterotic changes and chromosomal polymorphisms). Consequently, this totally spurious database could not produce valid correlations or predictions (see Section 5.2).

11.3 Hybrid zones

The comparative analyses used in this volume have placed a considerable amount of emphasis on hybridization between species, both in terms of the chromosomal and genetic impact of structural rearrangements and the role of reproductive isolation in speciation. Less emphasis has been placed on the study of hybrid zones, although many regard these areas as the most crucial part of species distributions. They are the site where reproductive isolating mechanisms are bought into play and where one can determine the resistance of chromosome races or species to introgression of chromosomal, genetical or morphometric markers.

It would seem a matter of logic that all hybrid zones between parapatric forms which inbreed to some degree, either F1, F2 or backcross, reflect an existing barrier to gene flow between the forms. One could justifiably ask why a hybrid zone would form at all if there was no barrier to gene flow? Reproductively unimpaired species would simply coalesce. It is only when levels of impairment are produced that a hybrid zone becomes recognizable and the greater the impact of reproductive isolation on gene flow the narrower the zone becomes. Presumably, species which have parapatric boundaries and F2 or backcross sterility, or inviability, would have a very narrow hybrid zone. Of course the size of this zone is also related to the mobility of the species and the size of the species distribution.

However, the analysis of hybrid zones is not a straightforward matter and the more detailed the information, the more complex the situation often becomes. In short, the analyses of hybrid zones can raise as many questions as they answer. Four examples will suffice to explain the problems here: the rodent *Peromyscus leucopus*, the frogs *Hyla cinerea* and *H. gratiosa*, and the grasshoppers *Podisma pedestris* and *Caledia captiva*.

The North American white-footed mouse, *Peromyscus leucopus*, is a geographically delimited species subdivided into two distinct chromosome races. Both of these have 2n = 48 chromosomes, but they differ by three fixed pericentric inversions. That is, the southwestern form has metacentric pairs 1, 2, 3, 5, 6, 9, 11, 12, 18, 21, 22, 23 with a degree of heterochromatin involvement in the morphology of all pairs but for 1, 2, 3, 5, 6 and 9. The northeastern race has acrocentric pairs 5 and 11 and pair 20 is biarmed (Baker *et al.*, 1983b). Stangl (1986) defined a narrow hybrid zone between these chromosome races and found that its distribution was congruent with that of an ecotone between great plains grasslands and eastern deciduous forests. Chromosome markers were asymmetrically distributed in the hybrid zone with northeastern markers being found at a higher frequency in the southwestern cytotype than the reverse. Electrophoretic, chromosomal and mtDNA analyses were made on 332 specimens taken from the hybrid zone and the same animals were run for the three techniques (Nelson *et al.*, 1987). Analysis of 38 presumptive loci revealed that there were no fixed allelic differences between the cytotypes, although the allelic frequency at *Adh* and *Pgm-1* varied across the zone. Significantly, the distribution of these loci was skewed across the zone and this matched the asymmetry of the chromosomal distribution. Forty-five per cent of alleles were unique to one side of the zone and not the other, thus showing an association with one parental cytotype. The cytotypes were most similar, having a Rogers similarity of 0.94. Nevertheless, all eastern and western populations clustered separately. A total of 38 uncommon alleles were found within the hybrid zone. The mtDNA analysis used eight restriction enzymes and corroborated the finding of both chromosomal and electrophoretic studies. These data not only showed that the two cytotypes had very different evolutionary histories, they also showed the asymmetry of the hybrid zone with a high incidence of northeastern markers in the southwestern cytotype.

Despite this wealth of information, the authors were forced to opt for the hybrid zone being an area of secondary contact rather than being primary (in the sense of stasipatric speciation), because it was difficult to envisage how strong selection could cause concordance between chromosomal, electrophoretic and mtDNA changes. In this respect the conclusions were reasonable, but the absence of reproductive data from structural hybrids was overpowering. But then even with these data being available could a different conclusion have been reached? My view would be that with a 'primary' hybrid zone in a stasipatric situation, clonal differentiation without geographic separation (if such a thing exists), could not be distinguished from a secondary contact zone which had been established between two recently differentiated chromosome races which were only distinguished by the chromosomal reproductive barrier which separated them. Indeed, the level of differentiation encountered in *P. leucopus* could be explained as an early secondary contact between otherwise undifferentiated chromosome races which subsequently established electrophoretic and mitochondrial divergence while in hybridiation parapatry. Alternatively, two allopatric chromosome races may have established electrophoretic and mtDNA differences in isolation and only recently formed hybridization parapatry. Even without reproductive data on the efficiency of chromosomal isolating mechanisms between the cytotypes, it is apparent that introgression is minimal and skewed. Thus, the data can support a number of possibilities for the evolution of this hybrid zone. The point I am making is unambiguous. Hybrid zone analysis, interesting though it may be, does not necessarily provide clear-cut conclusions to significant evolutionary problems.

Another hybridization study which was of considerable interest because of the questions it raised, was that between populations of *Hyla cinerea* and *H. gratiosa* in Alabama in North America. These species are sympatrically distributed over much of their distribution and generally do not hybridize; however, recent environmental disturbance has led to hybridization at a set of artificial ponds. Lamb and Avise (1986) made a mtDNA and allozyme analysis on 305 animals from a hybrid population. They were able to conclude that a behaviourally based asymmetry in the direction of interspecific hybridization was occurring, that is, between *H. cinerea* males and *H. gratiosa* females. All markers fulfilled the predictions on F1 backcross and later generation hybrids. The essentially artificial situation created by environmental disturbance led to the production of hybrids between species which did not normally hybridize when sympatric. When hybridization occurred, the directional introgression of the genes from one species into the other was demonstrated. One might question whether hybrid zones which have been interfered with by man's agricultural activities may not be similarly disrupted. A case which readily springs to mind is that of the grasshopper *Caledia captiva*. Here Shaw *et al.* (1980, 1985), found an unusual hybrid zone between the Torresian and Moreton chromosome races. These forms are distinguished by pericentric inversion and C-band differences between all elements in the complement. The chromosome races are separated by a narrow hybrid zone 200 km long, and from 200 to 800 metres wide (Shaw *et al.*, 1980; Marchant *et al.*, 1988). The situation is most unusual because there is a considerable level of reproductive isolation between the races (46% of F2 inviability has been attributed to inversion differences: Coates and Shaw, 1984). While this accounts for the narrow hybrid zone, it is difficult to explain why Torresian chromosomes introgress freely into the Moreton race, whereas no introgression of Moreton chromosomes occurs into the Torresian race outside of the hybrid zone. However, Moreton rDNA and mtDNA and certain allozyme markers are present in the Torresian zone (Arnold *et al.*, 1987; Marchant *et al.*, 1988).

Shaw *et al.* (1985) argued that over six generations the hybrid zone between the *Caledia* chromosome races remained in the same position. However, chromosomal genotypic frequencies were highly erratic during this period. These authors proposed that the metacentric karyotype was favoured on one side of the hybrid zone during major climatic changes and this accounted for fluctuations in the number of acrocentric chromosomes (favoured in dry years), compared to metacentrics which were favoured in more mesic years.

A contrasting view was presented by Marchant *et al.* (1988) who analysed populations 100 to 300 km away from the hybrid zone and found that there was evidence for the asymmetrical introgression of restriction fragment markers of the nuclear rRNA genes, mtDNA and four enzyme electromorphs to the north of the present hybrid zone. It was proposed that these markers were relics of ancient hybridization between the two chromosome races in an area where only the Torresian form is now found. That is, they are relics from the displacement of the Moreton by the chromosomally uniform Torresian form.

The time scale for these changes would be most interesting, for if the Torresian race is displacing the Moreton race, this could be interpreted as an ancient event. However, an alternative explanation is that the displacement is a relatively recent event and a consequence of man's agricultural activities which involved the clearing of forests and conversion of the countryside into grasslands (see aerial photograph Fig 11.17 in Shaw *et al.*, 1980). Thus, the *Caledia* hybrid zone may be collapsing; the Torresian form displacing the Moreton race by introgressive hybridization due to an adaptive advantage of the Torresian complement in the modified habitat. In this respect, the *Caledia* hybridizations may be analogous to those between *Hyla cinerea* and *H. gratiosa*.

One of the most intensively studied hybrid zones is that between chromosome races of the European alpine grasshopper *Podisma pedestris*. The races differ by an X–autosome fusion, producing a neo XY male sex chromosome system which characterizes one race; the other race has a single XO male sex chromosome system. The chromosomal difference between the chromosome races, although impinging on hybrid fertility, fails to provide reproductive isolation. Indeed, Hewitt *et al.* (1989) suggested that a series of allozyme, chromosomal, morphological and hybridization analyses have demonstrated this.

However, two of these studies provide significant results which do not necessarily agree with these conclusions. The electrophoretic analysis of 21 presumptive loci by Halliday *et al.* (1983) revealed that the genetic distance between the races is $D = 0.005$ which is well within the range of comparison between populations of the same karyotype. It is not surprising that the authors found that there was no correlation between chromosome change and gene frequency change. This raises the question as to whether the chromosomal hybrid zone is an ineffective barrier to gene flow as the authors suggest, or whether no genetic divergence has occurred between the races because they are of recent origin (only 10 000 years BP). The evidence suggests that there is an absence of genetic differentiation between chromosome races and populations. Gene flow is not estimated in this type of study, so conclusions which were reached suggesting that gene flow was not occurring across a hybrid zone are not supported by any data, they are simple supposition. One of the disturbing arguments raised by Barton and Hewitt (1981) was that the chromosome races had diverged at numerous loci which had minor cumulative effects on viability, so that hybrids were only half as viable as parental genotypes. This conclusion is not supported by any of the electrophoretic data and such differentiation is hard to reconcile with the short time that the chromosome races are thought to have been in existence.

In an attempt to resolve some of the problems associated with the evolution of this hybrid zone, Hewitt *et al.* (1989) mated individual virgin females of the two races with males of each race sequentially and in reciprocal order. The embryos were then karyotyped to determine the extent of each of the male's paternity. The excess of racial homozygosity demonstrated that homogamy was occurring. That is, females had a tendency to be fertilized by males of their own karyotype. Moreover, first sperm precedence was also demonstrated. One of the most interesting features was that the sex ratio was biased in favour of female embryo viability. The significant level of male infertility which was encountered was attributed to sperm or embryo failure. The number of infertile eggs produced in the crosses ranged from 30 to 60% of those analysed.

All of these data suggested that the impact of the X–autosome translocation (producing the neo XY sex chromosome system) on hybrid fertility may have been greater than was previously considered possible. Hewitt *et al.* (1989) pointed out that the preponderance of male sterility and sex ratio effects are another example of Haldane's rule (see Chapters 3 and 5). However, the effects on hybrid fertility and viability also suggest an association between chromosomally effected male fertility controls and meiotic drive mechanisms similar to those seen in *Drosophila* (see Chapters 3 and 6 *sensu* McKee, 1991), and these may provide an explanation for the homogamy and sex ratio distortions found in *Podisma*. These data also argue against the likelihood of 'inviability' genes.

It is of some significance that the difficulties encountered in the analysis of hybrid zones in *Podisma* have once again been at least partially explained by planned experimentation in the laboratory. The examples considered in this brief discussion on hybrid zones, suggest that the study of hybrid zones themselves may absorb a considerable amount of effort and produce what can only be regarded as a dubious if not arguable conclusion. It is my perception that the study of hybrid zones is not the key to our understanding of speciation processes. The most valuable form of study is that of controlled hybridization experimentation, and it is interesting to see that in both *Caledia captiva* and *Podisma pedestris* the authors turned to this form of analysis to resolve problems associated with their analysis of hybrid zones. The most meaningful reproductive data were direct results of carefully controlled hybridization studies, and in the case of *Caledia*, the inclusion of molecular analysis.

11.4 Punctuated equilibrium: a speciationist's view of evolution

The concept of punctuated equilibria was introduced by Eldredge (1971) and Eldredge and Gould (1972), and subsequently defended in a series of publications (Gould and Eldredge, 1977; Gould, 1982). It was formulated to account for the incompatibility of phyletic gradualism, which dominates evolutionary theory, with the palaeontological record. Morphological gaps in fossil sequences suggested a punctuated pattern of evolution with bursts of morphological change followed by prolonged periods of evolutionary stasis.

Gould and Eldredge (1977) equated the periods of morphological change to bursts of speciation. Rather than the widely held view of gradual change in large populations, they argued that speciation occurred rapidly in founding populations and was associated with colonizing radiations. This occurred so rapidly that considerable gaps would be present in the fossil record. Thus, 'accumulated speciation is the root of most evolutionary change and that what we have called anagenesis is no more than repeated cladogenesis filtered through the net of differential success at the species level' (Gould, 1982, p. 85).

Stasis was regarded by Gould and Eldredge as the key to the problem, for species seem to remain unchanged for millions of years, a view totally out of tune with phyletic gradualism. Thus, while punctuation was fully consistent with Mayrian allopatric speciation in peripheral isolates, stasis was problematical. For while this was explained away by gradualists as being due to directional selection, it is difficult to accept that selection of this type could operate for millions of years. Gould (1982) suggested that the answer to stasis could come from internal genetic regulatory mechanisms.

One of the major problems associated with the punctuated equilibrium concept is the equation between bursts of cladogenic activity and palaeontological detection of this. Fossils provide a relatively crude indication of what species have been present at any time in evolutionary history. This view can be supported if we consider extant speciating complexes from a fossil's perspective. For example, various Australian gekkonid lizards which have been analysed chromosomally and divided into chromosome races have subsequently been analysed morphologically and are now regarded as species complexes. Thus, *Diplodactylus vittatus* is now a complex of five chromosome races, four of which have been elevated to species status (King, 1977). The *Phyllodactylus marmoratus* complex has four chromosome races, two of which are now regarded as species (King and Rofe, 1976; King and King, 1977; Section 8.2.9). The *Gehyra australis* complex of seven chromosome races (King, 1983a), has been redefined as five species (King, 1983b). The *Gehyra variegata–punctata* complex was subdivided into at least ten chromosome races (King, 1979; Moritz, 1986 and unpublished), six of which are now regarded as species. Morphological differences between the entities in each of the speciating complexes are often subtle, being dependent on coloration, back pattern, differences in number and shape of scales, or number of preanal pores. The differences are not generally associated with osteological structures. Indeed, an osteological analysis of *Gehyra* species by Mitchell (1965) failed to detect any of the species complexes now known. Similarly, the chromosomally speciating rats of the *Rattus sordidus* complex (see Sections 7.3.2 and 8.2.3), which are distinguished by profound chromosomal differences, subtle differences in pelage and no electrophoretic or immunogenetic differences, are also osteologically indistinguishable (Watts and Aslin, 1981).

If it is assumed that chromosomally speciating complexes have played a significant evolutionary role in the past as they do today, it is possible to question how many species would be recognized as such if they were fossils? It is probable that palaeontologists would have been unable to make a distinction between species within complexes, because the only characters that they could assess are associated with skeletal structure. Thus, each species complex would have been regarded as a single species. Indeed, in vertebrates at least, palaeontologists could only distinguish isolated species from independent lineages of class 3 and 4 complexes (*sensu*

Chapter 8). Nevertheless, they may well be able to recognize the individual species in class 1 and 2 speciating complexes, the members of which have established numerous morphological adaptations which distinguish them from their congeners. If these findings apply to all fossilized animals and plants, and there is little doubt that this would be so, this would amount to many thousands of species. The fossil record is a coarse database which has particular problems in resolving subtle species differences. Models designed around bursts of cladogenic activity which cannot be readily resolved are therefore open to question. In the case outlined above, the problems with species identification are in the underestimation of species numbers. This can jeopardize a concept such as punctuated equilibrium just as much as overestimation of species diversity can (see Maynard Smith, 1987; Sheldon, 1987).

Concomitant with the view of punctuated equilibria being accompanied by bursts of founder-induced speciation is the probability that this was chromosomally induced speciation (see Gould, 1982). The Pavlovian response this elicited from the phyletic gradualists was an instant flailing around at anything in sight. Needless to say, a direct attack was launched on the concept of chromosomal speciation as a means of discrediting evolution by punctuated equilibria (see Charlesworth *et al.*, 1982). Some of the more bizarre claims made by these authors were dealt with in Section 11.2.

Gould (1982) regarded the possibility that reproductive isolation might be caused by chromosomal speciation, rather than as the accidental byproduct of adaptive divergence, as a two-edged sword for the concept of punctuated equilibrium. That is, while the rapidity of chromosomal speciation was favourable, the decoupling of morphological change from reproductive isolation created a profound dichotomy:

1 If the morphological adjustments that produce a new stable system occurred rapidly following reproductive isolation, then punctuated equilibrium was supported.
2 However, if morphological adaptations accumulated gradually (in geological perspective), with no tendency to any rapid initial setting and stabilization, then punctuated equilibrium was wrong.

The analyses of speciating complexes in Chapters 7 and 8, together with the above information on Australian gekkonids, provides a mixed bag for punctuated equilibria. Genetic distances between species and chromosome races within evolutionary lineages described in Chapter 8, show a linear increase in distance from the most recently formed to the most distantly related species in the sequence (see Fig. 8.11). This can be interpreted as a gradual accumulation of genetic differences with time within each species which is subsequent to the act of chromosomal differentiation. Much of the information from type 3 and 4 complexes indicates that morphological changes associated with chromosomal speciation are also subtle and accumulate gradually with time. This is not that surprising when it is considered that morphological change itself is not directly responsible for the primary speciation event. The chief advantage of chromosomal speciation is the breakdown of gene flow between parental and daughter species. This permits the rapid fixation of adaptive changes within the new species distribution. Such changes may involve minor adaptive modifications in coloration or back pattern (in gekkos), which can take advantage of differences in the habitat which has been colonized. In fact, it would seem unlikely that major macromutations of morphology could reach fixation, for these changes may force the species out of the niche that it had become well adapted to, or at least make it more difficult for it to survive in that niche. Nevertheless, if profound morphological change was adaptive, there are no intuitive reasons why such a change should not reach fixation. Although there appear to be very few examples of macromutations being established in chromosomally speciating complexes of this type, it is worth considering the formidable morphological changes associated with speciation by polyploidy in flowering plants (see Section 3.4.6).

In species complexes which have evolved by chromosomal speciation involving negatively heterotic changes, adaptive subtlety tends to prevail and, since most of the morphological changes do not involve major structural or osteological mutations (in higher vertebrates), support for macromutation and punctuated equilibria is not forthcoming. Nevertheless, the data derived from chromosomal analyses is not entirely negative in regard to this concept. The basic tenet of punctuated equilibria is that speciation is the agent responsible for evolutionary change, rather than gradual phyletic evolution. Chromosomal studies of both plants and animals indicate that many species have evolved by chromosomal speciation and that a large number of these are entities of chromosomally speciating lineages. That is, they are geographically aligned species which share sequentially derived chromosomal rearrangements with their direct ancestor. The fact that such speciating complexes stand out from other congeneric or confamilial relatives which may be chromosomally uniform, or have the odd difference present, suggests that bursts of relatively rapid cladogenic activity have occurred and that these are associated with colonizing radiations. The fact that macromutations do not generally distinguish entities in many of these complexes, would appear to be incidental to the fact of chromosomal speciation. How these speciating complexes would appear when telescoped into the fossil record must remain unresolved. When the desiderata of the propagandists is removed, evolution by bursts of speciation forming a punctuated equilibrium remains as an attractive hypothesis.

11.5 End view

We live in tumultuous times. The specialization and dedication of research workers to specific biochemical and molecular techniques, whether they are in the field of zoology, botany or genetics, has caused a minor revolution in our understanding of evolutionary issues and a questioning of established theories. And so should it be.

In some areas, however, the approach has become so introverted that researchers are more intent on developing the intricacies of their technique than ensuring the usable application of their findings. The born-again cladists who have not only constructed their own language, think with their own logic, and are obsessed with phylogenetic relationships to the exclusion of all else, now require their own species definitions.

One of the major problems that has developed in evolutionary biology is a failure to integrate specific findings into a common evolutionary perspective. This can be often seen in the very narrow scope of research articles, some of which appear to have been written for an audience restricted in their interests to a small group of North American rodents. A discovery of a particular chromosome structure may be described as if it was a first for cytogenetics, yet the same finding may have been made 20 years beforehand on plants and grasshoppers, and studied in detail, but this does not get a mention. There seems to be no questioning the fact that the channelling of research activities into restricted fields has been responsible for this lack of integration.

All of the biological sciences have felt the impact of molecular biology, both in terms of the expansion of new horizons and the possibility of answering unanswerable questions. An emphasis is being placed on cost-effective applied science and workers are being drawn away from fundamental evolutionary studies. The utilization of molecular techniques, to answer purely molecular questions, together with the breakdown in the integration of these findings with biological realities, is a disturbing trend. Students appear to be able to undertake specialized evolutionary research in molecular biology without any grounding in postgraduate or even undergraduate evolutionary theory, and their understanding of zoological or botanical evolution may be non-existent. How can the study of evolution progress under such a regime?

Despite these clouds on the horizon, the great leap forward in evolutionary biology in the last ten years has accompanied this specialization, and one cannot help but be optimistic for the future. Numerous evolutionary studies now utilize a suite of molecular, biochemical and cytogenetic techniques to answer basic evolutionary questions, and in this regard evolutionary advances have been highly significant. However, the key to our understanding of chromosomal speciation still lies in the analysis of hybridization and the coupling of this to sophisticated cytogenetic and molecular techniques which can break down the components responsible for the formation of reproductive isolating mechanisms between species. The evidence presented in this book demonstrates the fact of chromosomal speciation. In the future, studies will involve the unravelling of the intricate mechanisms responsible for this evolutionary process.

References

Adams, M., Baverstock, P. R., Tidemann, C. R. & Woodside, D. P. (1982). Large genetic differences between sibling species of bats, *Eptesicus*, from Australia. *Heredity*, **48**, 435–38.

Adams, M., Baverstock, P. R., Watts, C. H. S. & Reardon, T. (1987). Electrophoretic resolution of species boundaries in Australian Microchiroptera. I. *Eptesicus* (Chiroptera: Vespertilionidae). *Australian Journal of Biological Science*, **40**, 143–62.

Agassiz, L. (1857). *Essay on Classification*. Reprint 1962, ed. E. Lurie. Cambridge: Harvard University Press.

Ainsworth, C. C., Parker, J. S. & Horton, D. M. (1983). Chromosome variation and evolution in *Scilla autumnalis*. In *Kew Chromosome Conference II*, ed. P. E. Brandham & M. D. Bennett, pp. 261–68. London: Allen and Unwin.

Árnason, Ú., Lutley, R. & Sandholt, B. (1980). Banding studies on six killer whales: An account of C-band polymorphism and G-band patterns. *Cytogenetics and Cell Genetics*, **28**, 71–78.

Arnold, M. L. (1986). The heterochromatin of grasshoppers from the *Caledia captiva* species complex. III. Cytological organisation and sequence evolution in a dispersed highly repeated DNA family. *Chromosoma*, **94**, 183–88.

Arnold, M. L. & Shaw, D. D. (1985). The heterochromatin of grasshoppers from the *Caledia captiva* species complex. II. Cytological organisation of tandemly repeated DNA sequences. *Chromosoma*, **93**, 183–90.

Arnold, M. L., Shaw, D. D. & Contreras, N. (1987). Ribosomal RNA encoding DNA introgression across a narrow hybrid zone between two subspecies of grasshopper. *Proceedings of the National Academy of Sciences of the USA*, **84**, 3946–50.

Avise, J. C., Smith, M. H. & Selander, R. K. (1979). Biochemical polymorphism and systematics in the genus *Peromyscus*. VII. Geographic differentiation in members of the *truei* and *maniculatus* species groups. *Journal of Mammalogy*, **60**, 177–92.

Ayala, F. J. (1975). Genetic differentiation during the speciation process. *Evolutionary Biology*, **8**, 1–78.

Ayala, F. J. (1982). The genetic structure of species. In *Perspectives on Evolution*, ed. R. Milkman, pp. 60–82. Sunderland: Sinauer Associates.

Ayala, F. J., Powell, J. R., Tracey, M. L., Mourão, C. A. & Perez-Salas, S. (1972). Enzyme variability in the *Drosophila willistoni* group. IV. Genic variation in natural populations of *Drosophila willistoni*. *Genetics*, **70**, 113–39.

Ayala, F. J., Tracey, M. L., Hedgecock, D. & Richmond, R. C. (1974). Genetic differentiation during the speciation process in *Drosophila*. *Evolution*, **28**, 576–92.

Baker, R. J. (1981). Chromosome flow between chromosomally characterized taxa of a volant mammal, *Uroderma bilobatum* (Chiroptera: Phyllostomatidae). *Evolution*, **35**, 296–305.

Baker, R. J. & Bickham, J. W. (1986). Speciation by monobrachial centric fusions. *Proceedings of the National Academy of Sciences of the USA*, **83**, 8245–48.

Baker, R. J., Barnett, R. K. & Greenbaum, I. F. (1979). Chromosomal evolution in grasshopper mice (*Onychomys*: Critetidae). *Journal of Mammalogy*, 60, 297–306.

Baker, R. J., Koop, B. F. & Haiduk, M. W. (1983a). Resolving systematic relationships with G-bands: A study of five genera of South American cricetine rodents. *Systematic Zoology*, 32, 403–16.

Baker, R. J., Robbins, L. W., Stangl, F. B. & Birney, E. C. (1983b). Chromosomal evidence for a major subdivision in *Peromyscus leucopus*. *Journal of Mammalogy*, 64, 356–59.

Baker, R. J., Bickham, J. W. & Arnold, M. L. (1985). Chromosomal evolution in *Rhogeessa* (Chiroptera: Vespertilionidae): Possible speciation by centric fusions. *Evolution*, 39, 233–43.

Baldwin, L. & Macgregor, H. C. (1985). Centromeric satellite DNA in the newt *Triturus cristatus karelinii* and related species: Its distribution and transcription on lampbrush chromosomes. *Chromosoma*, 92, 100–07.

Baranov, V. S. (1980). Mice with Robertsonian translocations in experimental biology and medicine. *Genetica*, 52/53, 23–32.

Barrett, S. C. H. (1989). Mating system evolution and speciation in heterostylous plants. In *Speciation and its Consequences*, ed. D. Otte & J. A. Endler, pp. 257–83. Sunderland: Sinauer Associates.

Barsacchi-Pilone, G., Batistoni, R., Andronico, F., Vitelli, L. & Nardi, I. (1986). Heterochromatic DNA in *Triturus* (Amphibia, Urodela) 1. A satellite DNA component of the pericentric bands. *Chromosoma*, 5, 435–46.

Barton, N. H. (1980). The fitness of hybrids between two chromosomal races of the grasshopper *Podisma pedestris*. *Heredity*, 45, 49–61.

Barton, N. H. & Charlesworth, B. (1984). Genetic revolutions, founder effects, and speciation. *Annual Review of Ecology and Systematics*, 15, 133–64.

Barton, N. H. & Hewitt, G. M. (1981). Hybrid zones and speciation. In *Evolution and Speciation*, ed. W. R. Atchley and D. Woodruff, pp. 109–45. Cambridge: University Press.

Barton, N. H., Halliday, R. B. & Hewitt, G. M. (1983). Rare electrophoretic variants in a hybrid zone. *Heredity*, 50, 139–46.

Batistoni, R., Vignali, R., Negroni, A., Cremisi, F. & Barasacchi-Pilone, G. (1986). A highly repetitive DNA family is dispersed in small clusters throughout the *Triturus* genome. *Cell Biology International Reports*, 10, 486.

Baverstock, P. R. & Adams, M. (1987). Comparative rates of molecular, chromosomal and morphological evolution in some Australian vertebrates. In *Rates of Evolution*, ed. K. S. W. Campbell & M. F. Day, pp. 175–87. London: Allen and Unwin.

Baverstock, P. R., Gelder, M. & Jahnke, A. (1982). Cytogenetic studies of the Australian rodent, *Uromys caudimaculatus*, a species showing extensive heterochromatin variation. *Chromosoma*, 84, 517–33.

Baverstock, P. R., Watts, C. H. S., Gelder, M. & Jahnke, A. (1983a). G-banding homologies of some Australian rodents. *Genetica*, 60, 105–17.

Baverstock, P. R., Gelder, M. & Jahnke, A. (1983b). Chromosome evolution in Australian *Rattus* – G-banding and hybrid meiosis. *Genetica*, 60, 93–103.

Baverstock, P. R., Adams, M., Maxson, L. R. & Yosida, T. H. (1983c). Genetic differentiation among karyotypic forms of the black rat, *Rattus rattus*. *Genetics*, 105, 969–83.

Baverstock, P. R., Adams, M. & Watts, C. H. S. (1986). Biochemical differentiation among karyotypic forms of Australian *Rattus*. *Genetica*, 71, 11–22.

Bell, G. (1982). The *Masterpiece of Nature. The Evolution and Genetics of Sexuality*. London: Croom Helm.

Benado, M., Aguilera, M., Reig, O. A. & Ayala, F. J. (1979). Biochemical genetics of chromosome forms of Venezuelan spiny rats of the *Proechimys guairae and Proechimys trinitatus* superspecies. *Genetica*, **50**, 89–97.

Bengtsson, B. O. (1980). Rates of karyotypic evolution in placental mammals. *Hereditas*, **92**, 37–47.

Berg, R. (1941). Genetical analysis of two wild populations of *Drosophila melanogaster. Journal of General Biology*, **2**, 143–58.

Berland, H. M., Sharma, A., Cribiu, E. P., Darre, R., Boscher, J. & Popescu, C. P. (1988). A new case of Robertsonian translocation in cattle. *Journal of Heredity*, **79**, 33–36.

Bickham, J. W. & Baker, R. J. (1979). Canalization model of chromosomal evolution. *Bulletin of Carnegie Museum of Natural History*, **13**, 70–84.

Bickham, J. W. & Baker, R. J. (1980). Reassessment of the nature of chromosomal evolution in *Mus musculus. Systematic Zoology*, **29**, 159.

Bingham, P. M., Kidwell, M. G. & Rubin, G. M. (1982). The molecular basis of *P-M* hybrid dysgenesis: The role of the *P* element, a *P*-strain-specific transposon family. *Cell*, **29**, 995–1004.

Blair, W. F. (1950). Ecological factors in the speciation of *Peromyscus. Evolution*, **4**, 253–75.

Bock, I. R. (1984). Interspecific hybridization in the genus *Drosophila*. In *Evolutionary Biology*, vol. 18, ed. M. K. Hecht, B. Wallace & G. T. Prance, pp. 41–70. New York: Plenum Publishing Corporation.

Bock, W. J. (1986). Species concepts, speciation and macroevolution. In *Modern Aspects of Species*, ed. K. Iwatsuki, P. H. Raven & W. J. Bock, pp. 31–57. Tokyo: University of Tokyo Press.

Bock, W. J. (1989). Commentary on article. In *Evolutionary Biology at the Crossroads*, ed. M. K. Hecht, pp. 53–58. New York: Queens College Press.

Bonaminio, G. A. & Fechheimer, N. S. (1988). Segregation and transmission of chromosomes from a reciprocal translocation in *Gallus domesticus* cockerels. *Cytogenetics and Cell Genetics*, **48**, 193–97.

Bosma, A. A. (1978). The chromosomal G-banding pattern in the wart hog *Phacochoerus aethiopicus* (Suidae, Mammalia) and its implications for the systematic position of the species. *Genetica*, **49**, 15–19.

Bouvet, A. & Cribiu, E. P. (1990). Analysis of synaptonemal complexes behaviour in a bull carrying the 1;29 and 9;23 Robertsonian translocations. *Reproduction in Domestic Animals*, **25**, 215–19.

Bradshaw, W. N. & Hsu, T. C. (1972). Chromosomes of *Peromyscus* (Rodentia, Cricetidae). III. Polymorphisms in *Peromyscus maniculatus. Cytogenetics*, **11**, 436–51.

Bregliano, J. C. & Kidwell, M. G. (1983). Hybrid dysgenesis determinants. In *Mobile Genetic Elements*, ed. J. A. Shapiro, pp. 363–410. New York: Academic Press.

Briscoe, D. A., Calaby, J. H., Close, R. L., Maynes, G. M., Murtagh, C. E. & Sharman, G. B. (1982). Isolation, introgression and genetic variation in rock wallabies. In *Species at Risk*: Research in Australia, ed. R. H. Groves & W. D. L. Ride, pp. 73–87. Canberra: Australian Academy of Science.

Britten, R. J. & Davidson, E. H. (1971). Repetitive and non-repetitive DNA sequences and a speculation on the origins of evolutionary novelty. *Quarterly Review of Biology*, **46**, 111–38.

Britton-Davidian, J. (1990). Genic differentiation in *Mus musculus domesticus* populations from Europe, the Middle East and North Africa: Geographic patterns and colonization events. *Biological Journal of the Linnean Society*, **41**, 27–45.

Britton-Davidian, J., Nadeau, J. H., Croset, H. & Thaler, L. (1989). Genic differentiation and origin of Robertsonian populations of the house mouse (*Mus musculus domesticus* Rutty). *Genetical Research*, **53**, 29–44.

Brncic, D. (1954). Heterosis and the integration of the genotype in geographic populations of *Drosophila pseudoobscura*. *Genetics*, **39**, 77–88.

Brown, D. R., Koehler, C. M., Lindberg, G. L., Freeman, A. E., Mayfield, J. E., Meyers, A. M., Schutz, M. M. & Beitz, D. C. (1989). Molecular analysis of cytoplasmic genetic variation in Holstein cows. *Journal of Animal Science*, **67**, 1926–32.

Brown, G. G. & Simpson, M. V. (1981). Intra- and interspecific variation of the mitochondrial genome in *Rattus norvegicus* and *Rattus rattus*: Restriction enzyme analysis of variant mitochondrial DNA molecules and their evolutionary relationships. *Genetics*, **97**, 125–43.

Bruere, A. N. & Ellis, P. M. (1979). Cytogenetics and reproduction of sheep with multiple centric fusions (Robertsonian translocations). *Journal of Reproduction and Fertility*, **57**, 363–75.

Bruere, A. N., Scott, I. S. & Henderson, L. M. (1981). Aneupolid spermatocyte frequency in domestic sheep heterozygous for three Robertsonian translocations. *Journal of Reproduction and Fertility*, **63**, 61–66.

Brussard, P. F. (1984). Geographic patterns and environmental gradients: The central-marginal model in *Drosophila* revisited. *Annual Review of Ecology and Systematics*, **15**, 25–64.

Bryant, E. H. & Meffert, L. A. (1988). Effect of an experimental bottleneck on morphological integration in the house-fly. *Evolution*, **42**, 698–707.

Bucheton, A., Paro, R., Sang, H. M., Pelisson, A. & Finnegan, D. J. (1984). The molecular basis of IR hybrid dysgenesis in *Drosophila melanogaster*: Identification, cloning and properties of the *I* factor. *Cell*, **38**, 153–63.

Buckland, R. A. & Evans, H. J. (1978a). Cytogenetic aspects of phylogeny in the Bovidae. I. G-banding. *Cytogenetics and Cell Genetics*, **21**, 42–63.

Buckland, R. A. & Evans, H. J. (1978b). Cytogenetic aspects of phylogeny in the Bovidae. II. C-banding. *Cytogenetics and Cell Genetics*, **21**, 64–71.

Bullini, L. (1985). Speciation by hybridization in animals. *Bollettino di Zoologia*, **52**, 121–37.

Bunch, T. D. & Nadler, C. F. (1980). Giemsa-band patterns of the tahr and chromosomal evolution of the tribe Caprini. *Journal of Heredity*, **71**, 110–16.

Bunch, T. D., Foote, W. C. & Spillett, J. J. (1976). Translocations of acrocentric chromosomes and their implications in the evolution of sheep (*Ovis*). *Cytogenetics and Cell Genetics*, **17**, 122–36.

Bush, G. L. (1975). Modes of animal speciation. *Annual Review of Ecology and Systematics*, **6**, 339–64.

Bush, G. L. (1982). What do we really know about speciation? In *Perspectives on Evolution*, ed. R. Milkman, pp. 119–28. Sunderland: Sinauer Associates.

Bush, G. L., Case, S. M., Wilson, A. C. & Patton, J. L. (1977). Rapid speciation and chromosomal evolution in mammals. *Proceedings of the National Academy of Sciences of the USA*, **74**, 3942–46.

Butlin, R. K. (1987). Species, speciation, and reinforcement. *The American Naturalist*, **130**, 461–64.

Cabrero, J. & Camacho, J. P. M. (1987). Population cytogenetics of *Chorthippus vagans*. I. Polymorphisms for pericentric inversion and for heterochromatin deletion. *Genome*, **29**, 280–84.

Cain, A. J. (1954). *Animal Species and their Evolution*. New York: Harper and Row.

Capanna, E. (1982). Robertsonian numerical variation in animal speciation: *Mus musculus*, an emblematic model. In *Mechanisms of Speciation*, ed. C. Barigozzi, pp. 155–74. New York: Alan R. Liss.

Capanna, E., Civitelli, M. V. & Cristaldi, M. (1977). Chromosomal rearrangement, reproductive isolation and speciation in mammals. The case of *Mus musculus*. *Bollettino di Zoologia*, **44**, 213–46.

Capanna, E., Corti, M. & Nascetti, G. (1985). Role of contact areas in chromosomal speciation of the European long-tailed house mouse (*Mus musculus domesticus*). *Bollettino di Zoologia*, **52**, 97–119.

Carr, G. D., Robichaux, R. H., Witter, M. S. & Kyhos, D. W. (1989) Adaptive radiation of the Hawaiian Silversword alliance (Compositae – Madiinae): a comparison with Hawaiian picture-winged *Drosophila*. In *Genetics, Speciation and the Founder Principle*, ed. L. V. Giddings, K. Y. Kaneshiro & W. W. Anderson, pp. 79–97. New York: Oxford University Press.

Carson, H. L. (1949). Seasonal variation in gene arrangement frequencies over a three-year period in *Drosophila robusta* Sturtevant. *Evolution*, **3**, 322–29.

Carson, H. L. (1955). The genetic characteristics of marginal populations of *Drosophila*. *Cold Spring Harbor Symposia on Quantitative Biology*, **20**, 276–87.

Carson, H. L. (1968). The population flush and its genetic consequences. In *Population Biology and Evolution*, ed. R. C. Lewontin, pp. 123–37. New York: Syracuse University Press.

Carson, H. L. (1975). The genetics of speciation at the diploid level. *The American Naturalist*, **109**, 83–92.

Carson, H. L. (1982a). Speciation as a major reorganization of polygenic balances. In *Mechanisms of Speciation*, ed. C. Barigozzi, pp. 411–33. New York: Alan R. Liss.

Carson, H. L. (1982b). Evolution of *Drosophila* on the newer Hawaiian volcanoes. *Heredity*, **48**, 3–25.

Carson, H. L. (1983). Chromosomal sequences and inter-island colonizations in the Hawaiian *Drosophila*. *Genetics*, **103**, 465–82.

Carson, H. L. (1985). Unification of speciation theory in plants and animals. *Systematic Botany*, **10**, 380–90.

Carson, H. L. (1990). Increased genetic variance after a population bottleneck. *Tree*, **5**, 228–30.

Carson, H. L. & Bryant, P. J. (1979). Genetic variation in Hawaiian *Drosophila*. VI. Change in a secondary sexual characteristic as evidence of incipient speciation in *Drosophila*. *Proceedings of the National Academy of Sciences of the USA*, **76**, 1929–32.

Carson, H. L. & Templeton, A. R. (1984). Genetic revolutions in relation to speciation phenomena: The founding of new populations. *Annual Review of Ecology and Systematics*, **15**, 97–131.

Carson, H. L. & Wisotzkey, R. G. (1989). Increase in genetic variance following a population bottleneck. *The American Naturalist*, **134**, 668–73.

Carson, H. L., Clayton, F. E. & Stalker, H. D. (1967). Karyotypic stability and speciation in Hawaiian *Drosophila. Proceedings of the National Academy of Sciences of the USA*, **57**, 1280–85.

Cathcart, C. A. (1986). Mitochondrial DNA analysis in related species of *Petrogale* rock wallabies. *M.Sc. thesis*, Sydney: Macquarie University.

Catzeflis, F. M., Nevo, E., Ahlquist, J. E. & Sibley, C. G. (1989). Relationships of the chromosomal species in the Eurasian mole rats of the *Spalax ehrenbergi* group as determined by DNA–DNA hybridization, and an estimate of the Spalacid–Murid divergence time. *Journal of Molecular Evolution*, **29**, 223–32.

Chandley, A. C., Jones, R. C., Dott, H. M., Allen, W. R. & Short, R. V. (1974). Meiosis in interspecific equine hybrids. I. The male mule (*Equus asinus × E. caballus*) and hinny (*E. caballus × E. asinus*). *Cytogenetics and Cell Genetics*, **13**, 330–41.

Chapman, H. M. & Bruere, A. N. (1975). The frequency of aneuploidy in the secondary spermatocytes of normal and Robertsonian translocation-carrying rams. *Journal of Reproduction and Fertility*, **45**, 333–42.

Charlesworth, B. & Rouhani, S. (1988). The probability of peak shifts in a founder population. II. An additive polygenic trait. *Evolution*, **42**, 1129–45.

Charlesworth, B., Lande, R. & Slatkin, M. (1982). A neo-Darwinian commentary on macroevolution. *Evolution*, **36**, 474–98.

Chesser, R. K. & Baker, R. J. (1986). On factors affecting the fixation of chromosomal rearrangements and neutral genes: Computer simulations. *Evolution*, **40**, 625–32.

Christensen, K. & Pedersen, H. (1982). Variation in chromosome number in the blue fox (*Alopex lagopus*) and its effect on fertility. *Hereditas*, **97**, 211–15.

Cleland, R. E. (1957). Chromosome structure in *Oenothera* and its effect on the evolution of the genus. *Cytologia (Proceedings of the International Genetics Symposia)*, 1956, 5–19.

Cleland, R. E. (1962). The cytogenetics of *Oenothera. Advances in Genetics*, **11**, 147–237.

Close, R. L. & Lowry, P. S. (1990). Hybrids in marsupial research. *Australian Journal of Zoology*, **37**, 259–67.

Coates, D. J. & Shaw, D. D. (1984). The chromosomal component of reproductive isolation in the grasshopper *Caledia captiva*. III. Chiasma distribution patterns in a new chromosomal taxon. *Heredity*, **53**, 85–100.

Collins, M. & Rubin, G. M. (1984). Structure of chromosomal rearrangements induced by the *FB* transposable element in *Drosophila. Nature*, **308**, 323–27.

Coluzzi, M., Petrarca, V. & Di Deco, M. A. (1985). Chromosomal inversion intergradation and incipient speciation in *Anopheles gambiae. Bollettino di Zoologia*, **52**, 45–63.

Corneo, G. (1976). Do satellite DNAs function as sterility barriers in eukaryotes? *Evolutionary Theory*, **1**, 261–65.

Coyne, J. A. & Orr, H. A. (1989a). Patterns of speciation in *Drosophila. Evolution*, **43**, 362–81.

Coyne, J. A. & Orr, H. A. (1989b). Two rules of speciation. In *Speciation and its Consequences*, ed. D. Otte & J. A. Endler, pp. 180–207. Sunderland: Sinauer Associates.

Cracraft, J. (1983). Species concepts and speciation analysis. *Current Ornithology*, **1**, 159–87.

Craddock, E. M. & Carson, H. L. (1989). Chromosomal inversion patterning and population differentiation in a young insular species, *Drosophila silvestris. Proceedings of the National Academy of Sciences of the USA*, **86**, 4798–802.

Craddock, E. M. & Johnson, W. E. (1979). Genetic variation in Hawaiian *Drosophila*. V.

Chromosomal and allozymic diversity in *Drosophila silvestris* and its homosequential species. *Evolution*, **33**, 137–55.

Curtsinger, J. W. (1984). Components of selection in X chromosome lines of *Drosophila melanogaster*. Sex ratio modification by meiotic drive and viability selection. *Genetics*, **108**, 941–52.

da Cunha, A. B. & Dobzhansky, T. (1954). A further study of chromosomal polymorphism in *Drosophila willistoni* in relation to environment. *Evolution*, **8**, 119–34.

da Cunha, A. B., Dobzhansky, T., Pavlovsky, O. & Spassky, B. (1959). Genetics of natural populations. XXVIII. Supplementary data on the chromosomal polymorphism in *Drosophila willistoni* in relation to the environment. *Evolution*, **13**, 389–404.

Darlington, C. D. (1931). The cytological theory of inheritance in *Oènothera*. *Journal of Genetics*, **24**, 405–74.

Davidson, E. H. & Britten, R. J. (1979). Regulation of gene expression: Possible role of repetitive sequences. *Science*, **204**, 1052–59.

Davis, K. M., Smith, S. A. & Greenbaum, I. F. (1986). Evolutionary implications of chromosomal polymorphisms in *Peromyscus boylii* from southwestern Mexico. *Evolution*, **40**, 645–49.

Davisson, M. T., Poorman, P. A., Roderick, T. H. & Moses, M. J. (1981). A pericentric inversion in the mouse. *Cytogenetics and Cell Genetics*, **30**, 70–76.

de Boer, P., Searle, A. G., van der Hoeven, F. A., de Rooij, D. G. & Beechey, C. V. (1986). Male pachytene pairing in single and double translocation heterozygotes and spermatogenic impairment in the mouse. *Chromosoma*, **93**, 326.

DeSalle, R. (1984). Mitochondrial DNA evolution and phylogeny in the *planitibia* subgroup of Hawaiian *Drosophila*. *Ph.D. thesis*, St Louis: Washington University.

DeSalle, R., Giddings, L. V. & Templeton, A. R. (1986a). Mitochondrial DNA variability in natural populations of Hawaiian *Drosophila*. I. Methods and levels of variability in D. *silvestris* and D. *heteroneura* populations. *Heredity*, **56**, 75–85.

DeSalle, R., Giddings, L. V. & Kaneshiro, K. Y. (1986b). Mitochondrial DNA variability in natural populations of Hawaiian *Drosophila*. II. Genetic and phylogenetic relationships of natural populations of D. *silvestris* and D. *heteroneura*. *Heredity*, **56**, 87–96.

Dobzhansky, T. (1937). *Genetics and the Origin of Species*. 1st edn. New York: Columbia University Press.

Dobzhansky, T. (1950). Mendelian populations and their evolution. *The American Naturalist*, **84**, 401–18.

Dobzhansky, T. (1951). *Genetics and the Origin of Species*. 3rd edn. New York: Columbia University Press.

Dobzhansky, T. (1970). *Patterns of Species Formation. Genetics of the Evolutionary Process*. New York: Columbia University Press.

Dobzhansky, T. (1974). Genetic analysis of hybrid sterility within the species *Drosophila pseudoobscura*. *Hereditas*, **77**, 81–88.

Dobzhansky, T. & Epling, C. (1944). Contribution to the genetics, taxonomy and ecology of *Drosophila pseudoobscura* and its relatives. *Carnegie Institution of Washington Publication*, **554**, 183pp.

Dollin, A. E., Murray, J. D. & Gillies, C. B. (1989). Synaptonemal complex analysis of hybrid cattle. I. Pachytene substaging and the normal full bloods. *Genome*, **32**, 856–64.

Dollin, A. E., Murray, J. D. & Gillies, C. B. (1991). Synaptonemal complex analysis of hybrid cattle. II. *Bos indicus* × *Bos taurus* F₁ and backcross hybrids. *Genome*, **34**, 220–27.

Donoghue, M. J. (1985). A critique of the biological species concept and recommendations for a phylogenetic alternative. *Bryologist*, **88**, 172–81.

Doolittle, W. F. & Sapienza, C. (1980). Selfish genes, the phenotype paradigm and genome evolution. *Nature*, **284**, 601–03.

Dover, G. A. (1982). Molecular drive: A cohesive mode of species evolution. *Nature*, **299**, 111–17.

Dover, G. A. (1988). rDNA world falling to pieces. *Nature*, **336**, 623–24.

Dutrillaux, B. & Rumpler, Y. (1977). Chromosomal evolution in Malagasy lemurs. II. Meiosis in intra- and interspecific hybrids in the genus *Lemur*. *Cytogenetics and Cell Genetics*, **18**, 197–211.

Echelle, A. A. (1990). In defense of the phylogenetic species concept and the ontological status of hybridogenetic taxa. *Herpetologica*, **46**, 109–113.

Ehrlich, P. R. (1961). Has the biological species concept outlived its usefulness? *Systematic Zoology*, **10**, 167–76.

Eichenlaub-Ritter, U. & Winking, H. (1990). Non-disjunction, disturbances in spindle structure, and characteristics of chromosome alignment in maturing oocytes of mice heterozygous for Robertsonian translocations. *Cytogenetics and Cell Genetics*, **54**, 47–54.

Elder, F. F. B. & Hsu, T. C. (1988). Tandem fusion in the evolution of mammalian chromosomes. In *The Cytogenetics of Mammalian Autosomal Rearrangements*, ed. A. Daniel, pp. 481–506. New York: Alan R. Liss.

Eldredge, N. (1971). The allopatric model and phylogeny in Paleozoic invertebrates. *Evolution*, **25**, 156–67.

Eldredge, N. & Gould, S. J. (1972). Punctuated equilibria: An alternative to phyletic gradualism. In *Models of Paleobiology*, ed. T. J. M. Schopf, pp. 82–115. San Francisco: Freeman, Cooper and Company.

Eldridge, M. D. B., Dollin, A. E., Johnston, P. G., Close, R. L. & Murray, J. D. (1988). Chromosomal rearrangements in rock wallabies, *Petrogale* (Marsupialia, Macropodidae). I. The *Petrogale assimilis* species complex: G-banding and synaptonemal complex analysis. *Cytogenetics and Cell Genetics*, **48**, 228–32.

Eldridge, M. D. B., Close, R. L. & Johnston, P. G. (1990). Chromosomal rearrangements in rock wallabies, *Petrogale* (Marsupialia: Macropodidae). III. G-banding analysis of *Petrogale inornata* and *P. penicillata*. *Genome*, **33**, 798–802.

Endler, J. A. (1977). *Geographic Variations, Speciation, and Clines*. Princeton: University Press.

Endler, J. A. (1989). Conceptual and other problems in speciation. In *Speciation and its Consequences*, ed. D. Otte & J. A. Endler, pp. 625–48. Sunderland: Sinauer Associates.

Engels, W. R. & Preston, C. R. (1981). Identifying *P* factors in *Drosophila* by means of chromosome breakage hotspots. *Cell*, **26**, 421–28.

Engels, W. R. & Preston, C. R. (1984). Formation of chromosome rearrangements by *P* factors in *Drosophila*. *Genetics*, **107**, 657–78.

Engstrom, M. D. & Bickham, J. W. (1982). Chromosome banding and phylogenetics of the golden mouse, *Ochrotomys nuttalli*. *Genetica*, **59**, 119–26.

Evans, E. P., Lyon, M. F. & Daglish, M. (1967). A mouse translocation giving a metacentric marker chromosome. *Cytogenetics*, **6**, 105–19.

Ferris, S. D., Sage, R. D., Prager, E. M., Ritte, U. & Wilson, A. C. (1983). Mitochondrial DNA evolution in mice. *Genetics*, 105, 681–721.

Finch, R. A., Miller, T. E. & Bennett, M. D. (1984). 'Cuckoo' *Aegilops* addition chromosome in wheat ensures its transmission by causing breaks in meiospores lacking it. *Chromosoma*, 90, 84–88.

Forejt, J. (1974). Nonrandom association between a specific autosome and the X chromosome in meiosis of the male mouse: Possible consequence of homologous centromeres separation. *Cytogenetics and Cell Genetics*, 13, 369–83.

Forejt, J. (1979). Meiotic studies of translocations causing male sterility in the mouse. II. Double heterozygotes for Robertsonian translocations. *Cytogenetics and Cell Genetics*, 23, 163–70.

Forejt, J. (1982). X–Y involvement in male sterility caused by autosome translocations – a hypothesis. In *Genetic Control of Gamete Production and Function*, ed. P. G. Crosignani, B. L. Rubin & M. Fraccaro, pp. 135–51. New York, London: Academic Press.

Forejt, J. & Gregorová, S. (1977). Meiotic studies of translocations causing male sterility in the mouse. I. Autosomal reciprocal translocations. *Cytogenetics and Cell Genetics*, 19, 159–79.

Forejt, J., Gregorová, S. & Goetz, P. (1981). XY pair associates with the synaptonemal complex of autosomal male-sterile translocations in pachytene spermatocytes of the mouse (*Mus musculus*). *Chromosoma*, 82, 41–53.

Foster, G. G. & Whitten, M. J. (1991). Meiotic drive in *Lucilia cuprina* and chromosomal evolution. *The American Naturalist*, 137, 403–15.

Frost, D. R. & Hillis, D. M. (1990). Species in concept and practice: Herpetological applications. *Herpetologica*, 46, 87–104.

Fry, K. & Salser, W. (1977). Nucleotide sequences of HS-oc satellite DNA from kangaroo rat *Dipodomys ordii* and characterization of similar sequences in other rodents. *Cell*, 12, 1069–74.

Frykman, I. & Bengtsson, B. O. (1984). Genetic differentiation in *Sorex*. III. Electrophoretic analysis of a hybrid zone between two karyotypic races in *Sorex araneus*. *Hereditas*, 100, 259–70.

Frykman, I., Simonsen, V. & Bengtsson, B. O. (1983). Genetic differentiation in *Sorex*. I. Electrophoretic analysis of the karyotypic races of *Sorex araneus* in Sweden. *Hereditas*, 99, 279–92.

Futuyma, D. J. & Mayer, G. C. (1980). Non-allopatric speciation in animals. *Systematic Zoology*, 29, 254–71.

Gabriel-Robez, O., Ratomponirina, C., Dutrillaux, B., Carré-Pigeon, F. & Rumpler, Y. (1986). Meiotic association between the XY chromosomes and the autosomal quadrivalent of a reciprocal translocation in two infertile men, 46, XY, t(19;22) and 46, XY, t(17;21). *Cytogenetics and Cell Genetics*, 43, 154–60.

Gabriel-Robez, O., Jaafar, H. Ratomponirina, C., Boscher, J., Bonneau, J., Popescu, C. P. & Rumpler, Y. (1988). Heterosynapsis in a heterozygous fertile boar carrier of a 3;7 translocation. *Chromosoma*, 97, 26–32.

Galton, G. (1894). Discontinuity in evolution. *Mind NS*, 3, 362.

Garagna, S., Zuccotti, M., Searle, J. B., Redi, C. A. & Wilkinson, P. J. (1989). Spermatogenesis in heterozygotes for Robertsonian chromosomal rearrangements from natural populations of the common shrew, *Sorex araneus*. *Journal of Reproduction and Fertility*, 87, 431–38.

Garagna, S., Redi, C. A., Zuccotti, M., Britton-Davidian, J. & Winking, H. (1990). Kinetics of oogenesis in mice heterozygous for Robertsonian translocations. *Differentiation*, 42, 167–71.

George, M. & Ryder, O. A. (1986). Mitochondrial DNA evolution in the genus *Equus*. *Molecular Biology and Evolution*, 3, 535–46.

Georgiadis, N. J., Kat, P. W., Oketch, H. & Patton, J. (1990). Allozyme divergence within the Bovidae. *Evolution*, 44, 2135–49.

Ghiselin, M. T. (1974). A radical solution to the species problem. *Systematic Zoology*, 23, 536–44.

Gileva, E. A. (1987). Meiotic drive in the sex chromosome system of the varying lemming, *Dicrostonyx torquatus* Pall. (Rodentia, Microtinae). *Heredity*, 59, 383–89.

Gillespie, D., Donehower, L. & Strayer, D. (1982). Evolution of primate DNA organisation. In *Genome Evolution*, ed. G. A. Dover & R. B. Flavell, pp. 113–33. London: Academic Press.

Ginzburg, L. R., Bingham, P. M. & Yoo, S. (1984). On the theory of speciation induced by transposable elements. *Genetics*, 107, 331–41.

Goldschmidt, R. (1940). *The Material Basis of Evolution*. New Haven: Yale University Press.

Gould, S. J. (1977). *Ontogeny and Phylogeny*. Cambridge: Bellknap Press.

Gould, S. J. (1982). The meaning of punctuated equilibrium and its role in validating a hierarchical approach to macroevolution. In *Perspectives on Evolution*, ed. R. Milkman, pp. 83–104. Sunderland: Sinauer Associates.

Gould, S. J. & Eldredge, N. (1977). Punctuated equilibria: The tempo and mode of evolution reconsidered. *Paleobiology*, 3, 115–51.

Grant, V. (1957). The plant species in theory and practice. In *The Species Problem*, ed E. Mayr, pp. 39–80. American Association for the Advancement of Science, Publication No. 50.

Grant, V. (1963). *The Origin of Adaptations*. New York: Columbia University Press.

Grant V. (1971). *Plant Speciation*. New York: Columbia University Press.

Grant, V. (1985). *The Evolutionary Process: A Critical Review of Evolutionary Theory*. New York: Columbia University Press.

Grao, P., Coll, M. D., Ponsà, M. & Egozcue, J. (1989). Trivalent behaviour during prophase I in male mice heterozygous for three Robertsonian translocations: An electron-microscopic study. *Cytogenetics and Cell Genetics*, 52, 105–10.

Green, M. M. (1986). Genetic instability in *Drosophila melanogaster*: The genetics of an *MR* element that makes complete *P* insertion mutations. *Proceedings of the National Academy of Sciences of the USA*, 83, 1036–40.

Greenbaum, I. F. (1981). Genetic interactions between hybridizing cytotypes of the tent-making bat (*Uroderma bilobatum*). *Evolution*, 35, 306–21.

Greenbaum, I. F. & Reed, M. J. (1984). Evidence for heterosynaptic pairing of the inverted segment in pericentric inversion heterozygotes of the deer mouse (*Peromyscus maniculatus*). *Cytogenetics and Cell Genetics*, 38, 106–11.

Greenbaum, I. F., Baker, R. J. & Ramsey, P. R. (1978). Chromosomal evolution and the mode of speciation in three species of *Peromyscus*. *Evolution*, 32, 646–54.

Greilhuber, J. & Speta, F. (1976). C-banded karyotypes in the *Scilla hohenackeri* group, *Scilla persica* and *Puschkinia* (Liliaceae). *Plant Systematics and Evolution*, 126, 149–88.

Gropp, A. & Winking, H. (1981). Robertsonian translocations: Cytology, meiosis, segregation patterns and biological consequences of heterozygosity. *Symposia of the Zoological Society of London*, **47**, 141–81.

Gropp, A., Winking, H., Redi, C., Capanna, E., Britton-Davidian, J. & Noack, G. (1982). Robertsonian karyotype variation in wild house mice from Rhaeto-Lombardia. *Cytogenetics and Cell Genetics*, **34**, 67–77.

Gunn, S. J. & Greenbaum, I. F. (1986). Systematic implications of karyotypic and morphologic variation in mainland *Peromyscus* from the Pacific northwest. *Journal of Mammalogy*, **67**, 294–304.

Gustavsson, I. (1969). Cytogenetics, distribution and phenotypic effects of a translocation in Swedish cattle. *Hereditas*, **63**, 68–169.

Gustavsson, I. (1988). Reciprocal translocation in four boars producing decreased litter size. *Hereditas*, **109**, 159–68.

Gustavsson, I., Switonski, M., Larsson, K., Ploen, L. & Hojer, K. (1988). Chromosome banding studies and synaptonemal complex analyses of four reciprocal translocations in the domestic pig. *Hereditas*, **109**, 169–84.

Gustavsson, I., Switonski, M., Iannuzzi, L., Ploen, L. & Larson, K. (1989). Banding studies and synaptonemal complex analysis of an X–autosome translocation in the domestic pig. *Cytogenetics and Cell Genetics*, **50**, 188–94.

Haaf, T., Winking, H. & Schmid, M. (1989). Immunocytogenetics. III. Analysis of trivalent and multivalent configurations in mouse pachytene spermatocytes by human autoantibodies to synaptonemal complexes and kinetochores. *Cytogenetics and Cell Genetics*, **50**, 14–22.

Hafner, M. S., Hafner, J. C., Patton, J. L. & Smith, M. F. (1987). Macrogeographic patterns of genetic differentiation in the pocket gopher *Thomomys umbrinus*. *Systematic Zoology*, **36**, 18–34.

Hägele, K. (1984). Different hybrid effects in reciprocal crosses between *Chironomus thummi thummi* and *Ch. th. piger* including spontaneous chromosome abberations and sterility. *Genetica*, **63**, 105–11.

Haldane, J. B. S. (1922). Sex ratio and unisexual sterility in hybrid animals. *Journal of Genetics*, **12**, 101–09.

Hale, D. W. (1986). Heterosynapsis and suppression of chiasmata within heterozygous pericentric inversions of the Sitka deer mouse. *Chromosoma*, **94**, 425–32.

Halkka, L., Söderlund, V., Skarén, U. & Heikkilä, J. (1987). Chromosomal polymorphism and racial evolution of *Sorex araneus* L. in Finland. *Hereditas*, **106**, 257–75.

Hall, E. R. (1981). *The Mammals of North America*. 2nd edn. New York: John Wiley and Sons.

Hall, W. P. (1983). Modes of speciation and evolution in the sceloporine iguanid lizards. I. Epistemology of the comparative approach and introduction to the problem. In *Advances in Herpetology and Evolutionary Biology*, ed. A. G. J. Rhodin & K. Miyata, pp. 643–79. Cambridge: Museum of Comparative Zoology, Harvard University.

Hall, W. P. & Selander, R. K. (1973). Hybridization of karyotypically differentiated populations in the *Sceloporus grammicus* complex (Iguanidae). *Evolution*, **27**, 226–42.

Halliday, R. B., Barton, N. H. & Hewitt, G. M. (1983). Electrophoretic analysis of a chromosomal hybrid zone in the grasshopper *Podisma pedestris*. *Biological Journal of the Linnean Society*, **19**, 51–62.

Hamerton, J. L. (1971). *Human Cytogenetics*, vol. 1. New York: Academic Press.

Hammer, M. F. (1991). Molecular and chromosomal studies on the origin of *t* haplotypes in mice. *The American Naturalist*, **137**, 359–65.

Hansen, K. M. (1969). Bovine tandem fusion and infertility. *Hereditas*, **63**, 453–54.

Harris, M. J., Wallace, M. E. & Evans, E. P. (1986). Aneuploidy in the embryonic progeny of females heterozygous for the Robertsonian chromosome (9.12) in genetically wild Peru-Coppock mice (*Mus musculus*). *Journal of Reproduction and Fertility*, **76**, 193–203.

Hatch, F. T., Bodner, A. J., Mazrimas, J. A. & Moore, D. H. (1976). Satellite DNA and cytogenetic evolution. DNA quantity, satellite DNA and karyotypic variations in Kangaroo rats (genus *Dipodomys*). *Chromosoma*, **58**, 155–68.

Hausser, J., Catzeflis, F., Meylan, A. & Vogel, P. (1985). Speciation in the *Sorex araneus* complex (Mammalia: Insectivora). *Acta Zoologica Fennica*, **170**, 125–30.

Hayashi, J., Yonekawa, I., Gotoh, O., Tagashira, H., Moriwaki, K. & Yosida, T. H. (1979). Evolutionary aspects of variant types of rat mitochondrial DNAs. *Biochimica et Biophysica Acta*, **564**, 202–11.

Hayman, D. L. (1990). Marsupial cytogenetics. *Australian Journal of Zoology*, **37**, 331–49.

Hayman, D. L. & Martin, P. G. (1974). Mammalia. I. Monotremata and Marsupialia. vol. 4: Chordata 4. In *Animal Cytogenetics*, ed. B. John. Berlin, Stuttgart: Gebrüder Borntraeger.

Hecht, M. K. (1983). Microevolution, developmental processes, paleontology, and the origin of vertebrate higher categories. In *Modalites, Rythmes et Mecanismes de l'Evolution Biologique*, ed. J. Chaline, pp. 289–94. Paris: Centre Nationale de Research Scientifique.

Hecht, M. K. & Hoffman, A. (1986). Why not neo-Darwinism? A critique of paleobiological challenges. In *Oxford Surveys in Evolutionary Biology*, vol. 3, ed. R. Dawkins & M. Ridley, pp. 1–47. London: Oxford University Press.

Hedrick, P. W. (1981). The establishment of chromosomal variants. *Evolution*, **35**, 322–32.

Hedrick, P. W. & Levin, D. A. (1984). Kin-founding and the fixation of chromosomal variants. *The American Naturalist*, **124**, 789–97.

Hengeveld, R. (1988). Mayr's ecological species criterion. *Systematic Zoology*, **37**, 47–55.

Heth, G. & Nevo, E. (1981). Origin and evolution of ethological isolation in subterranean mole rats. *Evolution*, **35**, 259–74.

Hewitt, G. M. (1973). Variable transmission rates of a B-chromosome in *Myrmeleotettix maculatus* (Thunb.) (Acrididae: Orthoptera). *Chromosoma*, **40**, 83–106.

Hewitt, G. M. (1976). Meiotic drive for B-chromosome in the primary oocytes of *Myrmeleotettix maculatus* (Orthoptera: Acrididae). *Chromosoma*, **56**, 381–91.

Hewitt, G. M., Mason, P. & Nichols, R. A. (1989). Sperm precedence and homogamy across a hybrid zone in the alpine grasshopper *Podisma pedestris*. *Heredity*, **62**, 343–53.

Hickey, W. A. & Craig, G. B. (1966). Genetic distortion of sex ratio in a mosquito, *Aedes aegypti*. *Genetics*, **53**, 1177–96.

Holliday, R. (1964). A mechanism for gene conversion in fungi. *Genetical Research*, **5**, 282–304.

Hotta, Y. & Chandley, A. C. (1982). Activities of X-linked enzymes in spermatocytes of mice rendered sterile by chromosomal alterations. *Gamete Research*, **6**, 65–72.

Iannuzzi, L. & Di Berardino, D. (1985). Diagrammatic representation of RBA-banded chromosomes of swamp buffalo (*Bubalus bubalis* L.) and sex chromosome banding homologies with cattle (*Bos taurus* L.). *Caryologia*, **38**, 281–95.

Imai, H. T. (1983). Quantitative analysis of karyotype alteration and species differentiation in mammals. *Evolution*, **37**, 1154–61.

Jaafar, H., Gabriel-Robez, O. & Rumpler, Y. (1989). Pattern of ribonucleic acid synthesis in vitro in primary spermatocytes from mouse testis carrying an X-autosome translocation. *Chromosoma*, **98**, 330–34.

Johannisson, R., Löhrs, U., Wolff, H. H. & Schwinger, E. (1987). Two different XY-quadrivalent associations and impairment of fertility in men. *Cytogenetics and Cell Genetics*, **45**, 222–50.

John, B. (1976). *Population Cytogenetics*. Studies in Biology No. 70. London: Edward Arnold Ltd.

John, B. (1981). Chromosome change and evolutionary change: a critique. In *Evolution and Speciation*, ed. W. R. Atchley & D. S. Woodruff, pp. 23–51. Cambridge: University Press.

John, B. & Freeman, M. (1975). Causes and consequences of Robertsonian exchange. *Chromosoma* (Berlin), **52**, 123–36.

John, B. & Hewitt, G. M. (1968). Patterns and pathways of chromosome evolution within the Orthoptera. *Chromosoma*, **25**, 40–74.

John, B. & King, M. (1977a). Heterochromatin variation in *Cryptobothrus chrysophorus*. I. Chromosome differentiation in natural populations. *Chromosoma*, **64**, 219–39.

John, B. & King, M. (1977b). Heterochromatin variation in *Cryptobothrus chrysophorus*. II. C-banding. *Chromosoma*, **65**, 59–79.

John, B. & King, M. (1980). Heterochromatin variation in *Cryptobothrus chrysophorus*. III. Synthetic hybrids. *Chromosoma*, **78**, 165–86.

John, B. & King, M. (1982). Meiotic effects of supernumerary heterochromatin in *Heteropternis obscurella*. *Chromosoma*, **85**, 39–65.

John, B. & King, M. (1983). Population cytogenetics of *Atractomorpha similis*. I. C-band variation. *Chromosoma*, **88**, 57–68.

John, B. & Miklos, G. L. G. (1979). Functional aspects of satellite DNA and heterochromatin. *International Review of Cytology*, **58**, 1–113.

John, B. & Miklos, G. L. G. (1988). *The Eukaryote Genome in Development and Evolution*. London: Allen and Unwin.

John, B., Lightfoot, D. C. & Wiessman, D. B. (1983). The meiotic behaviour of natural F_1 hybrids between *Trimerotropis suffusa* Scudder and *T. cyaneipennis* Bruner (Orthoptera: Oedipodinae). *Canadian Journal of Genetics and Cytology*, **25**, 467–77.

John, B., Appels, R. & Contreras, N. (1986). Population cytogenetics of *Atractomorpha similis*. II. Molecular characterisation of the distal C-band polymorphisms. *Chromosoma*, **94**, 45–58.

Johnson, W. E., Carson, H. L., Kaneshiro, K. Y., Steiner, W. W. M. & Cooper, M. M. (1975). Genetic variation in Hawaiian *Drosophila*. II. Allozymic differentiation in the *D. planitibia* subgroup. In *Isozymes IV. Genetics and Evolution*, ed. C. L. Market, pp. 563–84. New York: Academic Press.

Johnston, F. P., Church, R. B. & Lin, C. C. (1982). Chromosome rearrangement between the Indian muntjac and Chinese muntjac is accompanied by a deletion of middle repetitive DNA. *Canadian Journal of Biochemistry*, **60**, 497–506.

Jones, R. N. (1991). B-chromosome drive. *The American Naturalist*, **137**, 430–42.

Jotterand-Bellomo, M. (1984). L'analyse cytogénétique de deux espèces de Muridae africains, *Mus oubanguii* et *Mus minutoides/musculoides*: polymorphisme chromosomique et ébauche d'une phylogénie. *Cytogenetics and Cell Genetics*, **38**, 182–88.

Jotterand-Bellomo, M. (1986). Le genre *Mus* africain, un exemple d'homogénéité caryotypique: étude cytogénétique de *Mus minutoides/musculoides* (Côte d'Ivoire), *M. setulosus* (République Centrafricaine), et de *M. mattheyi* (Burkina Faso). *Cytogenetics and Cell Genetics*, **42**, 99–104.

Jotterand-Bellomo, M. (1988). Chromosome analysis of five specimens of *Mus bufo-triton*

(Muridae) from Burundi (Africa): Three cytogenetic entities, a special type of chromosomal sex determination, taxonomy and phylogeny. *Cytogenetics and Cell Genetics*, **48**, 88–91.

Kaneshiro, K. Y. (1983). Sexual selection and direction of evolution in the biosystematics of Hawaiian *Drosophilidae*. *Annual Review of Entomology*, **28**, 161–78.

Karpechenko, G. D. (1927). Polyploid hybrids of *Raphanus sativus* L. × *Brassica oleracea* L. *Bulletin of Applied Botanical Genetics*, **17**, 305–10.

Kayano, H. (1957). Cytogenetic studies in *Lillium callosum*. III. Preferential segregation of a supernumerary chromosome in EMCs. *Proceedings of the Japan Academy*, **33**, 553–58.

Key, K. H. L. (1968). The concept of stasipatric speciation. *Systematic Zoology*, **17**, 14–22.

Key, K. H. L. (1974). Speciation in the Australian morabine grasshoppers: Taxonomy and ecology. In *Genetic Mechanisms of Speciation in Insects*, ed. M. J. D. White, pp. 43–56. Sydney: Australia and New Zealand Book Company.

Key, K. H. L. (1979). The genera *Culmacris* and *Stiletta* (Orthoptera: Eumastacidae: Morabinae). *Australian Journal of Zoology*, **25**, 31–108.

Key, K. H. L. (1981). Species, parapatry, and the morabine grasshoppers. *Systematic Zoology*, **30**, 425–58.

Kidwell, M. G. (1979). Hybrid dysgenesis in *Drosophila melanogaster*: The relationship between the P–M and I–R interaction systems. *Genetical Research*, **33**, 205–17.

Kimura, M. (1983). *The Neutral Theory of Molecular Evolution*. Cambridge: University Press.

Kimura, M. (1986). DNA and the neutral theory. *Philosophical Transactions of the Royal Society of London*, **312**, 343–54.

King, M. (1977). Chromosomal and morphometric variation in the gekko *Diplodactylus vittatus* (Gray). *Australian Journal of Zoology*, **25**, 43–57.

King, M. (1979). Karyotypic evolution in *Gehyra* (Gekkonidae: Reptilia). I. The *Gehyra variegata–punctata* complex. *Australian Journal of Zoology*, **27**, 373–93.

King, M. (1980). C-banding studies on Australian hylid frogs: Secondary constriction structure and the concept of euchromatin transformation. *Chromosoma*, **80**, 191–217.

King, M. (1981). Chromosome change and speciation in lizards. In *Evolution and Speciation*, ed. W. R. Atchley & D. S. Woodruff, pp. 262–85. Cambridge: University Press.

King, M. (1982). A case for simultaneous multiple chromosome rearrangements. *Genetica*, **59**, 53–60.

King, M. (1983a). Karyotypic evolution in *Gehyra* (Gekkonidae: Reptilia): III. The *Gehyra australis* complex. *Australian Journal of Zoology*, **31**, 723–41.

King, M. (1983b). The *Gehyra australis* species complex. (Sauria: Gekkonidae). *Amphibia–Reptilia*, **4**, 147–69.

King, M. (1984). Karyotypic evolution in *Gehyra* (Gekkonidae, Reptilia). IV. Chromosome change and speciation. *Genetica*, **64**, 101–14.

King, M. (1985). The canalization model of chromosomal evolution: A critique. *Systematic Zoology*, **34**, 69–75.

King, M. (1987). Chromosomal rearrangements, speciation and the theoretical approach. *Heredity*, **59**, 1–6.

King, M. (1990). Amphibia, vol. 4 Chordata 2. In *Animal Cytogenetics*, ed. B. John. Stuttgart, Berlin: Gebrüder Borntraeger.

King, M. (1991). The evolution of heterochromatin in the amphibian genome. In *Amphibian Cytogenetics and Evolution*, ed. D. M. Green & S. K. Sessions. San Diego: Academic Press.

King, M. & John, B. (1980). Regularities and restrictions governing C-band variation in Acridoid grasshoppers. *Chromosoma*, **76**, 123–50.

King, M. & King, D. (1977). An additional chromosome race of *Phyllodactylus marmoratus* (Gray) (Reptilia: Gekkonidae) and its phylogenetic implication. *Australian Journal of Zoology*, 25, 667–72.

King, M. & Rofe, R. (1976). Karyotypic variation in the Australian gekko *Phyllodactylus marmoratus* (Gray) (Gekkonidae: Reptilia). *Chromosoma*, 54, 75–87.

Kirby, G. W. M. (1979). Bali cattle in Australia. *World Animal Review*, 31, 1–7.

Lamb, T. & Avise, J. C. (1986). Directional introgression of mitochondrial DNA in a hybrid population of tree frogs: The influence of mating behaviour. *Proceedings of the National Academy of Sciences of the USA*, 83, 2526–30.

Lande, R. (1979). Effective deme sizes during long-term evolution estimated from rates of chromosomal rearrangement. *Evolution*, 33, 234–51.

Lande, R. (1984). The expected fixation rate of chromosomal inversions. *Evolution*, 38, 743–52.

Lande, R. (1985). The fixation of chromosomal rearrangements in a subdivided population with local extinction and colonization. *Heredity*, 54, 323–32.

Lansman, R. A., Avise, J. C., Aquadro, C. F., Shapira, J. F. & Daniel, S. W. (1983). Extensive genetic variation in mitochondrial DNAs among geographic populations of the deer mouse, *Peromyscus maniculatus. Evolution*, 37, 1–16.

Lara-Gongora, G. (1983). Two new species of the lizard genus *Sceloporus* (Reptilia, Sauria, Iguanidae) from the Adjusco and Ocuilan Sierras, Mexico. *Bulletin of the Maryland Herpetological Society*, 19, 1–14.

Larson, A., Prager, M. & Wilson, A. C. (1984a). Chromosomal evolution, speciation and morphological change in vertebrates: The role of social behaviour. *Chromosomes Today*, 8, 215–28.

Larson, A., Wake, D. B. & Yanev, K. P. (1984b). Measuring gene flow among populations having high levels of genetic fragmentation. *Genetics*, 106, 293–308.

Lay, D. M. & Nadler, C. F. (1972). Cytogenetics and origin of North African *Spalax* (Rodentia: Spalacidae). *Cytogenetics*, 11, 279–85.

Levin, D. A. (1978). Some genetic consequences of being a plant. In *Ecological Genetics: The Interface*, ed. P. Brussard, pp. 189–212. New York: Springer-Verlag.

Lewis, H. (1962). Catastrophic selection as a factor in speciation. *Evolution*, 16, 257–71.

Lewis, H. (1966). Speciation in flowering plants. *Science*, 152, 167–72.

Lewis, H. & Roberts, M. R. (1956). The origin of *Clarkia lingulata. Evolution*, 10, 126–38.

Lewontin, R. C. (1974). *The Genetic Basis of Evolutionary Change*. New York: Columbia University Press.

Lewontin, R. C. & Hubby, J. L. (1966). A molecular approach to the study of genic heterozygosity in natural populations of *Drosophila pseudoobscura. Genetics*, 54, 595–609.

Lifschytz, E. & Lindsley, D. (1972). The role of X-chromosome inactivation during spermatogenesis. *Proceedings of the National Academy of Sciences of the USA*, 69, 182–86.

Lim, J. K. (1981). Site-specific intrachromosomal rearrangements in *Drosophila melanogaster*: Cytological evidence for transposable elements. *Cold Spring Harbor Symposia on Quantitative Biology*, 45, 553–60.

Liming, S. & Pathak, S. (1981). Gametogenesis in a male Indian muntjac × Chinese muntjac hybrid. *Cytogenetics and Cell Genetics*, 30, 152–56.

Liming, S., Yingying, Y. & Xingsheng, D. (1980). Comparative cytogenetic studies on the red muntjac, Chinese muntjac, and their F1 hybrids. *Cytogenetics and Cell Genetics*, 26, 22–27.

Lindsley, D. L. & Tokuyasu, K. T. (1980). Spermatogenesis. *In Genetics and Biology of Drosophila*, vol. 2, ed. M. Ashburner & T. R. F. Wright, pp. 225–94. London: Academic Press.

Logue, D. N. & Harvey, M. J. A. (1978). Meiosis and spermatogenesis in bulls heterozygous for a presumptive 1/29 Robertsonian translocation. *Journal of Reproduction and Fertility*, **54**, 159–65.

Loh-Chung, Y., Lowensteiner, D., Wong, E. F. K., Sawada, I., Mazrimas, J. & Schmid, C. (1986). Localization and characterization of recombinant DNA clones derived from the highly repetitive DNA sequences in the Indian muntjac cells: Their presence in the Chinese muntjac. *Chromosoma*, **93**, 521–28.

Long, S. E. (1977). Cytogenetic examination of preimplantation blastocysts of ewes mated to rams heterozygous for the Massey I (t1) translocation. *Cytogenetics and Cell Genetics*, **18**, 82–89.

Loudenslager, E. J. (1978). Variation in the genetic structure of *Peromyscus* populations. I. Genetic heterozygosity – Its relationship to adaptive divergence. *Biochemical Genetics*, **16**, 1165–79.

Løvtrup, S. (1979). The evolutionary species: fact or fiction? *Systematic Zoology*, **28**, 386–92.

Luciani, J. M., Guichaoua, M. R., Mattei, A. & Morazzani, M. R. (1984). Pachytene analysis of a man with a 13q:14q translocation and infertility. *Cytogenetics and Cell Genetics*, **38**, 14–22.

Lucov, Z. & Nur, U. (1973). Accumulation of B chromosomes by preferential segregation in females of the grasshopper *Melanoplus femur-rubrum*. *Chromosoma*, **42**, 289–306.

Lyon, M. F. (1991). The genetic basis of transmission-ratio distortion and male sterility due to the *t* complex. *The American Naturalist*, **137**(3), 349–58.

Lyttle, T. W. (1981). Experimental population genetics of meiotic drive systems. III. Neutralization of sex-ratio distortion in *Drosophila* through sex chromosome aneuploidy. *Genetics*, **98**, 317–34.

Lyttle, T. W. (1982). A theoretical analysis of the effects of sex chromosome aneuploidy on X and Y chromosome meiotic drive. *Evolution*, **36**, 822–31.

Lyttle, T. W. (1989). Is there a role for meiotic drive in karyotype evolution? In *Genetics, Speciation and the Founder Principle*, ed. L. V. Giddings, K. Y. Kaneshiro & W. W. Anderson, pp. 149–64. New York: Oxford University Press.

Macgregor, H. C. & Kezer, J. (1971). The chromosomal localization of a heavy satellite DNA in the testis of *Plethodon c. cinereus*. *Chromosoma*, **33**, 167–82.

Macgregor, H. C. & Sherwood, S. (1979). The nucleolus organisers of *Plethodon* and *Aneides* located by *in situ* nucleic acid hybridisation with *Xenopus* ^{3}H-ribosomal RNA. *Chromosoma*, **72**, 271–80.

Mahadevaiah, S. & Mittwoch, U. (1986). Synaptonemal complex analysis in spermatocytes and oocytes of tertiary trisomic Ts(5)31 H mice with male sterility. *Cytogenetics and Cell Genetics*, **41**, 169.

Mahadevaiah, S. K., Setterfield, L. A. & Mittwoch, U. (1990). Pachytene pairing and sperm counts in mice with single Robertsonian translocations and monobrachial compounds. *Cytogenetics and Cell Genetics*, **53**, 26–31.

Mäkinen, A. & Lohi, O. (1987). The litter size in chromosomally polymorphic blue foxes. *Hereditas*, **107**, 115–19.

Mäkinen, A. & Remes, E. (1986). Low fertility in pigs with rcp (4q+;13q−) translocation. *Hereditas*, **104**, 223–29.

Mandahl, N. (1978). Variation in C-stained chromosome regions in European hedgehogs (Insectivora, Mammalia). *Hereditas*, **89**, 107–28.

Marchant, A. D., Arnold, M. L. & Wilkinson, P. (1988). Gene flow across a chromosomal tension zone. I. Relicts of ancient hybridization. *Heredity*, **61**, 321–28.

Mascarello, J. T. & Hsu, T. C. (1976). Chromosome evolution in woodrats, genus *Neotoma* (Rodentia: Cricetidae). *Evolution*, **30**, 152–69.

Matthews, K. A. (1981). Developmental stages of genome elimination resulting in transmission ratio distortion of the T-007 male recombination (MR) chromosome of *Drosophila melanogaster*. *Genetics*, **97**, 95–111.

Maynard-Smith, J. (1987). Darwinism stays unpunctured. *Nature*, **330**, 516.

Mayr, B., Krutzler, J., Auer, H., Kalat, M. & Schleger, W. (1987). NORs, heterochromatin, and R-bands in three species of Cervidae. *The Journal of Heredity*, **78**, 108–10.

Mayr, E. (1942). *Systematics and the Origin of Species*. New York: Columbia University Press.

Mayr, E. (1954). Change of genetic environment and evolution. In *Evolution as a Process*, ed. J. S. Huxley, A. C. Hardy & E. B. Ford, pp. 156–80. London: Allen and Unwin.

Mayr, E. (1963). *Animal Species and Evolution*. Cambridge: Harvard University Press.

Mayr, E. (1969). *Principles of Systematic Zoology*. New York: McGraw-Hill.

Mayr, E. (1970). *Populations, Species and Evolution*. Cambridge: Harvard University Press.

Mayr, E. (1982a). *The Growth of Biological Thought; Diversity, Evolution and Inheritance*. Cambridge: Harvard University Press.

Mayr, E. (1982b). Processes of speciation in animals. In *Mechanisms of Speciation*, ed. C. Barigozzi, pp. 1–20. New York: Alan R. Liss.

McClintock, B. (1956). Controlling elements and the gene. *Cold Spring Harbor Symposia on Quantitative Biology*, **21**, 197–216.

McClintock, B. (1978). Mechanisms that rapidly reorganize the genome. *Stadler Genetics Symposia*, **10**, 25–47.

McClintock, B. (1984). The significance of responses of the genome to challenge. *Science*, **226**, 792–801.

McKee, B. (1984). Sex chromosome meiotic drive in *Drosophila melanogaster* males. *Genetics*, **106**, 403–22.

McKee, B. (1991). X–Y pairing, meiotic drive, and ribosomal DNA in *Drosophila melanogaster* males. *The American Naturalist*, **137**, 332–39.

Melander, Y. & Hansen-Melander, E. (1980). Chromosome studies in African wild pigs (Suidae, Mammalia). *Hereditas*, **92**, 283–89.

Miklos, G. L. G. (1974). Sex chromosome pairing and male fertility. *Cytogenetics and Cell Genetics*, **13**, 558–77.

Miklos, G. L. G. & John, B. (1987). From genome to phenotype. In *Rates of Evolution*, ed. K. S. W. Campbell & M. F. Day, pp. 263–82. London: Allen and Unwin.

Miklos, G. L. G., Willcocks, D. A. & Baverstock, P. R. (1980). Restriction endonuclease and molecular analyses of three rat genomes with special reference to chromosome rearrangement and speciation problems. *Chromosoma*, **76**, 339–63.

Mitchell, F. J. (1965). Australian geckos assigned to the genus *Gehyra* Gray. (Reptilia, Gekkonidae). *Senckenbergiana Biologica*, **46**, 287–319.

Miyamoto, M. M., Tanhauser, S. M. & Laipis, P. J. (1989). Systematic relationships in the artiodactyl tribe Bovini (Family Bovidae), as determined from mitochondrial DNA sequences. *Systematic Zoology*, **38**, 342–49.

Mizuno, S. & Macgregor, H. C. (1974). Chromosomes, DNA sequences, and evolution in salamanders of the genus *Plethodon*. *Chromosoma*, **48**, 239–96.

Moran, C. (1981). Genetic demarcation of geographical distribution by hybrid zones. *Proceedings of the Ecological Society of Australia*, **11**, 67–73.

Moritz, C. (1983). Parthenogenesis in the endemic Australian lizard *Heteronotia binoei* (Gekkonidae). *Science*, **220**, 735–37.

Moritz, C. (1984). The evolution of a highly variable sex chromosome in *Gehyra purpurascens* (Gekkonidae). *Chromosoma*, **90**, 111–19.

Moritz, C. (1986). The population biology of *Gehyra* (Gekkonidae): Chromosome change and speciation. *Systematic Zoology*, **35**, 46–67.

Moriwaki, K., Sato, T. & Tsuchiya, K. (1971). Difference in amino acid composition of serum transferrin among various species of *Rattus*. *Annual Report of the National Institute of Genetics (Japan)*, **21**, 34–6.

Moriwaki, K., Yonekawa, H., Gotoh, O., Minezawa, M., Winking, H. & Gropp, A. (1984). Implications of the genetic divergence between European wild mice with Robertsonian translocations from the viewpoint of mitochondrial DNA. *Genetical Research*, **43**, 277–87.

Moses, M. J. (1977). The synaptonemal complex and meiosis. In *Molecular Human Cytogenetics*, vol. II, ed. R. S. Sparkes, D. Commings & C. F. Fox, pp. 101–25. New York: Academic Press.

Moses, M. J., Poorman, P. A., Roderick, T. H. & Davisson, M. T. (1982). Synaptonemal complex analysis of mouse chromosomal rearrangements. IV. Synapsis and synaptic adjustment in two paracentric inversions. *Chromosoma*, **84**, 457–74.

Moses, M. J., Dresser, M. E. & Poorman, P. A. (1984). Composition and role of the synaptonemal complex. *Symposia of the Society for Experimental Biology*, **38**, 245–70.

Mrongovius, M. J. (1979). Cytogenetics of the hybrids of three members of the grasshopper genus *Vandiemenella* (Orthoptera: Eumastacidae: Morabinae). *Chromosoma*, **71**, 81–107.

Murray, J. D. & Kitchin, R. M. (1976). Chromosomal variation and heterochromatin polymorphisms in *Peromyscus maniculatus*. *Experientia*, **32**, 307–08.

Nadeau, J. H. & Baccus, R. (1981). Selection components of four allozymes in natural populations of *Peromyscus maniculatus*. *Evolution*, **35**, 11–20.

Naveira, H. & Fontdevila, A. (1985). The evolutionary history of *Drosophila buzzatii*. IX. High frequencies of new chromosome rearrangements induced by introgressive hybridization. *Chromosoma*, **91**, 87–94.

Nei, M. (1972). Genetic distance between populations. *The American Naturalist*, **106**, 283–92.

Nelson, K., Baker, R. J. & Honeycutt, R. L. (1987). Mitochondrial DNA and protein differentiation between hybridizing cytotypes of the white-footed mouse, *Peromyscus leucopus*. *Evolution*, **41**, 864–72.

Nevo, E. (1978). Genetic variation in natural populations: Patterns and theory. *Theoretical Population Biology*, **13**, 121–77.

Nevo, E. (1982). Genetic structure and differentiation during speciation in fossorial gerbil rodents. *Mammalia*, **46**, 523–30.

Nevo, E. (1983). Population genetics and ecology: The interface. In *Evolution from Molecules to Man*, ed. D. S. Bendall, pp. 287–321. Cambridge: University Press.

Nevo, E. (1985a). Speciation in action and adaptation in subterranean mole rats: Patterns and theory. *Bollettino di Zoologia*, **52**, 65–95.

Nevo, E. (1985b). Genetic differentiation and speciation in spiny mice, *Acomys*. *Acta Zoologica Fennica*, **170**, 131–36.

Nevo, E. & Bar-El, H. (1976). Hybridization and speciation in fossorial mole rats. *Evolution*, **30**, 831–40.

Nevo, E. & Cleve, H. (1978). Genetic differentiation during speciation. *Nature*, **275**, 125.

Nevo, E. & Shaw, C. R. (1972). Genetic variation in a subterranean mammal, *Spalax ehrenbergi*. *Biochemical Genetics*, **7**, 235–41.

Nevo, E., Kim, Y. I., Shaw, C. R. & Thaeler, C. S. (1974). Genetic variation, selection and speciation in *Thomomys talpoides* pocket gophers. *Evolution*, **28**, 1–23.

Nevo, E., Dessauer, H. C. & Chuang, K. C. (1975). Genetic variation as a test of natural selection. *Proceedings of the National Academy of Sciences of the USA*, **72**, 2145–49.

Nevo, E., Bodmer, M. & Heth, G. (1976). Olfactory discrimination as an isolating mechanism in speciating mole rats. *Experientia*, **32**, 1511–12.

Nevo, E., Beiles, A. & Ben-Shlomo, R. (1984). The evolutionary significance of genetic diversity: Ecological, demographic and life history correlates. In *Evolutionary Dynamics of Genetic Diversity, Lecture Notes in Biomathematics* 53, ed. G. S. Mani, pp. 13–213. New York: Springer-Verlag.

Novitski, E. (1967). Nonrandom disjunction in *Drosophila*. *Annual Review of Genetics*, **1**, 71–86.

Nur, U. (1968). Synapsis and crossing-over within a paracentric inversion in the grasshopper *Camnula pellucida*. *Chromosoma*, **25**, 198–214.

Ohta, T. & Dover, G. A. (1984). The cohesive population genetics of molecular drive. *Genetics*, **108**, 501–21.

Orgel, L. E. & Crick, F. H. C. (1980). Selfish DNA: The ultimate parasite. *Nature*, **284**, 604–07.

Orgel, L. E., Crick, F. H. C. & Sapienza, C. (1980). Selfish DNA. *Nature*, **288**, 645–46.

Pardue, M. L. (1974). Localization of repeated DNA sequences in *Xenopus* chromosomes. *Cold Spring Harbor Symposia on Quantitative Biology*, **38**, 475–82.

Parker, J. S., Wilby, A. S. & Taylor, S. (1988). Chromosome stability and instability in plants. In *Kew Chromosome Conference III*, ed. P. E. Brandham, pp. 131–40. London: Her Majesty's Stationery Office.

Paterson, H. E. H. (1978). More evidence against speciation by reinforcement. *South African Journal of Science*, **74**, 369–71.

Paterson, H. E. H. (1982). Darwin and the origin of species. *South African Journal of Science*, **78**, 272–75.

Paterson, H. E. H. (1985). The recognition concept of species. In *Species and Speciation*, ed. E. S. Vrba, pp. 21–29. Pretoria: Transvaal Museum Monograph No. 4.

Paterson, H. E. H. (1986). Environment and species. *South African Journal of Science*, **82**, 62–65.

Paterson, H. E. H. (1987). On defining species in terms of sterility: Problems and alternatives. *Pacific Science*, **42**, 64–70.

Pathak, S. & Kieffer, N. (1979). Sterility in hybrid cattle. I. Distribution of constitutive heterochromatin and nucleolus organizer regions in somatic and meiotic chromosomes. *Cytogenetics and Cell Genetics*, **24**, 42–52.

Patterson, J. T. & Stone, W. S. (1952). *Evolution in the Genus* Drosophila. New York: Macmillan.

Patton, J. L. (1969). Chromosome evolution in the pocket mouse, *Perognathus goldmani* Osgood. *Evolution*, **23**, 645–62.

Patton, J. L. (1973). An analysis of natural hybridization between the pocket gophers, *Thomomys bottae and Thomomys umbrinus* in Arizona. *Journal of Mammalogy*, **54**, 561–84.

Patton, J. L. & Feder, J. H. (1978). Genetic divergence between populations of the pocket gopher, *Thomomys umbrinus* (Richardson). *Zeitschrift fuer Saeugetierkunde*, **43**, 17–30.

Patton, J. L. & Sherwood, S. W. (1982). Genome evolution in pocket gophers (genus *Thomomys*). I. Heterochromatin variation and speciation potential. *Chromosoma*, **85**, 149–62.

Patton, J. L. & Smith, M. F. (1981). Molecular evolution in *Thomomys* pocket gophers: Phyletic systematics, paraphyly, and rates of evolution. *Journal of Mammalogy*, **62**, 493–500.

Patton, J. L. & Smith, M. F. (1989). Population structure and the genetic and morphologic divergence among pocket gopher species (genus *Thomomys*). In *Speciation and its Consequences*, ed. D. Otte & J. A. Endler, pp. 284–304. Sunderland: Sinauer Associates.

Patton, J. L. & Yang, S. Y. (1977). Genetic variation in *Thomomys bottae* pocket gophers: Macrogeographic patterns. *Evolution*, **31**, 697–720.

Peacock, W. J., Dennis, E. S., Elizur, A. & Calaby, J. H. (1981). Repeated DNA sequences and kangaroo phylogeny. *Australian Journal of Biological Sciences*, **34**, 325–40.

Peters, G. B. (1982). The recurrence of chromosome fusion in inter-population hybrids of the grasshopper *Atractomorpha similis*. *Chromosoma*, **85**, 323–47.

Poorman, P. A., Moses, M. J., Davisson, M. T. & Roderick, T. H. (1981). Synaptonemal complex analysis of mouse chromosomal rearrangements. III. Cytogenetic observations on two paracentric inversions. *Chromosoma*, **83**, 419–29.

Popescu, C. P. & Boscher, J. (1986). A new reciprocal translocation in a hypoprolific boar. *Genetique Selection Evolution*, **18**, 123–30.

Porter, C. A. & Sites, J. W. (1985). Normal disjunction in Robertsonian heterozygotes from a highly polymorphic lizard population. *Cytogenetics and Cell Genetics*, **39**, 250–57.

Porter, C. A. & Sites, J. W. (1986). Evolution of *Sceloporus grammicus* complex (Sauria: Iguanidae) in central Mexico: Population cytogenetics. *Systematic Zoology*, **35**, 334–58.

Porter, C. A. & Sites, J. W. (1987). Evolution of *Sceloporus grammicus* complex (Sauria: Iguanidae) in central Mexico. II. Studies on rates of nondisjunction and the occurrence of spontaneous chromosomal mutations. *Genetica*, **75**, 131–44.

Prakash, S. (1973). Patterns of gene variation in central and marginal populations of *Drosophila robusta*. *Genetics*, **75**, 347–69.

Prakash, S. (1977). Further studies on gene polymorphism in the main body and geographically isolated populations of *Drosophila pseudoobscura*. *Genetics*, **85**, 713–19.

Ratomponirina, C., Brun, B. & Rumpler, Y. (1988). Synaptonemal complexes in Robertsonian translocation heterozygotes in lemurs. In *Kew Chromosome Conference III*, ed. P. E. Brandham, pp. 65–73. London: Her Majesty's Stationery Office.

Rattner, J. B. & Lin, C. C. (1985). Centromere organization in chromosomes of the mouse. *Chromosoma*, **92**, 325–29.

Redi, C. A. & Capanna, E. (1988). Robertsonian heterozygotes in the house mouse and the fate of their germ cells. In *The Cytogenetics of Mammalian Autosomal Rearrangements*, ed. A. Daniel, pp. 315–59. New York: Alan R. Liss.

Redi, C. A. Garagna, S., Mazzini, G. & Winking, H. (1986). Pericentromeric heterochromatin and A–T contents during Robertsonian fusion in the house mouse. *Chromosoma*, **94**, 31–35.

Redi, C. A., Garagna, S., Della Valle, G., Bottiroli, G., Dell'Orto, P., Viale, G., Peverali, F. A., Raimondi, E. & Forejt, J. (1990a). Differences in the organization and chromosomal allocation of satellite DNA between the European long-tailed house mice *Mus domesticus* and *Mus musculus*. *Chromosoma*, **99**, 11–17.

Redi, C. A., Garagna, S. & Capanna, E. (1990b). Nature's experiment with *in situ* hybridization? A hypothesis for the mechanism of Rb fusion. *Journal of Evolutionary Biology*, **3**, 133–37.

Redi, C. A., Garagna, S. & Zuccotti, M. (1990c). Robertsonian chromosome formation and fixation: the genomic scenario. *Biological Journal of the Linnean Society*, **41**, 235–55.

Rees, H., Jenkins, G., Seal, A. G. & Hutchinson, J. (1982). Assays of the phenotypic effects of changes in DNA amount. In *Genome Evolution*, ed. G. A. Dover & R. B. Flavell, pp. 287–97. London: Academic Press.

Reig, O. A., Aguilera, M., Barros, M. A. & Useche, M. (1980). Chromosomal speciation in a rassenkreis of Venezuelan spiny rats (genus *Proechimys*, Rodentia, Echimyidae). *Genetica*, **52/53**, 291–312.

Richardson, B. J., Baverstock, P. R. & Adams, M. (1986). *Allozyme Electrophoresis: A Handbook for Animal Systematics and Population Studies*. Sydney: Academic Press.

Robbins, L. W. & Baker, R. J. (1981). An assessment of the nature of chromosomal rearrangements in 18 species of *Peromyscus* (Rodentia: Cricetidae). *Cytogenetics and Cell Genetics*, **31**, 194–202.

Rofe, R. & Hayman, D. (1985). G-banding evidence for a conserved complement in the Marsupialia. *Cytogenetics and Cell Genetics*, **39**, 40–50.

Rogers, D. S., Greenbaum, I. F., Dunn, S. J. & Engstrom, M. D. (1984). Cytosystematic value of chromosomal inversion data in the genus *Peromyscus* (Rodentia: Cricetidae). *Journal of Mammalogy*, **65**, 457–65.

Roldan, E. R. S., Merani, M. S. & Von Lawzewitsch, I. (1984). Two abnormal chromosomes found in one cell line of a mosaic cow with low fertility. *Genetique Selection Evolution*, **16**, 135–42.

Rose, M. R. & Doolittle, W. F. (1983). Molecular biological mechanisms of speciation. *Science*, **220**, 157–62.

Rosen, D. E. (1978). Vicariant patterns and historical explanation in biogeography. *Systematic Zoology*, **27**, 159–88.

Rosenmann, A., Wahrman, J., Richler, C., Voss, R., Persitz, A. & Goldman, B. (1985). Meiotic association between the XY chromosomes and unpaired autosomal elements as a cause of human male sterility. *Cytogenetics and Cell Genetics*, **39**, 19–29.

Rouhani, S. & Barton, N. H. (1987). The probability of peak shifts in a founder population. *Journal of Theoretical Biology*, **126**, 51–62.

Rumpler, Y. & Dutrillaux, B. (1976). Chromosomal evolution in Malagasy lemurs. I. Chromosome banding studies in the genera *Lemur* and *Microcebus*. *Cytogenetics and Cell Genetics*, **17**, 268–81.

Ryder, O. A., Epel, N. C. & Benirschke, K. (1978). Chromosome banding studies of the Equidae. *Cytogenetics and Cell Genetics*, **20**, 323–50.

Ryder, O. A., Kumamoto, A. T., Durrant, B. S. & Benirschke, K. (1989). Chromosomal divergence and reproductive isolation in dik-diks. In *Speciation and its Consequences*, ed. D. Otte & J. A. Endler, pp. 208–25. Sunderland: Sinauer Associates.

Saïd, K., Jacquart, T., Montgelard, C., Sonjaya, H., Helal, A. N. & Britton-Davidian, J. (1986). Robertsonian house mouse populations in Tunisia: A karyological and biochemical study. *Genetica*, **68**, 151–56.

Sandler, L. R. G. & Novitski, E. (1957). Meiotic drive as an evolutionary force. *The American Naturalist*, **120**, 510–32.

Sannomiya, M. (1978). Relationship between crossing-over and chiasma formation in a translocation heterozygote of *Atractomorpha bedeli* (Acrididae, Orthoptera). *Heredity*, **40**, 305–08.

Saura, A., Lakovaara, S., Lokki, J. & Lankinen, P. (1973). Genic variation in central and marginal populations of *Drosophila subobscura. Hereditas*, **75**, 33–46.

Schmidtke, J., Brennecke, H., Schmid, M., Neitzel, H. & Sperling, K. (1981). Evolution of muntjac DNA. *Chromosoma*, **84**, 187–93.

Schroeder, G. L. (1968). Pericentric inversion polymorphism in *Trimerotropis helferi* (Orthoptera: Acrididae) and its effect on chiasma frequency. *Ph. D. Thesis*, Davis: University of California

Searle, J. B. (1984a). Nondisjunction frequencies in Robertsonian heterozygotes from natural populations of the common shrew, *Sorex araneus L. Cytogenetics and Cell Genetics*, **38**, 265–71.

Searle, J. B. (1984b). Hybridization between Robertsonian karyotypic races of the common shrew *Sorex araneus. Experientia*, **40**, 876–78.

Searle, J. B. (1985). Isoenzyme variation in the common shrew (*Sorex araneus*) in Britain, in relation to karyotype. *Heredity*, **55**, 175–80.

Searle, J. B. (1986). Meiotic studies of Robertsonian heterozygotes from natural populations of the common shrew, *Sorex araneus L. Cytogenetics and Cell Genetics*, **41**, 154–62.

Searle, J. B. (1988a). Selection and Robertsonian variation in nature: The case of the common shrew. In *The Cytogenetics of Mammalian Autosomal Rearrangements*, ed. A. Daniel, pp. 507–31. New York: Alan R. Liss.

Searle, J. B. (1988b). Karotypic variation and evolution in the common shrew, *Sorex araneus.* In *Kew Chromosome Conference III*, ed. P. E. Brandham, pp. 97–107. London: Her Majesty's Stationery Office.

Sene, F. M. & Carson, H. L. (1977). Genetic variation in Hawaiian *Drosophila*. IV. Allozymic similarity between *D. silvestris* and *D. heteroneura* from the island of Hawaii. *Genetics*, **86**, 187–98.

Setterfield, L. A. & Mittwoch, U. (1986). Reduced oocyte numbers in tertiary trisomic mice with male sterility. *Cytogenetics and Cell Genetics*, **41**, 177–80.

Sharman, G. B., Close, R. L. & Maynes, G. M. (1990). Chromosome evolution, phylogeny and speciation of rock wallabies (*Petrogale*: Macropodidae). *Australian Journal of Zoology*, **37**, 351–63.

Shaw, D. D. & Coates, D. J. (1983). Chromosomal variation and the concept of the co-adapted genome – a direct cytological assessment. In *Kew Chromosome Conference II*, ed. P. E. Brandham & M. P. Bennett, pp. 207–16. London: Allen and Unwin.

Shaw, D. D., Moran, C. & Wilkinson, P. (1980). Chromosomal reorganization, geographic differentiation and the mechanism of speciation in the genus *Caledia. Symposia of the Royal Entomological Society of London*, **10**, 171–94.

Shaw, D. D., Wilkinson, P. & Coates, D. J. (1983). Increased chromosomal mutation rate after hybridization between two subspecies of grasshoppers. *Science*, **220**, 1165–67.

Shaw, D. D., Coates, D. J., Arnold, M. L. & Wilkinson, P. (1985). Temporal variation in the chromosomal structure of a hybrid zone and its relationship to karyotypic repatterning. *Heredity*, **55**, 293–306.

Shaw, D. D., Coates, D. J. & Wilkinson, P. (1986). Estimating the genic and chromosomal components of reproductive isolation within and between subspecies of the grasshopper *Caledia captiva*. *Canadian Journal of Genetics and Cytology*, **28**, 686–95.

Sheldon, P. R. (1987). Parallel gradualistic evolution of Ordovician trilobites. *Nature*, **330**, 561–63.

Short, R. V., Chandley, A. C., Jones, R. C. & Allen, W. R. (1974). Meiosis in interspecific equine hybrids. II. The Przewalski horse/domestic horse hybrid. (*Equus przewalskii* × *E. caballus*). *Cytogenetics and Cell Genetics*, **13**, 465–78.

Simpson, G. G. (1961). *Principles of Animal Taxonomy. The Species and Lower Categories*. New York: Columbia University Press.

Sites, J. W. (1982). Morphological variation within and among three chromosome races of *Sceloporus grammicus* (Sauria: Iguanidae) in the north-central part of its range. *Copeia*, 1982, 920–41.

Sites, J. W. (1983). Chromosome evolution in the Iguanid lizard *Sceloporus grammicus*. I. Chromosome polymorphisms. *Evolution*, **37**, 38–53.

Sites, J. W. & Davis, S. K. (1989). Phylogenetic relationships and molecular variability within and among six chromosome races of *Sceloporus grammicus* (Sauria, Iguanidae), based on nuclear and mitochondrial markers. *Evolution*, **43**, 296–317.

Sites, J. W. & Greenbaum, I. F. (1983). Chromosome evolution in the Iguanid lizard *Sceloporus grammicus*. II. Allozyme variation. *Evolution*, **37**, 54–65.

Sites, J. W. & Moritz, C. (1987). Chromosomal evolution and speciation revisited. *Systematic Zoology*, **36**, 153–74.

Sites, J. W., Chesser, R. K. & Baker, R. J. (1988a). Population genetic structure and the fixation of chromosomal rearrangements in *Sceloporus grammicus* (Sauria, Iguanidae): A computer simulation study. *Copeia*, **4**, 1045–55.

Sites, J. W., Thompson, P. & Porter, C. A. (1988b). Cascading chromosomal speciation in lizards: A second look. *Pacific Science*, **42**, 89–104.

Sites, J. W., Camarillo, J. L., Gonzalez, A., Mendoza, F., Javier, L., Mancilla, M. & Lara-Gongora, G. (1988c). Allozyme variation and genetic divergence within and between three cytotypes of the *Sceloporus grammicus* complex (Sauria: Iguanidae) in central Mexico. *Herpetologica*, **44**, 297–307.

Slobodchikoff, C. N. (1976). *Concepts of Species*. Berkeley: University of California Press.

Smith, M. F. & Patton, J. L. (1988). Subspecies of pocket gophers: Causal bases for geographic differentiation in *Thomomys bottae*. *Systematic Zoology*, **37**, 163–78.

Smith, M. J., Hayman, D. L. & Hope, R. M. (1979). Observations on the chromosomes and reproductive systems of four Macropodine interspecific hybrids (Marsupialia: Macropodidae). *Australian Journal of Zoology*, **27**, 959–72.

Sokal, R. R. & Crovello, T. J. (1970). The biological species concept: A critical evaluation. *The American Naturalist*, **104**, 127–53.

Soma, H., Kada, H., Mtayoshi, K., Suzuki, Y., Meckvichal, C., Mahannop, A. & Vatanaromya, B. (1983). The chromosomes of *Muntiacus feae*. *Cytogenetics and Cell Genetics*, **35**, 156–58.

Speed, R. M. (1986). Oocyte development in XO foetuses of man and mouse: the possible role of heterologous X-chromosome pairing in germ cell survival. *Chromosoma*, **94**, 115–24.

Spencer, H. G., McArdle, B. H. & Lambert, D. M. (1986). A theoretical investigation of speciation by reinforcement. *The American Naturalist*, **128**, 241–62.

Spencer, H. G., Lambert, D. M. & McArdle, B. H. (1987). Reinforcement, species, and speciation: A reply to Butlin. *The American Naturalist*, **130**, 958–62.

Stack, S. & Anderson, L. (1988). An analysis of synaptic adjustment length and twisting of synaptonemal complexes in maize. In *Kew Chromosome Conference III*, ed. P. E. Brandham, pp. 299–312. London: Her Majesty's Stationery Office.

Stangl, F. B. (1986). Aspects of a contact zone between two chromosomal races of *Peromyscus leucopus* (Rodentia: Cricetidae). *Journal of Mammalogy*, **67**, 465–73.

Stewart-Scott, I. A. & Bruere, A. N. (1987). Distribution of heterozygous translocations and aneuploid spermatocyte frequency in domestic sheep. *Journal of Heredity*, **78**, 37–40.

Storr, G. M. (1987). The genus *Phyllodactylus* (Lacertilia: Gekkonidae) in Western Australia. *Records of the Western Australian Museum*, **13**, 275–84.

Sved, J. A. (1976). Hybrid dysgenesis in *Drosophila melanogaster*. A possible explanation in terms of spatial organization of chromosomes. *Australian Journal of Biological Sciences*, **29**, 375–88.

Switonski, M. (1980). Robertsonian translocation in Arctic fox (*Alopex lagopus*) and its effect on fertility. *4th European Conference on the Cytogenetics of Domestic Animals, Uppsala*, 45–49.

Szostak, J. W. & Wu, R. (1980). Unequal crossing over in the ribosomal DNA of *Saccharomyces cerevisiae*. *Nature*, **284**, 426–30.

Tease, C. & Fisher, G. (1988). Chromosome pairing in foetal oocytes of mouse inversion heterozygotes. In *Kew Chromosome Conference III*, ed. P. E. Brandham, pp. 293–98. London: Her Majesty's Stationery Office.

Temin, R. G., Ganetzky, B., Powers, P. A., Lyttle, T. W., Pimpinelli, S., Dimitri, P., Wu, C. I. & Hiraizumi, Y. (1991). Segregation distortion in *Drosophila melanogaster*: Genetic and molecular analyses. *The American Naturalist*, **137**, 287–331.

Templeton, A. R. (1980). The theory of speciation by the founder principle. *Genetics*, **92**, 1011–38.

Templeton, A. R. (1981). Mechanisms of speciation – A population genetic approach. *Annual Review of Ecology and Systematics*, **12**, 23–48.

Templeton, A. R. (1982). Genetic architectures of speciation. In *Mechanisms of Speciation*, ed. C. Barigozzi, pp. 105–21. New York: Alan R. Liss.

Templeton, A. R. (1987). Species and speciation. *Evolution*, **41**, 233–35.

Templeton, A. R. (1989). The meaning of species and speciation: A genetic perspective. In *Speciation and its Consequences*, ed. D. Otte and J. A. Endler, pp. 3–27. Sunderland: Sinauer Associates.

Thaeler, C. S. (1974). Four contacts between ranges of different chromosome forms of the *Thomomys talpoides* complex (Rodentia: Geomyidae). *Systematic Zoology*, **23**, 343–54.

Thompson, J. A. (1987). Evolution of gene structure in relation to function. In *Rates of Evolution*, ed. K. S. W. Campbell & M. F. Day, pp. 189–206. London: Allen and Unwin.

Thompson, P. & Sites, J. W. (1986). Comparision of population structure in chromosomally polytypic and monotypic species of *Sceloporus* (Sauria: Iguanidae) in relation to chromosomally-mediated speciation. *Evolution*, **40**, 303–14.

Thorpe, J. P. (1983). Enzyme variation, genetic distance and evolutionary divergence in relation to levels of taxonomic separation. In *Protein Polymorphism: Adaptive and Taxonomic Significance*, ed. G. S. Oxford & D. Rollinson, pp. 131–52. London: Academic Press.

Tikhonov, V. N. & Troshina, A. I. (1975). Chromosome translocations in the karyotypes of wild boars *Sus scrofa* L. of the European and the Asian areas of USSR. *Theoretical and Applied Genetics*, **45**, 304–08.

Turner, B. J., Grudzien, T. A., Adkisson, K. P. & Worrell, R. A. (1985). Extensive chromosomal divergence within a single river basin in the Goodeid fish, *Ilyodon furcidens*. *Evolution*, **39**, 122–34.

Valdez, R., Nadler, C. F. & Bunch, T. D. (1978). Evolution of wild sheep in Iran. *Evolution*, **32**, 56–72.

Van Valen, L. (1976). Ecological species, multispecies, and oaks. *Taxon*, **25**, 223–39.

Vetukhiv, M. A. (1953). Viability of hybrids between local populations of *Drosophila pseudo-obscura*. *Proceedings of the National Academy of Sciences of the USA*, **39**, 30–34.

Vigneault, G. & Zouros, E. (1986). The genetics of asymmetrical male sterility in *Drosophila mojavensis* and *Drosophila arizonensis* hybrids: Interactions between the Y-chromosome and autosomes. *Evolution*, **40**, 1160–70.

Vosa, C. G. (1973). Heterochromatin recognition and analysis of chromosome variation in *Scilla sibirica*. *Chromosoma*, **43**, 269–78.

Wahrman, J. & Goitein, R. (1972). Hybridization in nature between two chromosome forms of spiny mice. *Chromosomes Today*, **3**, 228–37.

Wahrman, J. & Gourevitz, P. (1973). Extreme chromosome variability in a colonizing rodent. *Chromosomes Today*, **4**, 399–424.

Wahrman, J., Goitein, R. & Nevo, E. (1969a). Geographic variation of chromosome forms in *Spalax*, a subterranean mammal of restricted mobility. In *Comparative Mammalian Cyto-genetics*, ed. K. Benirschke, pp. 30–48. New York: Springer-Verlag.

Wahrman, J., Goitein, R. & Nevo, E. (1969b). Mole rat *Spalax*: Evolutionary significance of chromosome variation. *Science*, **164**, 82–84.

Wahrman, J., Richler, C., Gamperl, R. & Nevo, E. (1985). Revisiting *Spalax*: Mitotic and meiotic chromosome variability. *Israel Journal of Zoology*, **33**, 15–38.

Wallace, B. (1953). On co-adaptation in *Drosophila*. *The American Naturalist*, **87**, 343–58.

Wallace, B. (1959). Influence of genetic systems on geographical distribution. *Cold Spring Harbor Symposia on Quantitative Biology*, **24**, 193–204.

Wallace, B. (1966). *Chromosomes, Giant Molecules, and Evolution*. New York: Norton.

Wallace, C. (1978). Chromosomal evolution in the antelope tribe Tragelaphini. *Genetica*, **48**, 75–80.

Walsh, J. B. (1982). Rate of accumulation of reproductive isolation by chromosome rearrangements. *The American Naturalist*, **120**, 510–32.

Warner, J. W. (1976). Chromosomal variation in the plains woodrat: Geographic distribution of three chromosomal morphs. *Evolution*, **30**, 593–98.

Watts, C. H. S. & Aslin, H. J. (1981). *The Rodents of Australia*. Sydney: Angus and Robertson Publishers.

Webb, G. C. (1976). Chromosome organization in the Australian plague locust, *Chortoicetes terminifera*. I. Banding relationships of the normal and supernumerary chromosomes. *Chromosoma*, **55**, 229–46.

Wendel, J. F., Edwards, M. D. & Stuber, C. W. (1987). Evidence for multilocus genetic control of preferential fertilisation in maize. *Heredity*, **58**, 297–301.

White, M. J. D. (1963). Cytogenetics of the grasshopper *Moraba scurra*. VIII. A complex spontaneous translocation. *Chromosoma*, **14**, 140–45.

White, M. J. D. (1968). Models of speciation. *Science*, **159**, 1065–70.

White, M. J. D. (1973). *Animal Cytology and Evolution*. 3rd edn. London: Cambridge University Press.

White, M. J. D. (1975). Chromosomal repatterning: Regularities and restrictions. *Genetics*, **79**, 63–72.

White, M. J. D. (1978a). *Modes of Speciation*. San Francisco: W. H. Freeman and Company.

White, M. J. D. (1978b). Chain processes in chromosomal speciation. *Systematic Zoology*, **27**, 285–98.

White, M. J. D. (1979). Speciation: Is it a real problem? *Scientia*, **114**, 455–68.

White, M. J. D. (1982). Rectangularity, speciation and chromosome architecture. In *Mechanisms of Speciation*, ed. C. Barigozzi, pp. 75–103. New York: Alan R. Liss.

White, M. J. D. & Andrew, L. E. (1960). Cytogenetics of the grasshopper *Moraba scurra*. V. Biometric effects of chromosomal inversions. *Evolution*, **14**, 284–92.

White, M. J. D., Blackith, R. E. & Cheney, J. (1967). Cytogenetics of the *viatica* group of morabine grasshoppers. I. The 'coastal' species. *Australian Journal of Zoology*, **15**, 263–302.

White, M. J. D., Contreras, N., Cheney, J. & Webb, G. C. (1977). Cytogenetics of the parthenogenetic grasshopper *Warramaba* (formerly *Moraba*) *virgo* and its bisexual relatives. II. Hybridization studies. *Chromosoma*, **61**, 127–48.

Wilby, A. S. & Parker, J. S. (1987). Population structure of hypervariable Y-chromosomes in *Rumex acetosa*. *Heredity*, **59**, 135–43.

Wilby, A. S. & Parker, J. S. (1988). Mendelian and non-Mendelian inheritance of newly-arisen chromosome rearrangements. *Heredity*, **60**, 263–68.

Wiley, E. O. (1978). The evolutionary species concept reconsidered. *Systematic Zoology*, **27**, 17–26.

Wiley, E. O. (1981). *Phylogenetics, The Theory and Practice of Phylogenetic Systematics*. New York: John Wiley and Sons.

Wilson, A. (1975). Relative rates of evolution of organisms and genes. *Stadler Genetics Symposia*, **7**, 117–34.

Wilson, A. C., Sarich, V. M. & Maxson, L. R. (1974). The importance of gene rearrangement in evolution: Evidence from studies on rates of chromosomal, protein and anatomical evolution. *Proceedings of the National Academy of Sciences of the USA*, **71**, 3028–30.

Wilson, A. C., Bush, G. L., Case, S. M. & King, M. C. (1975). Social structuring of mammalian populations and rate of chromosomal evolution. *Proceedings of the National Academy of Sciences of the USA*, **72**, 5061–65.

Winking, H., Dulic, B. & Bulfield, G. (1988). Robertsonian karyotype variation in the European house mouse, *Mus musculus*. Survey of present knowledge and new observations. *Zeitschrift fuer Saeugetierkunde*, **53**, 148–61.

Wójcik, J. M. (1986). Karyotypic races of the common shrew (*Sorex araneus* L.) from northern Poland. *Experientia*, **42**, 960–62.

Wood, R. J. & Newton, M. E. (1982). Meiotic drive and sex ratio distortion in mosquitoes. In *Recent Developments in the Genetics of Insect Disease Vectors*, ed. W. W. M. Steiner, pp. 130–52. Champaign, Illinois: Stripes Publishing Company.

Wood, R. J. & Newton, M. E. (1991). Sex-ratio distortion caused by meiotic drive in mosquitoes. *The American Naturalist*, **137**, 379–91.

Wright, S. (1931). Evolution in Mendelian populations. *Genetics*, **16**, 97–159.

Wright, S. (1932). The roles of mutation, inbreeding, crossbreeding, and selection in evolution. *Proceedings of the 6th International Congress of Genetics*, **1**, 356–66.

Wright, S. (1940). Breeding structure of populations in relation to speciation. *The American Naturalist*, **74**, 232–48.

Wright, S. (1941). On the probability of fixation of reciprocal translocations. *The American Naturalist*, **75**, 513–22.

Wright, S. (1977). *Evolution and the Genetics of Populations. Vol. 3. Experimental Results and Evolutionary Deductions*. Chicago: University of Chicago Press.

Wright, S. (1982a). The shifting balance theory and macroevolution. *Annual Review of Genetics*, **16**, 1–19.

Wright, S. (1982b). Character change, speciation, and the higher taxa. *Evolution*, **36**, 427–43.

Wurster, D. H. & Atkin, N. B. (1972). Muntjac chromosomes: A new karyotype for *Muntiacus muntjak. Experientia*, **28**, 972–73.

Wurster, D. H. & Benirschke, K. (1967). Chromosome studies in some deer, the springbok and pronghorn, with notes on placentation in deer. *Cytologia*, **32**, 273–85.

Wurster, D. H. & Benirschke, K. (1968). Chromosome studies in the superfamily Bovoicia. *Chromosoma*, **25**, 152–71.

Wurster, D. H. & Benirschke, K. (1970). Indian muntjac, *Muntiacus muntjak:* a deer with a low diploid chromosome number. *Science*, **168**, 1364–66.

Wurster-Hill, D. H. & Seidel, B. (1985). The G-banded chromosomes of Roosevelt's muntjac, *Muntiacus rooseveltorum. Cytogenetics and Cell Genetics*, **39**, 75–76.

Wyles, J. S. & Gorman, G. C. (1980). The albumin immunological and Nei electrophoretic distance correlation: A calibration for the Saurian genus *Anolis* (Iguanidae). *Copeia*, **1980**, 66–71.

Yamaguchi, O., Cardellino, R. A., & Mukai T. (1976). High rates of occurrence of spontaneous chromosome aberrations in *Drosophila melanogaster. Genetics*, **83**, 409–22.

Yonekawa, H., Moriwaki, K., Gotch, O., Miyashita, N., Migita, S., Bonhomme, F., Hjorth, J. P., Petras, M. L. & Tagashira, Y. (1982). Origins of laboratory mice deduced from restriction patterns of mitochondrial DNA. *Differentiation*, **22**, 222–26.

Yosida, T. H. (1980a). *Cytogenetics of the Black Rat. Karyotype Evolution and Species Differentiation*. Tokyo: University of Tokyo Press.

Yosida, T. H. (1980b). Segregation of karyotypes in the F_2 generation of the hybrids between Mauritius and Oceanian black rats with a note on their litter-size. *Proceedings of the Japan Academy Series B*, **56**, 557–61.

Yosida, T. H., Kato, H., Tsuchiya, K., Moriwaki, K., Ochiai, Y. & Monty, J. (1979). Mauritius type black rats with peculiar karyotypes derived from Robertsonian fission of small metacentrics. *Chromosoma*, **75**, 51–62.

Yunis, G. & Yasmineh, W. G. (1971). Heterochromatin, satellite DNA and cell function. *Science*, **174**, 1200–09.

Zong, E. & Fan, G. (1989). The variety of sterility and gradual progression to fertility in hybrids of the horse and donkey. *Heredity*, **62**, 393–406.

Zouros, E. (1974). Genic differentiation associated with the early stages of speciation in the *mulleri* subgroup of *Drosophila. Evolution*, **27**, 601–21.

Zouros, E. (1982). On the role of chromosomal inversions in speciation. *Evolution*, **36**, 414–16.

Zouros, E. (1986). A model for the evolution of asymmetrical male hybrid sterility and its implications for speciation. *Evolution*, **40**(6), 1171–84.

Zouros, E., Lofdahl, K. & Martin, P. A. (1988). Male hybrid sterility in *Drosophila*: Interactions between autosomes and sex chromosomes in crosses of *D. mojavensis* and *D. arizonensis*. *Evolution*, **42**, 1321–31.

Name index

Subject index

Printed in the United States
By Bookmasters